Pichia Protocols,
SECOND EDITION

METHODS IN MOLECULAR BIOLOGY™

John M. Walker, SERIES EDITOR

METHODS IN MOLECULAR BIOLOGY™

Pichia Protocols

SECOND EDITION

Edited by

James M. Cregg

Keck Graduate Institute of Applied Life Sciences,
Claremont, CA

HUMANA PRESS ✳ TOTOWA, NEW JERSEY

© 2007 Humana Press Inc.
999 Riverview Drive, Suite 208
Totowa, New Jersey 07512

www.humanapress.com

This publication is printed on acid-free paper. ∞
ANSI Z39.48-1984 (American Standards Institute)

Permanence of Paper for Printed Library Materials.

Cover illustration: Fig. 4, Chapter 12, "Selenomethionine Labeling of Recombinant Proteins," Anna M. Larsson and T. Alwyn Jones

Production Editor: Amy Thau

Cover design by Karen Schulz

For additional copies, pricing for bulk purchases, and/or information about other Humana titles, contact Humana at the above address or at any of the following numbers: Tel.: 973-256-1699; Fax: 973-256-8341; E-mail: humana@humanapr.com; or visit our Website: www.humanapress.com

Printed in the United States of America. 10 9 8 7 6 5 4 3 2 1

eISBN: 978-1-59745-456-8

Library of Congress Control Number: 2007932487

Preface

Pichia pastoris, as a recombinant protein production system, has continued to increase in popularity. In 1998, the year the first edition of *Pichia Protocols* was published, there were approx 500 published papers describing the use of this yeast for gene expression. As of the writing of this Preface, the number now stands at more than 1800 published reports (PubMed key word: *Pichia pastoris*). To be sure, part of the continued success of the system is simply a result of time and the effect of additional researchers finding and having success with the *P. pastoris* system. However, it is partly also due to efforts by our lab and others to increase the range of proteins that can be made in useful amounts in the system.

Pichia Protocols, Second Edition is intended mostly as an extension of our first edition, primarily focusing on new information that has come to light in recent years. However, key basic methods have been repeated here so that researchers are not absolutely dependent on having both editions to be able to utilize the system. In Chapter 1 of this edition, basics of the system are reviewed, with particular attention to contrasting the *P. pastoris* system with that of *Escherichia coli*, with which most researchers are familiar. Chapter 2 describes the many vectors and host strains now available for expression work, and Chapter 3 covers much of the information in the previous edition on transformation procedures for the yeast. With these three key chapters under their belts, researchers can get started with the system.

Chapters 4 to 7 discuss the use of fermentors to grow *P. pastoris* and provide good advice on how to go about the initial processing of material from these cultures. Chapters 8 to 10 present information on O- and N-linked glycosylation, and in particular, discuss the exciting work published in the last few years describing the modification of N-linked sugars on recombinant proteins secreted from this yeast. Chapters 11 to 13 present methods for labeling *P. pastoris*-expressed proteins for structural studies, in particular, labeling with selenomethionine, deuterium, and H^3 and N^{15}.

Chapter 14 repeats much of the basic information on classical genetic analysis with *P. pastoris* that was presented in the first edition; however, we have added an extensive discussion on how to formally analyze mutations in essential *P. pastoris* genes. In Chapter 15, the introduction of mutations in *P. pastoris* genes by the method of restriction enzyme-mediated integration (REMI) is presented. Finally, Chapters 16 to 18 are directed toward cell biologists using the

P. pastoris system to investigate protein interactions or to localize proteins in yeast cells.

An important advance in *P. pastoris* science that was a little late to make this volume is the sequencing of the genome of this yeast performed by Integrated Genomics (Chicago, IL). In the near future, researchers will be able to access the P. pastoris sequence through Integrated Genomics website (http://www.integratedgenomics.com).

I would like to thank the many authors and researchers involved in creating *Pichia Protocols, Second Edition* for their hard work in helping to put it together. I would especially like to thank my post-doc Dr. Bernard de la Cruz for meticulously going through each chapter and making sure all parts were present, accounted for, and formatted correctly. Without his efforts, this edition might still be sitting on my desk awaiting publication. As with the previous edition, I am indebted to Dr. John Walker, the series editor, for his work and patience, and to Mr. Tom Lanigan, President of Humana Press, who together with Dr. Walker provided much needed advice and suggestions.

James M. Cregg

Contents

vii

Contributors

RICK BARENT • *Biological Process Development Facility, Department of Chemical Engineering, University of Nebraska, Lincoln, NE*

KEITH BOOHER • *Section of Molecular Biology, Division of Biological Sciences, University of California San Diego, La Jolla, CA*

ROGER K. BRETTHAUER • *Department of Chemistry and Biochemistry, University of Notre Dame, Notre Dame, IN*

NICO CALLEWAERT • *Department of Biology, Institute of Microbiology, ETH Zürich, Zürich, Switzerland*

ROLAND CONTRERAS • *Fundamental and Applied Molecular Biology, Department of Molecular Biomedical Research, Ghent University, Ghent, Belgium*

JAMES M. CREGG • *Keck Graduate Institute of Applied Life Sciences, Claremont, CA*

WILLIAM A. DUNN, JR. • *Department of Anatomy and Cell Biology, University of Florida College of Medicine, Gainsville, FL*

SARAH A. FANDERS • *Department of Chemical Engineering, University of Nebraska, Lincoln, NE*

JEAN-CLAUDE FARRE • *Section of Molecular Biology, Division of Biological Sciences, University of California San Diego, La Jolla, CA*

TILLMAN U. GERNGROSS • *Thayer School of Engineering and the Department of Biological Sciences, Dartmouth College, Hanover, NH*

STEVEN GEYSENS • *Fundamental and Applied Molecular Biology, Department of Molecular Biomedical Research, Ghent University, Ghent, Belgium*

ERIN M. GIACCONE • *GlycoFi Inc., Lebanon, NH*

BENJAMIN S. GLICK • *Department of Molecular Genetics and Cell Biology, The University of Chicago, Chicago, IL*

MARK GOUTHRO • *Biological Process Development Facility, Department of Chemical Engineering, University of Nebraska, Lincoln, NE*

PETER J. HOTEZ • *Department of Microbiology and Tropical Medicine, The George Washington University, Washington, DC*

JICAI HUANG • *Biological Process Development Facility, Department of Chemical Engineering, University of Nebraska, Lincoln, NE*

MEHMET INAN • *Department of Chemical Engineering, University of Nebraska, Lincoln, NE*

T. ALWYN JONES • *Department of Cell and Molecular Biology, University of Uppsala, Uppsala, Sweden*

VLADIMIR KAIGORODOV • *Fundamental and Applied Molecular Biology, Department of Molecular Biomedical Research, Ghent University, Ghent, Belgium*

ANNA M. LARSSON • *Department of Cell and Molecular Biology, University of Uppsala, Uppsala, Sweden*

SEBASTIEN LEON • *Section of Molecular Biology, Division of Biological Sciences, University of California San Diego, La Jolla, CA*

HUIJUAN LI • *GlycoFi Inc., Lebanon, NH*

GEOFF P. LIN-CEREGHINO • *Department of Biological Sciences, University of the Pacific, Stockton, CA*

JOAN LIN-CEREGHINO • *Department of Biological Sciences, University of the Pacific, Stockton, CA*

MICHAEL M. MEAGHER • *Department of Chemical Engineering, University of Nebraska, Lincoln, NE*

ROBERT G. MIELE • *GlycoFi Inc., Lebanon, NH*

TERESA I. MITCHELL • *GlycoFi Inc., Lebanon, NH*

AMAR PATEL • *Section of Molecular Biology, Division of Biological Sciences, University of California San Diego, La Jolla, CA*

NAGANAND RAYAPURAM • *Section of Molecular Biology, Division of Biological Sciences, University of California San Diego, La Jolla, CA*

SANDRA E. RÍOS • *GlycoFi Inc., Lebanon, NH*

ERIC RODRIGUEZ • *Danone Vitapole, Cedex, France*

VIRGINIA P. ROXAS • *United States Army Medical Research Institute of Infectious Diseases, Fort Detrick, MD*

LAURA A. SCHRODER • *Department of Anatomy and Cell Biology, University of Florida College of Medicine, Gainsville, FL*

KANAE SHIRAHAMA-NODA • *Section of Molecular Biology, Division of Biological Sciences, University of California San Diego, La Jolla, CA*

LEONARD A. SMITH • *United States Army Medical Research Institute of Infectious Diseases, Fort Detrick, MD*

SURESH SUBRAMANI • *Section of Molecular Biology, Division of Biological Sciences, University of California San Diego, La Jolla, CA*

IVET SURIAPRANATA • *Section of Molecular Biology, Division of Biological Sciences, University of California San Diego, La Jolla, CA*

ILYA TOLSTORUKOV • *Keck Graduate Institute of Applied Life Sciences, Claremont, CA*

WOUTER VERVECKEN • *Fundamental and Applied Molecular Biology, Department of Molecular Biomedical Research, Ghent University, Ghent, Belgium*

JAMES W. WHITTAKER • *Department of Environmental and Biomolecular Systems, Oregon Health Sciences University, Beaverton, OR*

MINGDA YAN • *Section of Molecular Biology, Division of Biological Sciences, University of California San Diego, La Jolla, CA*

BIN ZHAN • *Department of Microbiology and Tropical Medicine, The George Washington University, Washington, DC*

WENHUI ZHANG • *Xoma Ltd., Berkeley, CA*

LAN ZHANG • *Section of Molecular Biology, Division of Biological Sciences, University of California San Diego, La Jolla, CA*

1

Introduction

Distinctions Between Pichia pastoris *and Other Expression Systems*

James M. Cregg

Abstract

The construction of *Pichia pastoris* expression strains and the general growth and manipulation of this yeast expression system are in many ways similar to those of bacterial expression systems, particularly *Escherichia coli*. Because of this, it is typically easy for researches experienced with bacterial systems to make the jump to this eukaryotic system. However, because the system is similar, users can be falsely fooled into assuming that the system is completely bacterial-like and may waste time and effort performing experiments that are unlikely to yield the desired results with this yeast. To aid in preventing *P. pastoris* users from falling into one or more or these traps, this introduction focuses directly on key ways that the *P. pastoris* expression system is different.

Key Words: *Pichia pastoris*; foreign protein expression; eukaryotic proteins.

1. General Comments

Pichia pastoris is a widely used yeast species for the production of recombinant proteins (*see 1* for a table of proteins made in *P. pastoris*). As a yeast, *P. pastoris* is a single-celled microorganism and therefore, like bacteria, relatively fast growing and inexpensive to manipulate in culture. However, unlike bacteria, yeasts are also eukaryotes with the same intracellular environment and many of the same post-translational protein processing capabilities as higher eukaryotes, including humans. Thus, eukaryotic proteins are more likely to be correctly processed, folded, and assembled into functional molecules when synthesized in a yeast.

Two factors have been most important to the success of the *P. pastoris* expression system relative to other yeast species. The first is that, unlike baker's yeast, *Saccharomyces cerevisiae*, and many other yeast species, *P. pastoris* does not like to ferment sugars or other carbon sources. The preference of this yeast for

From: *Methods in Molecular Biology, vol. 389:* Pichia *Protocols, Second Edition*
Edited by: J. M. Cregg © Humana Press Inc., Totowa, NJ

Fig. 1. Centrifuge bottles containing cultures of a recombinant production strain of *P. pastoris*. *Left bottle*: culture at a density of approx 1.0 OD600 units/mL. *Right bottle*: culture at a density of approx 400 OD600 units/mL.

respiratory growth greatly facilitates its culturing at high cell densities because it does not generate significant amounts of the toxic fermentative product, ethanol. It simply turns the carbon source into more biomass (*see* **Fig. 1**). Because the volumetric productivity of a given recombinant production process is generally proportional to the density of biomass generated, the ability to easily grow this organism at ultra-high cell densities is a major advantage. The second factor is related to the ability of *P. pastoris* to grow on methanol as a sole carbon and energy source. Like many alternative carbon source pathways, growth on methanol requires the induction of a specific set of metabolic enzymes. Two of these enzymes, alcohol oxidase (AOX) and dihydroxyacetone synthase (DHAS), although absent from cells grown on glucose or other carbon sources, each can constitute approx 30% of total protein in cells cultured on methanol *(2)*. The spectacular regulation of AOX and DHAS protein levels is controlled primarily at the level of transcription of their genes, *AOX1* and *DAS (3)*.

In addition, to recombinant protein production, *P. pastoris* is also used as a model to investigate certain problems in cell biology, including peroxisome biogenesis and assembly *(4)*, protein and organelle reprocessing by autophagy *(5)*, and Golgi organization and function *(6)*.

This introductory chapter reviews basic aspects of the expression system, mainly those that are either not intuitively obvious or significantly differ from procedures with other systems, and in particular *Escherichia coli*. For a more detailed description of this system, one can examine the first *Pichia* protocols book *(6)* or one of the many reviews written on the system *(7–18)*. Additional information and procedures can be found on the Invitrogen website *(22)* or on the authors' website *(1)*.

2. The Myth of Codon Optimization

One of the first considerations in expressing any gene in a foreign system is the structure of the gene itself (i.e., is it compatible with the host's translational machinery?). A number of reports on expression in *P. pastoris* have suggested that changing codons in a foreign gene to those of codons more frequently used by the yeast (as judged from the predicted amino acid sequences of highly expressed *P. pastoris* genes) significantly improves the levels of expression of those foreign genes *(19)*. In fact, the jury is still out on such a conclusion. To be sure, changing the A+T ratio of a foreign gene to one more similar to the *P. pastoris* genome can improve expression. It should be noted that when one changes to codons frequently utilized by the yeast, one is simultaneously also changing the A+T content of the gene to be more in line with *P. pastoris* expressed genes. However, the affect may be more related to the shift in A+T ratio than codon usage. In Sinclair and Choy *(20)*, investigators optimized codon usage of a gene. As a control they made a second gene in which the A+T ratio was changed to a similar extent but the specific codons utilized were not those thought optimal for *P. pastoris* to the extent possible. The result was that either modified gene improved expression to about the same degree, indicating that the A+T shift and not codon optimization was primarily responsible for the improved expression result. The conclusions were that optimization for A+T content is important and that it is not yet clear whether codon optimization itself actually improves expression of a gene is this system.

3. A Sensitive Assay is Critical

A major difference between the *P. pastoris* system and *E. coli* is the level of recombinant protein typically observed in early phases of development. With *E. coli*, one can typically count on 10% or greater of total protein being recombinant product. As a consequence, one can expect to observe the product as a band on a stained gel. This is typically not true for *P. pastoris*, especially in the early development phases where intracellular protein yields of less than 1% of total protein are common for secreted proteins and where proteinases may initially degrade most product before it can be visualized. Thus, one must have a sensitive assay for observing product in these early stages. This can be a functional (enzymatic) assay or a polyclonal antibody preparation against the protein product or the presence of an epitope tag fused to the product for visualization by a western blot-based technique. However, going into an expression effort with *P. pastoris* without such an assay is almost certainly a recipe for frustration and disappointment.

4. Secretion Using the α Factor Pre Pro Signal

P. pastoris offers the potential of secretion of a recombinant protein product, a significant advantage when possible as the yeast secretes only low levels of its own proteins. Thus, the recombinant protein is often virtually the only protein in the culture medium and the major purification step is simply the removal of the spent cells from the medium. In making the decision of whether to try and secrete a recombinant protein, the general rule of thumb is proteins that are normally made intracellularly in their native host should be produced intracellularly in *P. pastoris* and proteins that are normally secreted should be secreted. However, there have been a few successful exceptions when a normally intracellular foreign protein has been secreted from *P. pastoris* (*21*). Thus, you might get lucky secreting a normally intracellular recombinant protein from *P. pastoris*, but do not count on it.

A potential problem with secretion in this yeast is that the native signal sequence of a foreign may or may not function properly. Typically, the most convenient way of investigating this is to make two constructs. One in which sequences encoding native signal sequence of the recombinant protein remain. The other construct contains a foreign signal sequence fused in frame to the mature protein of the recombinant protein. The most popular foreign signal sequence to use for *P. pastoris* secretion is the *Saccharomyces cerevisiae* α mating factor pre pro (αMF) signal. This signal is the one most frequently utilized and has been remarkably successful in the secretion of a wide range of foreign proteins from *P. pastoris* cells. Furthermore, it has been incorporated into a wide range of *P. pastoris* vectors (*see* Chapter 2 and *[22]*). However, one must plan constructions properly and be prepared for potential complications with this signal.

The first complication with αMF vectors is that the only restriction site available in most for fusing sequences encoding the αMF signal to the recombinant protein in a manner that does not result in a final product with two or more nonauthentic amino acids added at the N-terminus is an *Xho*I site located just upstream (5') of the Kex2 dibasic endoproteinase cleavage site in the αMF DNA sequence (*see* **Fig. 2**). If a recombinant protein product with an authentic N-terminus is essential (as it typically is for most human pharmaceutical uses), you must take advantage of this site by preparing your recombinant gene insert with a 3'-*Xho*I (or compatible *Sal*I) restriction site and also include sequences that "build back" the *Kex*2 cleavage site of αMF. If a few additional N-terminal amino acids are no problem then any of the restriction sites located down stream of the Kex2 site are acceptable, as long as you make sure your construct keeps your gene in-frame with the signal sequence.

A second complication in using the αMF signal involves problems in its processing from the recombinant protein. The αMF signal has proven to be a potent

pPICZα A MCS

5′ end of *AOX1* mRNA   5′ *AOX1* priming site

```
811   AACCTTTTTT TTTATCATCA TTATTAGCTT ACTTTCATAA TTGCGACTGG TTCCAATTGA

871   CAAGCTTTTG ATTTTAACGA CTTTTAACGA CAACTTGAGA AGATCAAAAA ACAACTAATT

931   ATTCGAAACG ATG AGA TTT CCT TCA ATT TTT ACT GCT GTT TTA TTC GCA GCA
                 Met Arg Phe Pro Ser Ile Phe Thr Ala Val Leu Phe Ala Ala

983   TCC TCC GCA TTA GCT GCT CCA GTC AAC ACT ACA ACA GAA GAT GAA ACG GCA
      Ser Ser Ala Leu Ala Ala Pro Val Asn Thr Thr Thr Glu Asp Glu Thr Ala
```
                          α-factor signal sequence
```
1034  CAA ATT CCG GCT GAA GCT GTC ATC GGT TAC TCA GAT TTA GAA GGG GAT TTC
      Gln Ile Pro Ala Glu Ala Val Ile Gly Tyr Ser Asp Leu Glu Gly Asp Phe

1085  GAT GTT GCT GTT TTG CCA TTT TCC AAC AGC ACA AAT AAC GGG TTA TTG TTT
      Asp Val Ala Val Leu Pro Phe Ser Asn Ser Thr Asn Asn Gly Leu Leu Phe
```
                                                              Xho I*
```
1136  ATA AAT ACT ACT ATT GCC AGC ATT GCT GCT AAA GAA GAA GGG GTA TCT CTC
      Ile Asn Thr Thr Ile Ala Ser Ile Ala Ala Lys Glu Glu Gly Val Ser Leu
```
          Kex2 signal cleavage   *EcoR* I   *Pml* I   *Sfi* I   *BsmB* I *Asp*718 I
```
1187  GAG AAA AGA GAG GCT GAA GCT GAATTCAC GTGGCCCAG CCGGCCGTC TCGGATCGGT
      Glu Lys Arg Glu Ala Glu Ala Glu Ala
```
                          Ste13 signal cleavage
Kpn I *Xho* I   *Sac* II *Not* I   *Xba* I   *c-myc* epitope
```
1244  ACCTCGAGCC GCGGCGGCC GCCAGCTTTC TA GAA CAA AAA CTC ATC TCA GAA GAG
                                         Glu Gln Lys Leu Ile Ser Glu Glu
```
                              polyhistidine tag
```
1299  GAT CTG AAT AGC GCC GTC GAC CAT CAT CAT CAT CAT CAT TGA GTTTGTAGCC
      Asp Leu Asn Ser Ala Val Asp His His His His His His ***

1351  TTAGACATGA CTGTTCCTCA GTTCAAGTTG GGCACTTACG AGAAGACCGG TCTTGCTAGA
```
                          3′ *AOX1* priming site
```
1411  TTCTAATCAA GAGGATGTCA GAATGCCATT TGCCTGAGAG ATGCAGGCTT CATTTTTGAT
```
                              3′ polyadenylation site
```
1471  ACTTTTTTAT TTGTAACCTA TATAGTATAG GATTTTTTTT GTCATTTTGT TTCTTCTCGT
```

Fig. 2. Diagram of multiple cloning site (MCS) of vector pPICZα A showing point of insertion for recombinant genes.

secretion signal in that it will successfully direct secretion of most recombinant proteins into the culture medium. However, the proteolytic processing of the signal has been more problematic. For a significant number of recombinant proteins, processing at the αMF signal *Kex*2 site does not occur, leaving an N-terminus with various portions of the αMF signal still attached. Adding the αMF *Ste*13 cleavage sites (Glu-Ala-Glu-Ala) to the construct just after the *Kex*2 site (their normal position in αMF) often results in proper *Kex*2 processing at its dibasic cleavage site but not the Glu-Alas *(23)*. The impression is that the N-teminus of some recombinant proteins fold into a structure that prevents processing enzymes from reaching their processing site when located immediately adjacent to the N-terminus of the protein. Addition of the Glu-Alas extends the dibasic site so that *Kex*2 can efficiently process it but the Glu-Ala sequences then become unreachable by the *Ste*13 processing protein *(23)*. Unfortunately, when such a processing problem occurs, there is not a known uniform solution at hand. One must simply try a variety of other signal sequences to find one that both works in terms of secretion of the recombinant protein and is properly processed. One signal that has been purported to work in this situation is the plant PHA-E signal *(24)*. However, there are not a lot of examples of the use of this signal to secrete recombinant proteins and be properly processed.

5. Vector and Strain Selection

There are now a significant variety of vectors and strains to choose from for expression in *P. pastoris* (*see* Chapter 2). Invitrogen sells some of the basic vectors and strains including the popular pPICZ, Zeocin resistance selection vector series *(22)*. Additional auxotrophically marked *P. pastoris* host strains and complementing biosynthetic-gene containing vectors are available for my lab *(1)*.

With regard to marked strains, it is usually best to plan for your final expression strain to be prototrophic. The alternative of leaving a nutritional marker in the final strain is likely to create a problem for culturing in high-density fermentors as feeding nutrients to these cultures while at the same time fully inducing expression from an *AOX1* promoter construct is difficult.

A second consideration is whether or not to use a protease-defective strain of *P. pastoris* as a host. Although a number of examples now exist in the literature demonstrating convincingly that a protease-defective strain (e.g., a *pep4* mutant) can significantly improve levels of intact recombinant protein *(25)*, these examples still reflect a minority (~10%) of all recombinant proteins. Furthermore, there are distinct disadvantages to using one of the *pep4* defective strains as a host. These strains are not nearly as healthy as *PEP4* proficient strains (e.g., slower growth on all carbon sources, more difficult to transition to methanol and lower levels of expression, more rapid decline and death on agar medium plates, lower DNA-mediated transformation frequencies, and a greatly reduced ability mate with other *P. pastoris* strains). Thus, it is not a good idea to utilize a *P. pastoris pep4* mutant strain as a first choice prophylactically. For intracellular proteins, there has yet to be published an example where a *pep4* host improved expression. For secreted proteins, if proteolytic sensitivity is not a problem for your protein, then use of a *pep4* mutant host will likely reduce your expression levels and lead to a production system that is unnecessarily difficult to handle.

6. Finding the Best Expression Strain

The *P. pastoris* system differs from bacterial and many other expression systems in that autonomously replicating vectors are generally not used for expression and vectors are integrated into the genome of the host. Like *S. cerevisiae*, such integration events happen during transformation and because the organism has a dominant homologous recombination system, virtually all integration events occur between regions of significant (>0.5 kb) homology between one or more sequences on the vector and the host genome (e.g., between the *AOX1* promoter sequences) *(26)*. An advantage of homologous recombination is that one can know with certainty where an expression vector has integrated without too much work. In addition, and as opposed to higher eukaryotic genomes, the yeast genome seems to have few regions where transcription is silent or repressed. Thus, where a vector is integrated does not seem to play a major role in the level of transcription.

One of the most useful strategies for increasing expression of a specific gene in *P. pastoris* is by increasing the number of expression vectors or expression cassettes in a host strain *(27)*. To be sure, this strategy is not universal in that a number of genes have now been identified whose expression is not increased with increasing copies of an expression vector *(28)*. However, this strategy is successful frequently enough that it should be at the top of your list of strategies to increase expression further. Three basic methods have been described to accomplish this feat. The first is to construct a single vector with multiple headto-tail copies of an expression cassette (i.e., a fragment containing the promoter-foreign gene-transcriptional terminator) *(29)*. The second is to simply construct multiple expression vectors each with a different selectable marker gene and to transform each vector into a suitable *P. pastoris* host. The third and by far the most popular is to construct a single expression cassette into a vector that has, as its selectable marker, one of several drug resistant marker genes *(27)*. These marker genes include the bacterial Zeocin (Zeo^R) *(30)*, Blasticidin (Bla^R) *(22)*, or Geneticin ($G418^R$) *(27)* resistance genes or the *P. pastoris* formaldehyde dehydrogenease (*FLD1*) gene *(31)*. All four of these genes work in a similar manner. *P. pastoris* competent cells are transformed with an appropriate vector and then spread on an agar plate with medium containing a high level of the appropriate antibiotic (or in the case of vectors with the *FLD1* gene, formaldehyde). An important point in using any of these multicopy selection systems with a high level of drug is that multicopy vector containing transformants still represent a minority (~10%) of transformed colonies. Thus, to be successful in finding one or more multicopy strains, a researcher must collect on the order of 100 high-level drug resistant transformed colonies to have a high probability of identifying at least one or a few that have greater than 5 copies of the expression vector.

7. Conclusion

Success in using the *P. pastoris* expression system requires a little planning and work, along with knowledge of how the system works. The above represents the wisdom of more than 20 yr of working with the system and as a consultant with many companies and individuals that have used the system. It is a relatively short and quick read that will keep the inexperienced out of most of the potential traps that can be a source of frustration and a major waste of time and effort.

Acknowledgments

I thank Ms. Deborah Flynn for help in writing this and other chapters in this book and for making sure they were all formatted correctly. The writing of this manuscript was supported in part by grants from the US Department of Energy, Office of Biosciences (DE-FG02-03ER15407), and the National Institutes of Health (DK067371) to J.M.C.

References

1. Keck Graduate Institute, Faculty and Research, James M. Cregg, Resources. http://faculty.kgi.edu/cregg/index.htm. Last accessed on May 8, 2007.
2. Couderc, R. and Barratti, J. (1980) Oxidation of methanol by the yeast *Pichia pastoris*: purification and properties of alcohol oxidase. *Agric. Biol. Chem.* **44,** 2279–2289.
3. Tschopp, J. F., Brust, P. F., Cregg, J. M., Stillman, C. and Gingeras, T. R. (1987) Expression of the *lacZ* Gene from Two Methanol Regulated Promoters in *Pichia pastoris*. *Nuc. Acids Res.* **15,** 3859–3876.
4. Subramani, S., Koller, A., and Snyder, W. B. (2000) Import of peroxisomal matrix and membrane proteins. *Annu. Rev. Biochem.* **69,** 399–418.
5. Farre, J. C. and Subramani, S. (2004) Peroxisome turnover by micropexophagy. *Trends Cell Biol.* **14,** 515–523.
6. Soderholm, J., Bhattacharyya, D., Strongin, D., et al. (2004) The transitional ER localization mechanism of Pichia pastoris Sec12. *Dev. Cell* **6,** 649–659.
7. Romanos, M. A., Scorer, C. A., and Clare, J. J. (1992) Foreign gene expression in yeast: a review. *Yeast* **8,** 423–488.
8. Higgins, D. R. and Cregg, J. M. (eds.) (1998) *Methods in Molecular Biology*: Pichia *Protocols,* vol. 103, Humana Press, Totowa, NJ.
9. Cregg, J. M., Vedvick, T. S., and Raschke, W. C. (1993) Recent Advances in the Expression of Foreign Genes in *Pichia pastoris*. *Bio/Technology* **11,** 905–910.
10. Romanos, M. (1995) Advances in the use of *Pichia pastoris* for high-level expression. *Curr. Opin. Biotech.* **6,** 527–533.
11. Nico-Farber, K., Harder, W., AB, G., and Veenhuis, M. (1995) Review: Methylotrophic yeasts as factories for the production of foreign proteins. *Yeast* **11,** 1331–1344.
12. Cregg, J. M. (1999) Expression in the methylotrophic yeast *Pichia pastoris*, in *Gene Expression Systems: Using Nature for the Art of Expression* (Fernandez, J. and Hoeffler, J., eds.), Academic Press, San Diego, CA, pp. 157–191.

13. Lin Cereghino, G. P. and Cregg, J. M. (1999) Applications of yeast in biotechnology: protein production and genetic analysis. *Curr. Opin. Biotechnol.* **10,** 422–427.
14. Lin Cereghino, J. and Cregg, J. M. (2000) Heterologous protein expression in the methylotrophic yeast *Pichia pastoris. FEMS Microbiol. Rev.* **24,** 45–66.
15. Cregg, J. M., Lin Cereghino, J., Shi, J., and Higgins, D. R. (2000) Recombinant protein expression in *Pichia pastoris. Mol. Biotechnol.* **16,** 23–52.
16. Lin Cereghino, G. P., Sunga, A. J., Lin Cereghino, J., and Cregg, J. M. (2001) Expression of foreign genes in the yeast *Pichia pastoris*, in *Genetic Engineering: Principles and Methods,* vol. 23. (Setlow, J., ed.), Kluwer Academic/Plenum Publishers, New York, NY, pp. 157–169.
17. Lin Cereghino, G. P., Lin Cereghino, J., Ilgen, C., and Cregg, J. M. (2002) Production of recombinant proteins in fermenter cultures of the yeast *Pichia pastoris. Curr. Op. Biotech.* **13,** 329–332.
18. Ilgen, C., Cereghino, J. L., and Cregg, J. M. (2004) Chapter 7: *Pichia pastoris*, in *Production of Recombinant Proteins: Microbial and Eukaryotic Expression Systems* (Gellissen, G., ed.), Wiley-VCH Verlag, Weinheim, Germany, pp. 143–162.
19. Gurkan, C., and Ellar, D. J. (2003) Expression of the Bacillus thuringiensis Cyt2Aa1 toxin in *Pichia pastoris* using a synthetic gene construct. *Biotechnol. Appl. Biochem.* **38,** 25–33.
20. Sinclair, G. and Choy, F. Y. (2002) Synonymous codon usage bias and the expression of human glucocerebrosidase in the methylotrophic yeast, *Pichia pastoris. Protein Expr. Purif.* **26,** 96–105.
21. Su, D. and Wilson, J. E. (2002) Purification of the Type II and Type III isozymes of rat hexokinase, expressed in yeast. *Protein Expr. Purif.* **24,** 83–89.
22. http://www.invitrogen.com. Last accessed May 8, 2007.
23. Vedvick, T., Buckholz, R. G., Engel, M., Urcan, M., Kinney, J., Provow, S., Siegel, R. S., and Thill, G. P. (1991) High-level secretion of biologically active aprotinin from the yeast *Pichia pastoris. J. Ind. Microbiol.* **7,** 197–202.
24. Raemarkers, R. J., de Muro, L., Gatehouse, J. A., Fordham-Skelton, A. P. (1999) Functional phytohemagglutinin (PHA) and *Galanthus nivalis* agglutinin (GNA) expressed in *Pichia pastoris*: correct N-terminal processing and secretion of heterologous proteins expressed using the PHA-E signal peptide. *Eur. J. Biochem.* **265,** 394–403.
25. White, C. E., Hunter, M. J., Meininger, D. P., White, L. R., and Komives, E. A. (1996) Large scale expression, purification and characterization of the smallest active fragment of thrombomodulin: the roles of the sixth domain and of methionine-388. *Protein Eng.* **8,** 1177–1187.
26. Cregg, J. M., Barringer, K. L., Hessler, A. Y., and Madden, K. R. (1985) *Pichia pastoris* as a host system for transformations. *Mol. Cell Biol.* **5,** 3376–3385.
27. Scorer, C. A., Clare, J. J., McCombie, W. R., Romanos, M. A., and Sreekrishna, K. (1994) Rapid selection using G418 of high copy number transformants of *Pichia pastoris* for high-level foreign gene expression. *Bio/Technology* **12,** 181–184.
28. Thill, G. P., Davis, G. R., Stillman, C., et al. (1991) Positive and negative effects of multi-copy integrated expression vectors on protein expression in *Pichia pastoris*, in *Proceedings of the 6th International Symposium on Genetics of Microorganisms,*

vol. II (Heslot, H., Davies, J., Florent, J., Bobichon, L., Durand, G., and Penasse, L., eds.) Societe Francaise de Microbiologie, Paris, pp. 477–490.

29. Brierley, R. A. (1998) Secretion of recombinant human insulin-like growth factor (IGF-1), in *Methods in Molecular Biology:* Pichia *Protocols*, vol. 103, (Higgins, D. R., and Cregg, J. M., eds.), Humana Press, Totowa, NJ, pp. 149–177.

30. Higgins, D. R., Busser, K., Comiskey, J., Whittier, P. S., Purcell, T. J., and Hoeffler, J. P. (1998) Small vectors for expression based on dominant drug resistance with direct multicopy selection, in *Methods in Molecular Biology:* Pichia *Protocols*, vol. 103, (Higgins, D. R., and Cregg, J. M., eds.), Humana Press, Totowa, NJ, pp. 41–53.

31. Sunga, A. J. and Cregg, J. M. (2004) The *Pichia pastoris* formaldehyde dehydrogenase gene (*FLD1*) as a marker for selection of multicopy expression strains of *P. pastoris. Gene* **330,** 39–47.

2

Vectors and Strains for Expression

Joan Lin-Cereghino and Geoff P. Lin-Cereghino

Abstract

Selection of both an appropriate expression vector and corresponding strain is crucial for successful expression of heterologous proteins in *Pichia pastoris*. This chapter explores both the standard and new vector/strain options available for protein expression in this yeast. Incorporated into expression vectors are selectable markers based on biosynthetic pathway genes, dominant drug resistance, or the *P. pastoris* formaldehyde dehydrogenase gene (*FLD1*). Novel strains available for expression include those that increase secretion of heterologous protein by overexpressing eukaryotic protein disulfide isomerase, and those that decrease hyperglycosylation or provide human-type glycosylation. This chapter also discusses methods to create multicopy strains that will potentially provide optimized expression of recombinant proteins in *P. pastoris*.

Key Words: Expression vectors; plasmids; expression strains; heterologous protein expression in *Pichia pastoris*.

1. Introduction

Fundamental to the generation of heterologous proteins in *Pichia pastoris* is construction of expression strains. This requires the selection of both an appropriate expression vector and a corresponding strain. Although many of the several hundred reports describing the use of this methylotrophic yeast utilize commonly available vectors and strains *(1)*, several new alternatives have been developed in recent years. New tools include selectable marker/auxotrophic strains based on the biosynthetic genes—*ARG4, ADE1, URA3,* and *URA5* from *P. pastoris* that allow for the expression of several heterologous genes simultaneously *(2,3)*; a new dominant drug resistance marker based on the Blasticidin gene of *Aspergillus terreus*; and a novel selectable marker system based on the *P. pastoris* formaldehyde dehydrogenase gene (*FLD1*) for DNA-mediated transformation and selection of multicopy expressions strains *(4)*.

From: *Methods in Molecular Biology, vol. 389:* Pichia *Protocols, Second Edition*
Edited by: J. M. Cregg © Humana Press Inc., Totowa, NJ

Additional strains include mutated *P. pastoris* strains that increase secretion of hetero-logous protein, by overexpressing eukaryotic protein disulfide isomerase (*PDI*) *(5)* and strains which decrease hyperglycosylation *(6)* or reconstruct an in vivo human glycosylation pathway *(7–9)*. This chapter explores the variety of vector and strain options for optimizing expression of heterologous proteins in *Pichia pastoris*.

2. Materials

2.1. Strains

All *P. pastoris* strains are derivatives of the wild-type strain Y-11430 from Northern Regional Research Laboratories, (NRRL; Peoria, IL). A listing of *P. pastoris* strains available for expression is shown in **Table 1**. Strain X-33, also a wild type, is GS115 (*his4*) transformed with the wild type *P. pastoris HIS4* gene. These strains require no supplementation for growth on minimal media.

2.1.1. Strains for Use With Biosynthetic Markers

Many expression strains are derived from the *his4* auxotrophic strain GS115 to allow for transformation with *HIS4*-based vectors. Several other strains are available which have different combinations of one of more auxotrophic mutations in the *ARG4, ADE1, HIS4, URA3,* and *URA5* biosynthetic genes of *P. pastoris (2,3)*. These strains allow for the selection of expression vectors containing the corresponding selectable marker gene upon transformation. Although the selec-tion of more than one expression cassette requires sequential transformations, the availability of multiple markers allows the concomitant expression of several heterologous genes.

2.1.2. Strains for Use With FLD Expression Vectors

The strain MS105 is also a *his4* auxotroph; however, it has an additional mutation in the *P. pastoris* formaldehyde dehydrogenase gene rendering the strain defective in the ability to grow on methanol as a carbon source or methy-lamine as a nitrogen source *(10)*. Additionally, *P. pastoris fld1* mutants have increased sensitivity to formaldehyde relative to wild-type cells. Vectors con-taining the *FLD1* marker can then be used to select multicopy expression strains by screening for transformed strains with high levels of resistance to formaldehyde *(4)*.

2.1.3. Methanol Utilization Mutants

Most *P. pastoris* strains grow on methanol at the same rate as wild-type strains (Mut[+], methanol utilization plus phenotype). However, because strains with *AOX* mutations are sometimes capable of producing higher levels of heterologous proteins *(11–13)*, strains with deletions in one or both *AOX* genes

Table 1
Selected *P. pastoris* Strains

Strain	Genotype	Reference
Y-11430	Wild type	Northern Regional Research Laboratories Peoria, IL
X-33	Wild type	Invitrogen
Auxotrophic strains		
GS115	*his4*	*37*
GS190	*arg4*	*38*
JC220	*ade1*	*37*
JC254	*ura3*	*38*
GS200	*arg4 his4*	*37*
JC227	*ade1 arg4*	*2*
JC304	*ade1 his4*	*2*
JC305	*ade1 ura3*	*2*
JC306	*arg4 ura3*	*2*
JC307	*his4 ura3*	*2*
JC300	*ade1 arg4 his4*	*2*
JC301	*ade1 his4 ura3*	*2*
JC302	*ade1 arg4 ura3*	*2*
JC303	*arg4 his4 ura3*	*2*
JC308	*ade1 arg4 his4 ura3*	*2*
YJN165	*ura5*	*3*
Protease-deficient strains		
KM71	*his4 arg4 aox1Δ::ScARG4*	*39*
MC100-3	*his4 arg4 aox1Δ::ScARG4 aox2Δ::Pphis4*	*15*
SMD1168	*his4 Δpep4::URA3 ura3*	*40*
SMD1165	*prb1 his4*	*40*
SMD1163	*pep4 prb1 his4*	*40*
SMD1168 *kex1::SUC2*	*Δpep4::URA3 Δ kex1::SUC2 his4 ura3*	*18*
Other strains		
GS241	*fld1*	*10*
MS105	*his4 fld1*	*10*

are available. An additional benefit of using these strains is that the large amounts of methanol routinely used for large-scale fermentations of Mut$^+$ strains are not required. KM71 (*his4 arg4 aox1Δ::ScARG4*) is a strain where a partially deleted *AOX1* has been replaced with the *Saccharomyces cerevisiae ARG4* gene *(14)*. This strain (Mut^s, methanol utilization slow phenotype) grows slowly on methanol because it must rely on the weaker *AOX2* for methanol metabolism. Another strain, MC100-3 (*his4 arg4 aox1Δ::ScARG4 aox2Δ::Pphis4*) is deleted for both *AOX1* and *AOX2* and therefore is totally incapable of growing on methanol (Mut⁻, methanol utilization minus phenotype) *(15)*. Both Mut^s and Mut⁻ strains, although compromised in the ability to metabolize methanol, retain the ability to induce high-level expression from the *AOX1* promoter *(11)*.

2.1.4. Protease Deficient Strains

Strains deficient in vacuolar proteases are sometimes effective at reducing protein degradation of recombinant proteins *(16,17)*. The protease-deficient strains SMD1163 (*his4 pep4 prb1*), SMD1165 (*his4 prb1*), and SMD1168 (*his4 pep4*) are particularly useful for expression of heterologous proteins in fermenter cultures because the combination of high-cell density and lysis of a small percentage of cells results in a relatively high concentration of vacuolar proteases such as proteinase A (*pep4*) and proteinase B (*prb1*). A derivative of SMD1168 in which the gene encoding the carboxypeptidase Kex1 has been disrupted with *S. cerevisiae SUC2* (SMD1168 *kex1::SUC2*), is also available. This strain was developed to inhibit carboxy-terminal proteolysis of lysines and arginines and led to purification of intact human endostatin after 40 h of fermentation *(18)*. It should be noted, however, that protease-deficient strains typically exhibit slower growth rates, lower transformation efficiencies, and lower viability and therefore should not be used unless other causes for low yields have been explored.

2.1.5. Glycosylation Mutants

In comparison with *S. cerevisiae*, *P. pastoris* recombinant proteins tend to be hyperglycosylated less frequently and do not have hyper-immunogenic terminal α-1,3-linked mannosylation *(19,20)*. However, differences in protein-linked carbohydrate synthesis between *P. pastoris* and humans have made *P. pastoris* synthesized proteins unsuitable for human pharmaceutical uses *(21)*. Therefore, two approaches to engineer human-type N-glycans on yeast recombinant proteins have been taken. Coexpression of endoplasmic reticulum-targeted *Trichoderma reesei* 1,2-α-D-mannosidase reduced by more than 85% the number of α-1,2-linked mannose residues on *P. pastoris* produced influenza virus haemagglutinin and *Trypanosoma cruzi* trans-sialidase. Additionally, the human-type glycan, $Man_5GlcNAc_2$, was the major oligosaccharide on the trans-sialidase *(6)*. Furthermore, in the "GlycoSwitch" system, strains have been created to produce uniform, small asparagines-linked glycans on any glycoprotein of interest. Any GS115 strain of interest can be converted to a GlycoSwitch strain yielding predominantly $Man_8GlcNAc_2$, $Man_5GlcNAc_2$, or $GlcNAcMan_5GlcNAc_2$ oligosaccharides, all of which have the same structure as processing intermediates of the mammalian N-glycosylation pathway.

In another effort to "humanize" *P. pastoris*, the strain YSH44 was constructed by eliminating the endogenous yeast glycosylation pathway and by using five eukaryotic proteins (mannosidases I and II, N-acetylglucosaminyl transferases I and II, and uridine 5′-diophosphate-*N*-acetyl glucosamine transporter) to establish a synthetic glycosylation pathway that generates the human oligosaccharide $GlcNAc_2Man_3GlcNAc_2$ *(7)*. Recent advances by GlycoFi (Lebanon, NH) report further development of engineered strains blocked in dolichol oligosaccharide assembly. Deletion of the *PpALG3* gene encoding Dol-P-Man:$Man_5GlcNAc_2$-PP-Dol mannosyltransferase results in a strain which synthesizes homogeneous $GlcNAc_2Man_3GlcNAc_2$ N-glycans *(7)*.

2.1.6. PDI Overexpression Strains

Overexpression of *PDI* protein in *P. pastoris* has been found to increase production of disulfide rich proteins such as Pfs25H, a *Plasmodium falciparum* transmission blocking vaccine candidate *(5)*. *PDI* enzyme forms and arranges disulfide bonds during protein folding in the endoplasmic reticulum. In shake-flask and 5L bioreactor expression of secreted protein, *PDI* overexpression increased the secretion of Pfs25H two- to fourfold depending on the induction temperature.

2.2. Vectors

2.2.1. General Considerations

Although there exists a wide variety of host strains for heterologous expression, there are an even greater number of expression vectors available. All expression vectors are *E. coli/P. pastoris* shuttle vectors, containing an origin of replication for plasmid maintenance in *E. coli* and markers functional in one or both organisms. Most expression vectors have an expression cassette consisting of a multiple cloning site (MCS) for insertion of a foreign coding sequence, flanked by promoter and termination sequences derived from *AOX1*. However, optimal expression results are more likely if the ATG of the heterologous coding sequence is inserted into the first restriction site in most MCSs *(22)*. In addition, for secretion of heterologous proteins, vectors that include sequences designed to make in-frame fusions between foreign proteins and various secretion signals (such as the *S. cerevisiae* α-mating factor [MF] prepro peptide or the *P. pastoris* acid phosphatase secretion signal) can be constructed *(23)*. A listing of commonly used expression vectors is shown in **Table 2**.

2.2.2. Promoters

Most expression vectors rely on the *AOX1* promoter to drive expression, but there are times where this promoter may not be ideal. Use of methanol induction protocols may not be appropriate for the food industry because the

Table 2
Common *P. pastoris* Expression Vectors

Vector name	Selectable markers	Features	Reference
Vectors for intracellular expression			
pHIL-D2	*HIS4*	*Not*I sites for *AOX1* gene replacement	Invitrogen
pAO8015	*HIS4*	*Bam*HI and *Bgl*II sites flanking expression cassette for generation of multicopy expression vector	*41*
pPIC3.5K	*HIS4* and kan^r	Multiple cloning site for insertion of foreign genes; G418 selection for multicopy strains	*42*
pPICZ	ble^r	Multiple cloning site for insertion of foreign genes; potential for fusion of foreign protein to His$_6$ and *myc* epitope tags; Zeocin selection for multicopy strains	Invitrogen
pPIC6	bsd^r	Similar to pPICZ except blasticidin resistance used for direct selection of multicopy strains	Invitrogen
pGAPZ	ble^r	Expression controlled by constitutive *GAP* promoter: multiple cloning site for insertion of foreign genes; Zeocin selection for multicopy strains; potential for fusion of foreign protein to His$_6$ and *myc* epitope tags	Invitrogen, *26*
pFLD	ble^r	Similar to pGAPZ except expression controlled by *FLD1* promoter for inducible expression with methylamine	Invitrogen, *10*
pJL-IX	*FLD1*	*AOX1* promoter with *FLD1* as a selectable marker	*4*
pBLHIS-IX	*HIS4*	*AOX1p* vector series each with one of four biosynthetic gene markers: *HIS4, ARG4, ADE1, URA3*	*2*
pBLARG-IX	*ARG4*		
pBLADE-IX	*ADE1*		
pBLURA-IX	*URA3*		
Vectors for secreted expression			
pHIL-S1	*HIS4*	*AOX1p* fused to *PHO1* secretion signal; *Xho*I, *Eco*RI, and *Bam*HI site available for insertion of foreign genes	Invitrogen
pPIC9K	*HIS4* and kan^r	*AOX1p* fused to α-MF prepro signal sequence; *Eco*RI, *Not*I, *Sna*BI and *Avr*II sites available for insertion of foreign genes; G418 selection for multicopy strains	*42*

(Continued)

Table 2 *(Continued)*

Vector name	Selectable markers	Features	Reference
pPICZα	ble^r	*AOX1p* fused to α-MF prepro signal sequence; multiple cloning site for insertion of foreign genes; potential for fusion of foreign protein to His$_6$ and *myc* epitope tags; Zeocin selection for multicopy strains	Invitrogen
pPIC6α	bsd^r	Similar to pPICZα except blasticidin resistance used for direct selection of multicopy strains	Invitrogen
pGAPZα	ble^r	*GAPp* fused to α-MF prepro signal sequence multiple cloning site for insertion of foreign genes; potential for fusion of foreign protein to His$_6$ and *myc* epitope tags; Zeocin selection for multicopy strains	Invitrogen, **26**
pFLDα	ble^r	Similar to pGAPZα except expression controlled by *FLD1* promoter for inducible expression with methylamine	Invitrogen, **10**
pJLl-SX	*FLD1*	*AOX1p* fused to α-MF prepro signal sequence; *FLD1* as selectable marker	**4**
pBLHIS-SX	*HIS4*	Series of vectors with *AOX1p* fused to α-MF prepro signal each with one of four biosynthetic gene markers: *HIS4, ARG4, ADE1, URA3*	**2**
pBLARG-SX	*ARG4*		
pBLADE-SX	*ADE1*		
pBLURA-SX	*URA3*		

petroleum-related compounds used are health and fire hazards. Alternative promoters include the glucose-inducible strong constitutive *P. pastoris* glyceraldehyde-3-phosphate dehydrogenase *(GAP)* gene promoter, and the moderately expressing promoters from the *P. pastoris* genes for *PEX8* and *YPT1 (24–26)*. The *FLD1* gene encodes a glutathione-dependent formaldehyde dehydrogenase, a key enzyme required for the methanol and methylated amine metabolism. The *FLD1* gene promoter yields expression levels similar to the *AOX1* promoter using a β-lactamase reporter gene; however, P_{FLD1} offers flexible induction options as it can be induced with either methanol as a sole carbon source (and ammonium sulfate as a nitrogen source) or methylamine as a sole nitrogen source (and glucose as a carbon source) *(10)*. Vectors based on the *FLD1* gene, pFLD and pFLDα, are available from Invitrogen (http://www.invitrogen.com).

2.2.3. Secretion Signals

Heterologous proteins expression in *P. pastoris* can be produced either intracellularly or extracellularly. Because this yeast secretes low levels of endogenous proteins, secreted recombinant protein often constitutes a majority of the total protein in the medium. Therefore, directing a heterologous protein to the culture medium serves as substantial first step in purification. However, because of protein stability and folding requirements, the option of secretion is usually reserved for foreign proteins that are normally secreted by their native hosts. Using the appropriate *P. pastoris* expression vector, researchers can clone a foreign gene in frame with sequences encoding either the native secretion signal, the *S. cerevisiae,* α-MF prepro peptide, or the *P. pastoris* acid phosphatase (*PHO1*) signal.

Although several other different secretion signal sequences, including PHA-E from the plant lectin *Phaseolus vulgaris* agglutinin *(27)*, native signal sequences present on heterologous proteins, or synthetic leaders for more efficient processing and secretion *(28,29)* have been used successfully, results for a wide range of proteins have either been variable or unavailable. The *S. cerevisiae* α-MF prepro peptide, which is readily available on vectors such as the pPICZα A, B, C series from Invitrogen (*see* **Fig. 1**), has been used with the most consistent success. To construct an expression cassette with the added *S. cerevisiae* α-MF prepro peptide, researchers need only to clone their heterologous gene in frame with the secretion signal using restriction enzymes represented in the multiple cloning site of the appropriate vector. In some cases, α-MF is a better secretion signal for expression in *P. pastoris* than the leader sequence of the native heterologous protein, as was demonstrated with the expression of the lipase *Lip1* from *Candida rugosa (30)*.

2.2.4. Selectable Markers

For cloning purposes, all *P. pastoris* expression vectors are designed for transformation into both *E. coli* and *P. pastoris*. Maintenance in bacteria requires the presence of a selectable antibiotic resistance gene, which in the case of *P. pastoris* expression vectors is usually the *bla* gene encoding ampicillin resistance.

In **Subheading 2.1.1.**, *P. pastoris* biosynthetic pathway genes from the yeasts, *S. cerevisiae* and *P. pastoris,* were described. A total of five biosynthetic markers are readily available in expression vectors. They include: *ADE1* (PR-amidoimidazolesuccinocarboxamide synthase)*, HIS4* (histidinol dehydrogenase)*, ARG4* (arginosuccinate lyase), *URA3* (orotidine-5′-phosphate decarboxylase), and *URA5* (orotate-phosphoribosyltransferase) from *P. pastoris (2,3,21)*.

A number of expression vectors, available from Invitrogen, rely on a single resistance marker for selection in both bacteria and yeast. These vectors take advantage of either the *Sh ble* gene from *Streptoalloteichus hindustanus,* that confers resistance to the bleomycin-related drug Zeocin *(31)*, or the Blasticidin

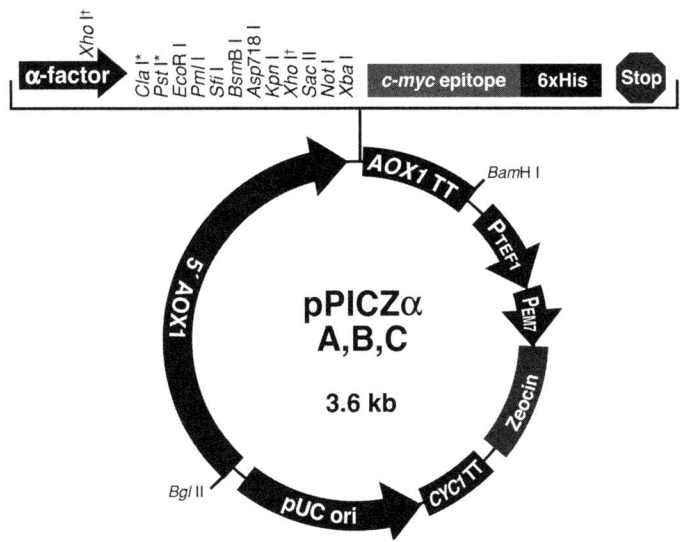

Fig. 1. *Pichia pastoris* vector pPICZαA, B, C. Multiple cloning site (MCS) is flanked by *S. cerevisiae* α-MF secretion signal, *c-myc*, and 6xHIS tags for immunological detection and purification. The *TEF*1 promoter and *CYC1*promoter are from *S. cerevisiae. EM7* promoter is from *E. coli. Bam*HI and *Bgl*II sites allow for cloning of multiple head-to-tail expression cassettes. (Reprinted with permission from Invitrogen Corporation © 2006.)

S deaminase gene, *BSD,* from *Aspergillus terreus*, which confers resistance to the nucleoside antibiotic blasticidin S HCl isolated from *Streptomyces griseochromogenes (32)*. Use of these resistance markers reduces by almost half the size of the expression vectors. At the present time, vectors containing the Zeocin resistance gene, such as the pPICZαA, B, C series (*see* **Fig. 1**), appear to be more popular than the *Bsd^r* vectors. Although the blasticidin resistance vectors appear to be comparable with Zeocin resistance vectors, there are relatively few reports of their use.

A novel set of expression vectors using the *P. pastoris* gene for formaldehyde dehydrogenase (*FLD1*) as selectable markers allow researchers to make expression vectors composed almost entirely of *P. pastoris* DNA (except for the heterologous gene to be expressed) *(4)*. These vectors allow for selection of multicopy transformants and are devoid of prokaryotic antibiotic resistance genes or other bacterial sequences.

3. Methods

3.1. Integration of Expression Vectors Into the P. pastoris *Genome*

Integration of expression vectors into the *P. pastoris* maximizes the stability of expression strains. The easiest integration method is to restrict the expression vector at a unique site in either the biosynthetic marker gene if present (e.g., a *P. pastoris HIS4* gene), or in the *AOX1* promoter fragment. The resulting linear DNA can then be cleaned of salts and transformed into the appropriate strain. Homologous recombination events stimulated by the free DNA termini result in single crossover-type integration events and efficient recombination of expression vectors at a predetermined genomic locus.

3.2. Construction of Multicopy Expression Strains

There are several well-established methods for creating multicopy expression strains. One method is simply to look for naturally occurring multicopy strains by screening large numbers of transformants for multicopies of the heterologous gene through DNA hybridization techniques *(33)*, or by screening for high product expression levels using sodium dodecyl sulfate-polyacrylamide gel electrophoresis (SDS-PAGE), immunoblotting, or colony immunoblotting *(33–35)*.

A second method involves the construction of a vector with several head-to-tail copies of an expression cassette *(16)*. The advantage to this approach, particularly in the production of human pharmaceuticals, is that the exact number of expression cassettes is known and can be verified by DNA sequencing. Generation of this type of vector relies on having an expression cassette flanked by restriction sites that have complementary ends (e.g., *Bam*HI-*Bgl*II, *Sal*I–*Xho*I combinations) allowing for the repeated reinsertion of additional expression cassettes into a single cleavage site.

Other approaches utilize vectors that confer resistance to kanamycin and G418 (*Tn903kan*[r]), Zeocin (*Sh ble*), or *BSD*. Use of the kanamycin resistance gene requires initial selection of yeast transformants using a biosynthetic marker such as *HIS4* followed by replica-plating to G418 plates. By this approach, strains carrying up to 30 copies of an expression cassette have been isolated *(36)*.

3.2.1. Generation of Multicopy Strains Using Zeocin or Blasticidin Resistance

One of the most popular methods for generation of multicopy strains is through the use of vectors with the bacterial *Sh ble* gene, which confers resistance to the antibiotic Zeocin *(31)*. Unlike G418 selection, strains transformed with expression cassettes containing the Zeocin marker can be selected directly by resistance to the drug. Additionally, enrichment for multicopy expression transformation occurs simply by plating on increased concentrations of Zeocin in the selection plates. Selection for multicopy blasticidin transformants is very similar.

However, with any of the methods, most transformants still only contain a single vector copy, even at high levels of drug concentration. Therefore, a large number of transformants (50–100) must be subjected to further analysis of copy number and/or expression level to identify multicopy expression candidates. The multicopy strains, once isolated, are stable with standard microbial handling procedures and do not require continued drug selection on plates or in liquid medium.

To generate multicopy Zeocin or blasticidin resistant transformants, expression vectors are transformed into *P. pastoris* by electroporation. Electroporated yeast cells are then immediately resuspended in 1.0 mL of a 1:1 mixture of YPD and 1M sorbitol, and incubated for 3 to 5 h at 30°C in a shaking incubator (200 rpm). Transformants can then be selected on YPD plates containing increasing concentrations of either Zeocin (100–2000 µg/mL) or blasticidin (50–500 µg/mL). After incubation at 30°C for 3 to 5 d, transformants appearing on selective plates should be restreaked and analyzed further for copy number and heterologous protein expression.

3.2.2. Generation of Multicopy Strains Using Resistance to Formaldehyde

Recently, two vectors, pJL IX and pJL SX (*see* **Fig. 2**), were described that allow for selection using the *P. pastoris* *FLD1* gene for formaldehyde dehydrogenase in the strains MS105 (*his4 fld1*) or GS241 (*fld1*). Transformants are inititally selected on minimal medium plates containing 0.25% methylamine chloride in place of NH_4SO_4 as nitrogen source. Populations of these transformants are then enriched for multicopy strains by selection for increased resistance to formaldehyde. As with the drug resistance selections, it is necessary to examine from 50 to 100 high-formaldehyde resistant strains to find a few that contain multiple copies of the vector. An additional advantage to these *FLD1*-based vectors is that they have been created in such a way that, after transformation into yeast, they are completely devoid of antibiotic resistance or other sequences of bacterial origin. Therefore these vectors do not pose a potential biohazard problem *(4)*.

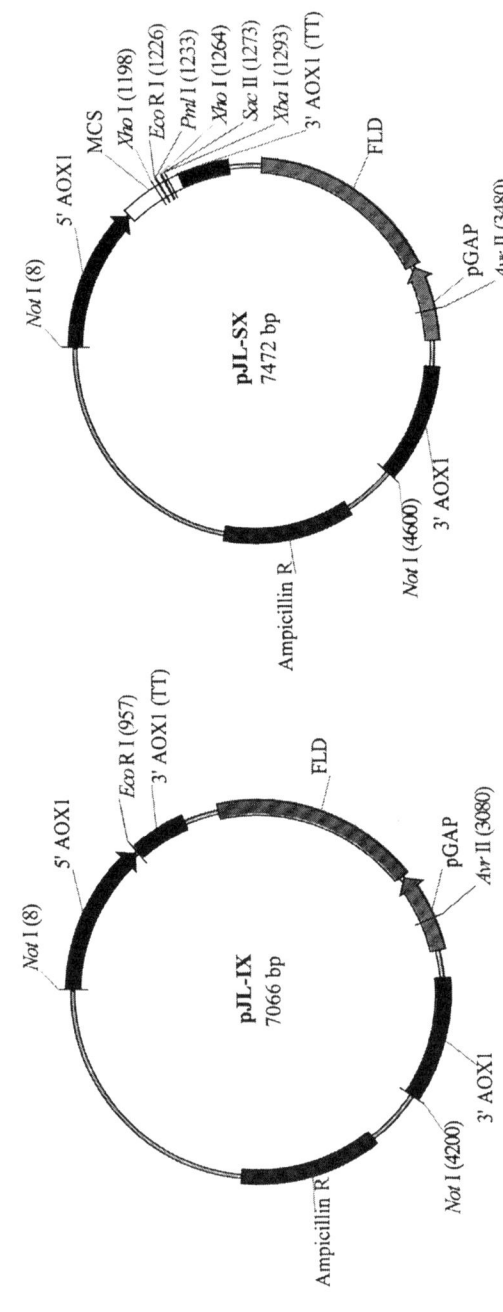

Fig. 2. FLD vectors for intracellular or secreted expression. *NotI* sites allow for transformation of expression cassette without any bacterial or antibiotic resistance sequences. (Plasmid diagrams courtesy of A. J. Sunga.)

22

References

1. Keck Graduate Institute, Faculty and Research, James M. Cregg, Resources. http://faculty.kgi.edu/cregg/index.htm. Last accessed May 8, 2007.
2. Lin Cereghino, G. P., Lin Cereghino, J., Sunga, A. J., et al. (2001) New selectable marker/auxotrophic host strain combinations for molecular genetic manipulation of *Pichia pastoris*. *Gene* **263,** 159–169.
3. Nett, J. H. and Gerngross, T. U. (2003) Cloning and disruption of the *PpURA5* gene and construction of a set of integration vectors for the stable genetic modification of *Pichia pastoris*. *Yeast* **20,** 1279–1290.
4. Sunga, A. J. and Cregg, J. M. (2004) The *Pichia pastoris* formaldehyde dehydrogenase gene (*FLD1*) as a marker for selection of multicopy expression strains of *P. pastoris*. *Gene* **330,** 39–47.
5. Tsai, C. W., Duggan, P. F., Shimp, R. L., Miller, L. H., and Narum, D. L. (2006) Overproduction of *Pichia pastoris* or *Plasmodium falciparum* protein disulfide isomerase affects expression, folding and O-linked glycosylation of a malaria vaccine candidate expressed in *P. pastoris*. *J. Biotechnol.* **121,** 458–470.
6. Callewaert, N., Laroy, W., Cadirgi, H., et al. (2001) Use of HDEL-tagged Trichoderma reesei mannosyl oligosaccharide 1,2-alpha-D-mannosidase for N-glycan engineering in *Pichia pastoris*. *FEBS Letters* **503,** 173–178.
7. Bobrowicz, P., Davidson, R. C., Li, H., et al. (2004) Engineering of an artificial glycosylation pathway blocked in core oligosaccharide assembly in the yeast *Pichia pastoris*: production of complex humanized glycoproteins with terminal galactose. *Glycobiology* **14,** 757–766.
8. Choi, B. K., Bobrowicz, P., Davidson, R. C., et al. (2003) Use of combinatorial genetic libraries to humanize N-linked glycosylation in the yeast *Pichia pastoris*. *Proc. Natl. Acad. Sci.* **100,** 5022–5027.
9. Hamilton, S. R., Bobrowicz, P., Bobrowica, B., et al. (2003) Production of complex human glycoproteins in yeast. *Science* **301,** 1244–1246.
10. Shen, S., Sulter, G., Jeffries, T. W., and Cregg, J. M. (1998) A strong nitrogen source-regulated promoter for controlled expression of foreign genes in the yeast *Pichia pastoris*. *Gene* **216,** 93–102.
11. Chiruvolu, V., Cregg, J. M., and Meagher, M. M. (1997) Recombinant protein expression in an alcohol oxidase-defective strain of *Pichia pastoris* in feedbatch fermentations. Enzyme Microbiol Technol. *Enzyme Microb. Technol.* **21,** 277–283.
12. Cregg, J. M., Tschopp, J. F., Stillman, C., et al. (1987) High level expression and efficient assembly of hepatitis B surface antigen in the methylotrophic yeast, *Pichia pastoris*. *Bio/Technology* **5,** 479–485.
13. Tschopp, J. F., Sverlow, G., Kosson, R., Craig, W., and Grinna, L. (1987) High level secretion of glycosylated invertase in the methylotrophic yeast, *Pichia pastoris*. *Bio/Technology* **5,** 1305–1308.
14. Cregg, J. M. and Madden, K. R. (1987) Development of yeast transformation systems and construction of methanol-utilization-defective mutants of *Pichia pastoris* by gene disruption, in). *Biological Research on Industrial Yeasts* (Stewart, G. G., Russell, I., Klein, R. D. and Hiebsch, R. R., eds.), CRC Press, Boca Raton, FL, pp. 1–18.

15. Cregg, J. M., Madden, K. R., Barringer, K. J., Thill, G. P., and Stillman, C. A. (1989) Functional characterization of the two alcohol oxidase genes from the yeast *Pichia pastoris*. *Mol. Cell. Biol.* **9,** 1316–1323.

16. Brierley, R. A. (1998) Secretion of recombinant human insulin-like growth factor I (IGF-I). *Methods Mol. Biol.* **103,** 149–177.

17. White, C. E., Hunter, M. J., Meininger, D. P., White, L. R., and Komives, E. A. (1995) Large scale expression, purification and characterization of the smallest active fragment of thrombomodulin: the roles of the sixth domain and of methionine-388. *Protein Eng.* **8,** 1177–1187.

18. Boehm, T., Pirie-Shepard, S., Trinh, L. B., Shiloach, J., and Folkman, J. (1999) Disruption of the *KEX1* gene in *Pichia pastoris* allows expression of full-length murine and human endostatin. *Yeast* **15,** 563–567.

19. Montesino, R., Garcia, R., Qunitero, O., and Cremata, J. A. (1998) Variation in N-linked oligosaccharide structures on heterologous proteins secreted by the methylotrophic yeast *Pichia pastoris*. *Protein Expr. Purif.* **14,** 197–207.

20. Verostek, M. F. and Trimble, R. B. (1995) Mannosyltransferase activities in membranes from various yeast strains. *Glycobiology* **5,** 671–681.

21. Romanos, M. A., Scorer, C. A., and Clare, J. J. (1992) Foreign gene expression in yeast: a review. *Yeast* **8,** 423–488.

22. Ellis, S. B., Brust, P. F., Koutz, P. J., Waters, A. F., Harpold, M. M., and Gingeras, T. R. (1985) Isolation of alcohol oxidase and two other methanol regulatable genes from the yeast *Pichia pastoris*. *Mol. Cell. Biol.* **5,** 1111–1121.

23. Lin Cereghino, J. and Cregg, J. M. (2001) Heterologous protein expression in the methylotrophic yeast *Pichia pastoris*. *FEMS Microbiology Reviews* **24,** 45–66.

24. Liu, H., Tan, X., Russell, K. A., Veenhuis, M., and Cregg, J. M. (1995) *PER3*, a gene required for peroxisome biogenesis in *Pichia pastoris*, encodes a peroxisomal membrane protein involved in protein import. *J. Biol.Chem.* **270,** 10,940–10,951.

25. Sears, I. B., O'Connor, J., Rossanese, O. W., and Glick, B. S. (1998) A versatile set of vectors for constitutive and regulated gene expression in *Pichia pastoris*. *Yeast* **14,** 783–790.

26. Waterham, H. R., Digan, M. E., Koutz, P. J., Lair, S. V., and Cregg, J. M. (1997) Isolation of the *Pichia pastoris* glyceraldehyde-3-phosphate dehydrogenase gene and regulation and use of its promoter. *Gene* **186,** 37–44.

27. Raemaeker, R. J. M., deMuro, L., Gatehouse, J. A., and Fordham-Skelton, A. P. (1999) Functional phytohemagglutinin (PHA) and *Galanthus nivalis* agglutinin (GNA) expressed in *Pichia pastoris* correct N-terminal processing and secretion of heterologous proteins expressed using the PHA-E signal peptide. *Eur. J. Biochem.* **65,** 394–403.

28. Kjeldsen, T., Pettersson, A. F., and Hatch, M. (1999) Secretory expression and characterization of insulin in *Pichia pastoris*. *Biotechnol. Appl. Biochem.* **29,** 79–86.

29. Martinez-Ruiz, A., Martinez del Pozo, A., Lacadena, J., et al. (1998) Secretion of recombinant pro- and mature fungal alpha-sarcin ribotoxin by the methylotrophic yeast *Pichia pastoris*: the Lys-Arg motif is required for maturation. *Protein Expr. Purif.* **12,** 315–322.

30. Brocca, S., Schmidt-Dannert, C., Lotti, M., Alberghina, L., and Schmid, R. D. (1998) Design, total synthesis, and functional overexpression of the *Candida rugosa lip1* gene coding for a major industrial lipase. *Protein Sci.* **7,** 1415–1422.

31. Higgins, D. R., Busser, K., Comiskey, J., Whittier, P. S., Purcell, T. J., and Hoeffler, J. P. (1998) Small vectors for expression based on dominant drug resistance with direct multicopy selection. *Method Mol. Biol.* **103,** 41–53.

32. Kimura, M., Takatsuki, A., and Yamaguchi, I. (1994) Blasticidin S deaminase gene from *Aspergillus terreus* (BSD): a new drug resistance gene for transfection of mammalian cells. *Biochim. Biophys. ACTA* **1219,** 653–659.

33. Romanos, M. A., Clare, J. J., Beesley, K. M., et al. (1991) Recombinant *Bordetella pertussis* pertactin (P69) from the yeast *Pichia pastoris*: high-level production and immunological properties. *Vaccine* **9,** 901–906.

34. Clare, J. J., Rayment, F. B., Ballantine, S. P., Sreekrishna, K., and Romanos, M. A. (1991) High-level expression of tetanus toxin fragment C in *Pichia pastoris* strains containing multiple tandem integrations of the gene. *Bio/Technol* **9,** 455–460.

35. Wung, J. L. and Gascoigne, N. R. (1996) Antibody screening for secreted proteins expressed in *Pichia pastoris*. *BioTechniques* **21,** 808–812.

36. Clare, J. J., Romanos, M. A., Rayment, F. B., et al. (1991) Production of mouse epidermal growth factor in yeast: high-level secretion using *Pichia pastoris* strains containing multiple gene copies. *Gene* **105,** 205–212.

37. Cregg, J. M., Barringer, K. J., Hessler, A. Y., and Madden, K. R. (1985) *Pichia pastoris* as a host system for transformations. *Mol.Cell. Biol.* **5,** 3376–3385.

38. Cregg, J. M., Shen, S., Johnson, M., and Waterham, H. R. (1998) Classical genetic manipulation, in *Pichia Protocols*, (Higgins, D. R., and Cregg, J. M. eds.), Humana Press, Totowa, NJ.

39. Tschopp, J. F., Brust, P. F., Cregg, J. M., Stillman, C. A., and Gingeras, T. R. (1987) Expression of the *lacZ* gene from two methanol-regulated promoters in *Pichia pastoris*. *Nuc. Acids Res.* **15,** 3859–3876.

40. Gleeson, M. A. G., White, C. E., Meninger, D. P., and Komives, E. A. (1998) Generation of protease-deficient strains and their use in heterologous protein expression. *Methods Mol. Biol.* **103,** 81–94.

41. Cregg, J. M., Vedvick, T. S., and Raschke, W. C. (1993) Recent advances in the expression of foreign genes in *Pichia pastoris*. *Bio/Technol* **11,** 905–910.

42. Scorer, C. A., Clare, J. J., McCombie, W. R., Romanos, M. A., and Sreekrishna, K. (1994) Rapid selection using G418 of high copy number transformants of *Pichia pastoris* for high-level foreign gene expression. *Bio/Technol* **12,** 181–184.

3

DNA-Mediated Transformation

James M. Cregg

Abstract

Several methods for DNA-mediated transformation of *Pichia pastoris* have been developed which vary in type of DNA that is transformable (e.g., linear versus circular) efficiency, cost, and labor and each is described in detail. As in *Saccharomyces cerevisiae*, gene replacement (also known as gene knock-out) methods provide a unique tool to investigate the function of specific *P. pastoris* genes. After construction, the function of the deleted gene is investigated from the phenotype of the mutant strain. In *S. cerevisiae*, an efficient polymerase chain reaction (PCR)-based method for the construction of gene replacement fragments has been developed. Modifications of this PCR method have been developed to adapt this approach to *P. pastoris*.

Key Words: *Pichia pastoris*; DNA-mediated transformation; gene knock-out; gene replacement; selectable markers.

1. Introduction
1.1. General Characteristics of the Methods

The key to the molecular-genetic manipulation of any organism is the ability to introduce and maintain DNA sequences of interest. For *Pichia pastoris*, the fate of introduced DNAs is generally similar to those described for *Saccharomyces cerevisiae*. Vectors can be maintained as autonomously replicating elements or integrated into the *P. pastoris* genome, if desired *(1–4)*. As in *S. cerevisiae*, integration events occur primarily by homologous recombination between sequences shared by the transforming vector and *P. pastoris* genome *(1)*. Thus, the controlled integration of vector sequences at preselected positions in the genome via gene targeting and gene replacement (gene knock-out) strategies are readily performed in *P. pastoris (2)*.

Four methods for introducing DNAs into *P. pastoris* have been described and vary with regard to convenience, transformation frequencies, and other

From: *Methods in Molecular Biology, vol. 389:* Pichia *Protocols, Second Edition*
Edited by: J. M. Cregg © Humana Press Inc., Totowa, NJ

Table 1
General Characteristics of *P. pastoris* Transformation Methods

Method	Transformation frequency (per µg)	Convenience factor	Multicopy integration?
Spheroplast	10^5	Low	Yes
Electroporation	10^5	High	Yes
PEG$_{1000}$*	10^3	High	No
LiCl*	10^2	High	No

*See **Note 11**.

characteristics (**Table 1**). With any of the four transformation procedures, it is possible to introduce vectors as autonomous elements or to integrate them into the *P. pastoris* genome. The spheroplast generation-polyethylene glycol-CaCl$_2$ (spheroplast) method is the best characterized of the techniques and yields a high frequency of transformants (~10^5/µg) but is laborious and results in transformed colonies that must be recovered from agar embedding (*1*). The other three methods utilize intact or whole cells (i.e., they do not involve spheroplasting) and, therefore are more convenient and result in transformants on the surface of agar plates which are easily picked or replica plated for further analysis. Of the whole-cell methods, electroporation yields transformants at frequencies comparable with those from spheroplasting and is the method of choice for most researchers. However, for laboratories that do not have access to an electroporation instrument, either of the other whole-cell procedures based on polyethylene glycol or alkali cations generates adequate numbers of transformants for most types of experiments and without the labor of spheroplasting. This chapter describes procedures for each of these four methods. However, to aid in understanding these methods, general features of *P. pastoris* vectors and host strains are described first.

1.2. P. pastoris Vectors

All *P. pastoris* vectors are of the shuttle type, (i.e., composed of sequences necessary to selectively grow and maintain them in either *Escherichia coli* or *P. pastoris* hosts). *P. pastoris* transformations most commonly involve either an auxotrophic mutant host and vectors containing a complementary biosynthetic gene or a dominant drug resistance gene and are described in Chapter 2 of this volume.

Autonomous replication of plasmids in *P. pastoris* requires the inclusion of a *P. pastoris*-specific autonomous replication sequence (PARS) (*1*). PARSs will maintain vectors as circular elements at an average copy number of approx 10/cell. However, relative to *S. cerevisiae*, *P. pastoris* appears to be particularly recombinogenic and, even with a PARS, any vector that also contains more than approx 0.5 kb of *P. pastoris* DNA will integrate into the *P. pastoris* genome at some point during the first ~100 generations after transformation. Thus, when cloning *P. pastoris* genes by functional complementation, it is important to recover the complementing vector from the yeast as soon as possible. Replication elements analogous to the *S. cerevisiae* 2μ circle plasmid or centromeres have not been described for *P. pastoris*.

1.3. Gene Replacement

In *P. pastoris*, the frequency of gene replacement events is highly dependent on the length of the terminal fragments responsible for proper targeting replacement vectors. The frequency of gene replacement events can be greater than 50% of the total transformant population when the targeting fragments are each greater than 1 kb but drops precipitously to less than 0.1% when their total length is less than 0.5 kb. Gene replacements can also be performed with linear vectors composed of only one terminal targeting fragment with the other targeting fragment located internally, although the frequency of replacement events which such vectors is substantially reduced.

Gene replacements have been performed with linear vectors that lack a selectable marker gene via cotransformation *(2)*. This is an especially useful technique in situations where the number of genome manipulations to be performed is greater than the number of selectable markers available for the host strain. For cotransformations, *P. pastoris* cells are simply transformed with a mixture of two DNA vectors: an autonomously replicating vector that contains a selectable marker, and an approx 10-fold excess of a linear gene replacement vector. Transformants are selected for the presence of the marker gene phenotype and then screened for ones that also receive the nonselected gene replacement vector. Typically, less than 1% of the transformants will have undergone a gene replacement event with the nonselected vector, a frequency sufficient to identify cotransformants by the phenotype conferred by the replacement event. After identifying a proper cotransformant, the autonomous vector is cured from the strain by growing it in a nonselective medium.

A polymerase chain reaction (PCR)-based method has made the construction of yeast gene replacement strains even easier because typically no recombinant DNA steps are required in the construction of the gene replacement fragment. The general procedure is to utilize a DNA fragment (or a plasmid containing a DNA fragment), which contains a selectable marker gene and a set of oligonucleotides that will amplify that selectable marker fragment and simultaneously add long (i.e., 50 bp or more) terminal sequences to the marker gene fragment that are complementary to the 5′- and 3′-flanking regions of the gene one wishes to delete. The PCR product is then transformed into a suitable strain of yeast, most often by electroporation. In *P. pastoris*, because the frequency of recombination with these relatively short complementary regions is significantly lower than in *S. cerevisiae*, "tricks" are useful to enhance the frequency of replacement events with these PCR products. One trick is to carry out the gene replacement transformation using a diploid strain of *P. pastoris* and then to subject the population of transformants to sporulation to haploidize them *(5)*. The haploid spore products are then screened for ones with the appropriate knock-out phenotype. For reasons that are not known, the frequency of gene replacement events is much higher with a diploid host relative to a haploid host. A second trick is to carry out a second round of PCR on the gene replacement fragment to lengthen the 5′ and 3′-complementary termini to approx 100 to 200 bp each *(6)*. The fragment with lengthened complementary regions is then transform into a suitable haploid *P. pastoris* strain and screened for the gene replacement event by phenotype as per usual.

2. Materials

2.1. Culture Medium

Prior to all transformation procedures, *P. pastoris* strains are cultured in YPD medium (1% yeast extract, 2% peptone, and 2% dextrose). Growth of *P. pastoris* in liquid media and on agar media plates is at 30°C unless otherwise specified.

2.2. Spheroplast Procedure

2.2.1. Stock Solutions

Prepare and autoclave the following:

1. 1 L H_2O.
2. $2M$ sorbitol (1 L).
3. 250 mM ethylene diamine tetraacetic acid (EDTA) (pH 8.0) (100 mL).
4. $1M$ Tris-HCl (pH 7.5) (100 mL).
5. 100 mM $CaCl_2$ (100 mL).
6. 10× yeast nitrogen base without amino acids (YNB) (6.7 g/100 mL).
7. 20% glucose (100 mL).
8. Regeneration agar: 55 g sorbitol, 6 g agar, and 240 mL H_2O (*see* **Notes 1** and **2**).

Prepare and filter sterilize the following solutions:

1. 100 mM Na citrate (pH 5.8) (100 mL).
2. 40% polyethylene glycol, 3350 average molecular weight (PEG_{3350}) (100 mL).
3. $0.5M$ dithiothreitol (DTT)(10 mL).
4. Prepare 1 mL of a 4 mg/mL solution of Zymolyase T100 (ICN, Costa Mesa, CA) and dispense in 100 µL aliquots into minicentrifuge tubes.

All of the above solutions can be stored at room temperature except for DTT and Zymolyase which must be stored at –20°C (*see* **Note 3**). All are stable for at least 1 yr.

2.2.2. Sterile Working Solutions

1. 1 *M* sorbitol (200 mL).
2. SCE: 1 *M* sorbitol, 10 m*M* Na citrate (pH 5.8), and 10 m*M* EDTA (pH 8.0) (100 mL).
3. CaS: 1 *M* sorbitol and 10 m*M* $CaCl_2$ (100 mL).
4. Polyethylene glycol (PEG)-CaT : 20% PEG, 10 m*M* $CaCl_2$ and 10 m*M* Tris-HCl, pH 7.5 (100 mL).
5. SOS: 1 *M* sorbitol, 0.3× YPD medium, and 10 m*M* $CaCl_2$ (20 mL).
6. SED: 1 *M* sorbitol, 25 m*M* EDTA (pH 8.0) and 50 m*M* DTT (10 mL). Prepare fresh and hold on ice until use.
7. Regeneration medium agar: 240 mL of regeneration agar (from stock remelted via autoclave, microwave, or boiling water bath), 30 mL 10X YNB, and 30 mL 20% glucose. Hold in a 45°C water bath until use. Prepare fresh the day of the transformation.

2.3. Electroporation

1. H_2O (1 L).
2. 1 *M* sorbitol (100 mL).
3. YND agar plates: 0.67% YNB, 2% glucose, and 2% agar (0.5 L/~25 plates).
4. 1*M* DTT (2.5 mL).
5. YPD medium (100 mL) with 20 mL 1 *M* HEPES buffer (pH 8.0).
6. Electroporation instrument (e.g., BTX Electro Cell Manipulator 600, BTX, San Diego, CA; Bio-Rad Gene Pulser, Bio-Rad, Hercules, CA; Electroporator II, Invitrogen, San Diego, CA).
7. Sterile electroporation cuvettes.

All solutions should be autoclaved except the DTT and HEPES solutions, which should be filter sterilized.

2.3.1. Rapid Heat-Shock/Electroporation

1. BEDS solution (9 mL): 10 m*M* bicine-NaOH (pH 8.3), 3% ethylene glycol, and 5% dimethyl sulfoxide (DMSO).
2. 1 *M* sorbitol supplemented with 1 mL 1.0*M* DTT.

2.4. Polyethylene Glycol Procedure

1. Buffer A: 1 *M* sorbitol, 10 m*M* Bicine (pH 8.35) and 3% ethylene glycol (100 mL).
2. Buffer B: 40% PEG 1000 average mol wt (PEG_{1000}) (*see* **Note 4**), and 0.2 *M* Bicine (pH 8.35) (50 mL).
3. Buffer C: 0.15 *M* NaCl, and 10 m*M* Bicine (pH 8.35) (50 mL). Filter sterilize and store at –20°C until use.

4. DMSO (*see* **Note 5**). Store at –70°C.
5. YND agar plates: 0.67% YNB, 2% glucose, and 2% agar (0.5 L/~25 plates).

2.5. Alkali Cation Procedure

Prepare and autoclave the following solutions:

1. H_2O (1 L).
2. TE buffer: 10 mM Tris-HCl (pH 7.4) and 1 mM EDTA (pH 8.0) (100 mL).
3. LiCl buffer: 0.1M LiCl, 10 mM Tris-HCl (pH 7.4) and 1 mM EDTA (pH 8.0) (100 mL).
4. PEG + LiCl buffer: 40% PEG_{3350}, 0.1M LiCl, 10 mM Tris-HCl (pH 7.4) and 1 mM EDTA (pH 8.0) (100 mL).
5. YND agar plates: 0.67% YNB, 2% glucose, and 2% agar (0.5 L/~25 plates).

3. Methods

3.1. DNA Preparation

For highest transformation frequencies, *E. coli-P. pastoris* shuttle vector DNAs should be pure and dissolved in water or Tris-EDTA (TE) buffer. Most standard procedures for purification of plasmid DNAs, such as those involving CsCl ethidium bromide centrifugation or a commercial plasmid preparation kit (e.g., QIAGEN; Hilden, Germany), work well. However, plasmids prepared by "mini-prep" methods, such as the alkaline lysis procedure, have lower transformation frequencies but are adequate for most purposes.

For gene targeting and gene replacement constructions, vectors should be digested with a restriction enzyme(s) that cuts within *P. pastoris* DNA sequences in the vector such that a minimum of 200 bp of *P. pastoris* DNA are available at each terminus to direct targeted integration. Prior to transformation, linear vectors should be extracted with phenol-chloroform-isoamyl alcohol (PCA) (25:24:1), alcohol precipitated, and dissolved in water or TE buffer. It is typically not necessary to separate a gene replacement fragment from the remaining nontransforming fragment. However, vector fragments purified from agarose gels via standard electroelution procedures or commercially available kits (e.g., QIAEX gel extraction kit; QIAGEN, Hilden, Germany) can be transformed into *P. pastoris*.

3.2. Spheroplast Procedure

This procedure is a modified version of that described by Hinnen et al. *(7)* and Cregg et al. *(1)*. The general characteristics of the method are summarized in **Table 1**.

3.2.1. Preparation of Spheroplasts

1. Inoculate a 10-mL YPD culture with a single *P. pastoris* colony of the strain to be transformed from a fresh agar plate and grow overnight with shaking. This culture can be stored at 4°C for several days.

2. Inoculate three 200-mL YPD cultures into 500-mL baffled culture flasks with 5, 10, and 20 µL of the above culture and incubate overnight with shaking.
3. In the morning, select the culture that has an OD_{600} of between 0.2 and 0.3 (*see* **Note 6**).
4. Wash the culture by centrifugation at 2000g in a 50-mL conical tube at room temperature once with 10 mL of water, once with 10 mL of freshly prepared SED, once with 10 mL of 1M sorbitol, and once with 10 mL of SCE.
5. Add between 1 and 10 µL of Zymolyase T100 and incubate at 30°C without shaking. To monitor spheroplasting, remove 100 µL aliquots of the cells before adding the Zymolyase and at times of 5, 10, 15, 20, 30, and 45 min after addition of the enzyme, and dispense them into a series of glass tubes containing 900 µL of 1% sodium dodecyl sulfate (SDS). After addition of each sample to the SDS solution, mix and visually examine the solution for cell lysis as judged by decreased turbidity and increased viscosity. The optimal time for spheroplasting is between 15 and 30 min. Adjust the amount of Zymolyase in future transformations accordingly (*see* **Note 7**). Proceed with the next step in the procedure as soon as spheroplasts appear ready.

3.2.2. Transformation

1. Wash the spheroplast preparation by centrifugation at 1500g for 10 min once in 10 mL of 1M sorbitol and once in 10 mL of CaS. Spheroplasts are very fragile. Therefore decant supernatants carefully and resuspend gently by tapping the side of the tube or by gentle pipetting.
2. Centrifuge the preparation and resuspend spheroplasts in 1.0 mL of CaS.
3. Dispense 100 µL aliquots of spheroplasts into 5 mL polypropylene snap-top Falcon (or similar) tubes. Add DNAs to each tube and incubate at room temperature for 20 min.
4. Add 1 mL of PEG-CaT to each tube and incubate an additional 15 min at room temperature.
5. Centrifuge samples at 1500g for 10 min, carefully decant PEG-CaT, and gently resuspend spheroplasts in 200 µL of SOS. Incubate samples at room temperature for 30 min and then add 800 µL of 1M sorbitol.

3.2.3. Plating

1. Prepare plates containing a 10 mL bottom layer of regeneration medium agar (one for each transformation sample, plus at least two additional bottom agar plates for viability testing).
2. Dispense a 10 mL aliquot of molten 45°C regeneration medium agar into a 50 mL polypropylene tube for each transformation sample. Gently mix transformation samples of between 10 µL and 0.5 mLwith the regeneration agar and pour over the bottom agar layer. After the agar solidifies (~10 min), incubate the plates for 4 to 7 d.
3. Monitor the quality of the spheroplast preparation and regeneration conditions as follows. Remove a 10 µL aliquot of one of the transformation samples and add to

a tube containing 990 μL of 1*M* sorbitol (10^{-2} dilution). Mix, remove a 10 μL sample of the 10^{-2} dilution, and dilute again by addition to a second tube containing 990 μL of 1*M* sorbitol (10^{-4} dilution). Spread 100 μL of each dilution on a YPD plate to determine the concentration of remaining whole (unspheroplasted) cells. To determine the concentration of spheroplasts with the potential to regenerate into viable cells, add a second 100 μL aliquot of each dilution to a tube containing 10 mL of molten regeneration medium agar supplemented with 50 μg/mL of the missing nutrient (e.g., histidine). Gently mix and pour over a bottom agar plate. Incubate these control plates with the transformation plates as described in **step 2**. Good spheroplast preparations and regeneration reagents will produce greater than 1×10^7 colonies/mL (i.e., >100 colonies on the 10^{-4} spheroplast regeneration plate) and less than 1×10^4 colonies/mL unspheroplasted whole cells (i.e., <10 colonies on the 10^{-2} remaining whole-cell plate).

3.2.4. Recovery of Transformants

Most transformant colonies will be embedded within the top agar. To recover individual embedded transformants, dig them out of the agar with an inoculation loop and streak for single colonies on the surface of an agar plate containing the selective medium. To recover embedded cells from a large proportion of the embedded colonies in a transformation plate for further analysis (e.g., to screen for ones that have undergone a gene replacement event), the following procedure can be used.

1. Scrape the colony-containing top agar layer into a sterile 50-mL centrifuge tube using a spatula or forceps (sterilized with 70% ethanol). Add 20 mL of sterile water and mix vigorously to break up the agar and release embedded cells.
2. Filter the suspension through 4 folds of sterile cheesecloth. Rinse cells from the agar by addition of approx 20 mL of water to the agar. Centrifuge the filtrate at 2000*g* for 5 min and decant.
3. Suspend the cells (and a small amount of remaining agar) in 5 mL of sterile water and vortex vigorously to disperse the cells.
4. Spread dilutions of the cells on selective medium agar plates at a concentration that will generate from100 to 500 colonies/plate and incubate for 2 to 3 d (*see* **Note 6**).
5. Replica plate onto an agar medium appropriate for identifying transformants that have undergone a gene replacement event. For example, when looking for transformants with a gene replacement at *AOX1* and, as a result, have acquired a methanolutilization-slow (Mut^s) phenotype, replica plate colonies from YND plates to two sets of plates, one containing YNB plus methanol and the other containing YND. After 1 or 2 d incubation, compare the two sets of plates and select colonies from the YND plate that are not growing well on the YNB plus methanol plate.

3.3. Electroporation

This procedure is a modified version of that described by Becker and Guarente *(8)*. The characteristics of the method are listed in **Table 1**. Parameters

Table 2
Parameters for Electroporation Using Selected Instruments

Instrument	Cuvette gap (mm)	Sample volume (µL)	Charging voltage (V)	Capacitance (µF)	Resistance (Ω)	Field strength (kV/cm)	Pulse length (~ms)	Ref.
ECM600* (BTX)	2	40	1500	Out	129	7500	5	*6*
Electroporator II (Invitrogen)	2	80	1500	50	200	7500	10	*9*
Gene-Pulser (Bio-Rad)	2	40	1500	25	200	7500	5	*10*
Cell-Porator (BRL)	1.5	20	480	10	Low	2670	NS	*11,12*

*For ECM600, select 2.5 kV/RESISTANCE high voltage.
Abbr: NS = not specified.

for electroporation with four different instruments are shown in **Table 2** (*see* **Note 8**).

3.3.1. Preparation of Competent Cells

1. Inoculate a 10 mL YPD culture with a single fresh *P. pastoris* colony of the strain to be transformed from an agar plate and grow overnight with shaking.
2. In the morning, use the overnight culture to inoculate a 500-mL YPD culture in a 2.8-L Fernback culture flask to a starting OD_{600} of 0.1 and grow to an OD_{600} of 1.0 (*see* **Note 6**).
3. Harvest the culture by centrifugation at 2000g at 4°C, and suspend the cells in 100 mL of YPD medium plus HEPES.
4. Add 2.5 mL of 1M DTT and gently mix.
5. Incubate at 30°C for 15 min.
6. Bring to 500 mL with cold water. Wash by centrifugation at 4°C once in 250 mL of cold water, once in 20 mL of cold 1M sorbitol and resuspend in 0.5 mL of cold 1M sorbitol. (Final volume including cells will be 1.0 to 1.5 mL.)
7. For highest frequencies, transform the cells directly without freezing as described below.
8. To freeze competent cells, distribute in 40 µL aliquots to sterile 1.5-mL minicentrifuge tubes, and place the tubes in a –70°C freezer until use (*see* **Note 9**).

3.3.2. Rapid Heat-Shock/Electroporation (*9*)

1. Grow 5 mL culture of *P. pastoris* cells in YPD overnight with shaking.
2. The next morning, dilute the overnight culture to an A_{600} 0.15 to 0.20 in a volume of 50 mL YPD medium in a flask large enough to provide good aeration.

3. Grow to an A_{600} of 0.8 to 1.0 with shaking (4–5 h).
4. Centrifuge cells at 500*g* for 5 min at room temperature then decant supernatant.
5. Suspend cells in 9 mL of ice-cold BEDS solution supplemented with DTT.
6. Incubate the cell suspension for 5 min at with shaking at 30°C.
7. Centrifuge cells at 500*g* for 5 min at room temperature and resuspend in 1 mL of BEDS (without DTT).
8. Immediately perform electroporation as described in **Subheading 3.3.3** below, or freeze cells in small aliquots at –80°C.

3.3.3. Electroporation

1. Mix up to 10 µg of DNA sample in no more than 5 µL total volume of water or TE buffer to a tube containing 40 µL of frozen or fresh competent cells and transfer to a 2-mm gap electroporation cuvette held on ice (*see* **Notes 8** and **9**).
2. Pulse cells according to the parameters suggested for yeast by the manufacturer of the specific electroporation instrument being used (**Table 2**).
3. Immediately add 1 mL of cold 1*M* sorbitol and transfer the cuvette contents to a sterile 1.5-mL minicentrifuge tube.
4. Spread selected aliquots onto agar plates containing YND or other selective medium and incubate for 2 to 4 d (*see* **Note 10**).

3.4. PEG Procedure

This procedure is a modified version of that described in Klebe et al. *(9)*. The characteristics of the method are listed in **Table 1** (*see* **Note 11**).

3.4.1. Preparation of Competent Cells

1. Inoculate a 10 mL YPD culture with a single fresh *P. pastoris* colony of the strain to be transformed from an agar plate and grow overnight with shaking.
2. In the morning, use the overnight culture to inoculate a 100 mL YPD culture to a starting OD_{600} of 0.1 and grow to an OD_{600} of 0.5 to 0.8 (*see* **Note 6**).
3. Harvest the culture by centrifugation at 2000*g* at room temperature and wash the cells once in 50 mL of Buffer A.
4. Resuspend cells in 4 mL of Buffer A and distribute in 0.2-mL aliquots to sterile 1.5-mL minicentrifuge tubes. Add 11 µL of DMSO to each tube, mix, and quickly freeze cells in a liquid nitrogen bath.
5. Store frozen competent cells at –70°C.

3.4.2. Transformation

1. Add up to 50 µg of DNA sample (in no more than 20 µL total volume) directly to a still-frozen tube of competent cells (*see* **Note 12**). Carrier DNA (40 µg of denatured and sonified salmon sperm DNA) should be included with sub-µg DNA samples for maximum transformation frequencies.

2. Incubate samples in a 37°C water bath for 5 min. Gently mix samples once or twice during this incubation period.
3. Remove tubes from bath and add 1.5 mL of Buffer B to each. Mix contents thoroughly.
4. Incubate tubes in a 30°C water bath for 1 h.
5. Centrifuge sample tubes at 2000g for 10 min at room temperature and gently resuspend cells in 1.5 mL of Buffer C.
6. Centrifuge samples a second time and resuspend cells gently in 0.2 mL of Buffer C.
7. Spread contents of each tube on an agar plate containing selective growth medium and incubate plates for 3 to 4 d (*see* **Note 10**).

3.5. Alkali Cation Procedure

This procedure is essentially the same as described by Ito et al. *(10)* except that LiCl is used instead of Li acetate. The characteristics of the method are listed in **Table 1** (*see* **Note 11**). Note that, unlike the other transformation methods, transformation frequencies are very low for both PARS- and nonPARS-containing circular vectors.

3.5.1. Preparation of Competent Cells

1. Inoculate a 10 mL YPD culture with a single fresh *P. pastoris* colony of the strain to be transformed from an agar plate and grow overnight with shaking.
2. In the morning, use the overnight culture to inoculate a 50 mL YPD culture to a starting OD_{600} of 0.1 and grow to an OD_{600} of 0.5 to 0.8 (*see* **Note 6**).
3. Harvest the culture, and wash the cells once with 10 mL of H_2O by centrifugation at 2000g at room temperature, once with 10 mL of TE buffer, and once with 20 mL of LiCl buffer.
4. Incubate the cells for one hour at 30°C. For highest transformation frequencies, transform the cells directly without freezing as described below.
5. To freeze competent cells, distribute in 0.2-mL aliquots to sterile 1.5-mL minicentrifuge tubes. Add 11 µL of DMSO to each tube, mix, and quickly freeze cells in a liquid nitrogen bath. Store frozen competent cells at −70°C.

3.5.2. Transformation

1. For each transformation sample, add the following to a sterile 1.5-mL centrifuge tube: 0.1 mL of competent cells, 0.1 to 20 µg of vector DNA in no more than a 20 µL volume. For maximum transformation frequencies with sub-µg amounts of DNA, add 10 µg of a carrier DNA (e.g., denatured and sonified salmon sperm DNA).
2. Incubate samples at 30°C for 30 min.
3. Add 0.7 mL of PEG + LiCl solution and briefly vortex to mix.
4. Heat shock at 37°C for 5 min.

5. Centrifuge samples at 2000*g* and resuspend in 0.1 mL of H$_2$O.
6. Spread samples onto selective medium plates and incubate for 3 d (*see* **Note 10**).

3.6 Construction of Gene Replacement Strains Using PCR-Generated Fragments

3.6.1. Diploid Host Strain Method

In *P. pastoris* and other yeast species, it has been shown that the frequency of gene replacement events with DNA fragments containing short (50–100 bp) terminal complementary regions is significantly higher when performed in diploid strains relative to standard haploid strains *(5)*. To perform such a gene replacement, the researcher must construct a suitably marked diploid host strain as well as amplify an appropriate DNA fragment to transform into the host.

3.6.1.1. DESIGN OF THE OLIGONUCLEOTIDES

To design the transforming DNA fragment, the researcher must first select a marker gene. Either a biosynthetic gene, such as *ARG4*, or a dominant drug resistance gene, such as the commonly used Zeocin-resistance gene, can be used as a marker. With biosynthetic genes two caveats should be noted. The first is the diploid host strain must be mutant in both alleles of that gene. For example, if one wishes to use an *ARG4* gene, then the diploid host must be *arg4/arg4* (more on the construction of such a strain below). Second, we prefer to utilize *S. cerevisiae* biosynthetic genes for gene replacement constructions when possible as experience has shown that these genes lack sufficient homology with the *P. pastoris* genome to recombine with it and thus, gene replacement events do not occur at a significant frequency. It is important with any marker gene that one includes sufficient 5′- and 3′-sequence flanking the ORF of the gene so that a promoter and transcriptional termination signals are likely to be present (*see* **Note 13**).

3.6.1.2. CONSTRUCTION OF THE DIPLOID TRANSFORMATION STRAIN

The diploid host strain for the gene replacement is constructed using the general procedures described in detail in Chapter 14 on genetic manipulation of *P. pastoris*. As mentioned in **Subheading 3.6.1.1.**, if a biosynthetic gene is being used as the selectable marker for the gene replacement fragment, the diploid strain must have the same mutant allele; for example, for an *ARG4* marker construction, an *arg4/arg4* diploid strain must be used. This can be done by mating *his4 arg4* and *ade1 arg4* strains (*see* **Note 14**).

3.6.1.3. SPORULATION AND ANALYSIS FOR THE GENE REPLACEMENT PRODUCT

Once the appropriate diploid strain and PCR-generated gene replacement fragments have been constructed, the fragment is transformed into the strain utilizing the electroporation protocol described in **Subheading 3.3.** above. (It is likely that the spheroplast-based procedure described in **Subheading 3.2** would also work but has yet to be tried.) After selection for Arg+ transformants, an entire population of several hundred transformants is collected, and subjected to sporulation by spreading a sample of the entire culture onto a sporulation medium plate (*see* Chapter 14). Spores are then spread on YPD plates to generate colonies and the resulting colonies are then replica plated onto a selective medium (e.g., medium with methanol as carbon source for selection of *pex* mutants) and a nonselective medium (e.g., medium with glucose as carbon source) and colonies that have the proper phenotype for the desired gene replacement mutant are identified (for *pex* mutants that phenotype is an inability to grow on methanol).

3.6.2. Multiple PCR-Round Sequence Extension Method

An alternative "trick" to improve the frequency of gene-replacement events in a transformant population is to increase the length of the complementary segments on a gene-replacement fragment by performing a second round of PCR, utilizing the product of the first round as template (*6*). A second set of primers for PCR are designed that overlap the termini of the first set by 15 to 20 bases and extend the flanking complementary regions by 100 to 150 bases on each terminus. After PCR, the lengthened product is transformed into a suitable haploid or diploid strain of *P. pastoris* by electroporation as per usual and transformants are screened for ones with the expected phenotype for the gene replacement as usual.

4. Notes

1. Spheroplast transformation reagents can be purchased from Invitrogen (San Diego, CA).
2. Some lots of sorbitol are much better than others for regeneration of *P. pastoris* spheroplasts. As an alternative to sorbitol, 0.6*M* KCl can be substituted (13.4 g KCl, 6 g agar, and 240 mL H_2O).
3. Zymolyase lots vary with regard to suitability for use in *P. pastoris* transformations. Since the activity in the dry powder is stable for years at –20°C, it is recommended that good lots be set aside exclusively for transformations. As an alternative spheroplasting agent, the snail gut preparation Glusulase (DuPont, Wilmington, DE) can be used. Glusulase is stable for periods of at least a year stored at 4°C. To spheroplast with Glusulase, follow the same procedure and add approx 20 µL of Glusulase/40 to 60 OD_{600} units of cells in SCE.
4. The purity of the PEG_{1000} is critical for transformation. Known sources of high quality PEG_{1000} are Sigma Chemical Company (St. Louis, MO) and Carl Roth GmbH (Karlsruhe, Germany; United States distributor is Atomergic Chemetals Corp., New York, NY).

5. DMSO must be from a fresh unopened bottle or from a stock of DMSO prepared from a new unopened bottle and stored at –70°C until use.

6. One OD_{600} unit of *P. pastoris* culture equals approx 5×10^7 cells.

7. For highest transformation frequencies, it is necessary to establish the optimal spheroplasting conditions (i.e., the optimal amount of Zymolyase or Glusulase, and incubation time) which vary with different lots of enzyme and other reagents and equipment used for cell growth and spheroplasting. Another method of establishing these conditions is to divide the washed cells into equal portions before spheroplasting and add enzyme to only one tube while holding the other on ice. Spheroplasting is monitored by incubating the tube with enzyme at 30°C and removing samples at selected times to SDS solution. Samples in SDS are then examined spectrophotometrically at OD_{800} and the time required for 70% of the cells to become sensitive to lysis with SDS determined quantitatively. Subsequently, enzyme is added to the remaining tube which is then incubated at 30°C for the length of time determined for optimal (70%) spheroplasting.

8. In general, procedures described for *S. cerevisiae* also work well for *P. pastoris*. Thus, if your instrument is not listed, use the protocol recommended for electroporation of *S. cerevisiae* with your instrument. See technical literature provided by manufacturers for information on the use of specific electroporation devices.

9. If using Invitrogen electroporation cuvettes, the volume of cells should be 80 μL. If using BRL 1.5-mm gap chambers, the volume of cells should be 20 μL.

10. *P. pastoris* cells are flocculent (i.e., tend to grow in multi-cell clumps). As a result, transformant colonies are frequently composed of more than one transformed strain. To avoid the problem of mixed colonies, pick and restreak them for single colonies on selective medium at least once before proceeding with further analysis. When looking for gene replacement transformants by replica plate screening for a recessive phenotype (e.g., *AOX1* gene replacements with a recessive methanol-utilization slow or Mut^s phenotype), colonies should be recovered from the transformation plates, suspended in water, and replated on selective medium before screening.

11. Multicopy vector integration events are much less frequent with the PEG_{1000} or alkali cation methods than with either the spheroplast or electroporation methods (Dr. Koti Sreekrishna, personal communication).

12. Cell competence decreases very rapidly after cells thaw even if the cells are held on ice. Therefore, adding DNA to frozen samples is critical. To transform large numbers of samples, it is convenient to process them in groups of about six at a time.

13. As an example, we utilized the *S. cerevisiae ARG4* gene in a gene replacement construct designed to delete the *PEX14* ORF and designed oligonucleotide primers that were complementary to 20 bases outside the *ARG4* ORF starting with nucleotide –375 to –352 (with the A in the methionine initiator ATG codon as +1) at the 5′-end and +327 to +346 (relative to the translational stop codon) on the 3′-end *(5)*. In addition, these oligos had ~70 nucleotides complementary to sequences just flanking the *PEX14* gene to be knocked-out. For example, the 5′-oligo had sequences from –75 to –1 (relative to the *PEX14* methionine initiator codon) fused with one of the *ARG4* primers and the 3′-oligo had sequences from +26 to +102

(relative to the *PEX14* translational stop codon) fused with the other *ARG4* primer. After PCR, the resulting fragment of 2256 bp (2104 of *ARG4* gene plus 75 bp of 5′ and 77 bp of 3′ primers from the *PEX14* locus) was ready for electroporation into *P. pastoris*.

14. Suitable strains and marker gene containing vectors can be obtained by going to the website http://faculty.kgi.edu/cregg/index.htm and requesting the desired strains from the author.

References

1. Cregg, J. M., Barringer, K. J., Hessler, A. Y., and Madden K. R. (1985) *Pichia pastoris* as a host system for transformations. *Mol. Cell. Biol.* **5,** 3376–3385.
2. Cregg, J. M., Madden, K. R., Barringer, K. J., Thill, G. P., and Stillman, C. A. (1989) Functional characterization of the two alcohol oxidase genes from the yeast *Pichia pastoris*. *Mol. Cell. Biol.* **9,** 1316–1323.
3. Liu, H., Tan, X., Wilson, K., Veenhuis, M., and Cregg, J. M. (1995) *PER3*, a gene required for peroxisome biogenesis in *Pichia pastoris*, encodes a peroxisomal membrane protein involved in protein import. *J. Biol. Chem.* **270,** 10,940–10,951.
4. Waterham, H. R., de Vries, Y., Russell, K. A., Xie, W., Veenhuis, M., and Cregg, J. M. (1996) The *Pichia pastoris PER6* gene product is a peroxisomal integral membrane protein essential for peroxisome biogenesis and has sequence similarity to PAF-1. *Mol. Cell. Biol.* **16,** 2527–2536.
5. Johnson, M. A., Snyder, W. B., Lin-Cereghino, J., Veenhuis, M., Subramani, S., and Cregg, J. M. (2001) *Pichia pastoris* Pex14p, a phosphorylated peroxisomal membrane protein, is part of a PTS-receptor docking complex and interacts with many peroxins. *Yeast* **18,** 621–641.
6. Snyder, W. B., Koller, A., Choy, A. J., et al. (1999) Pex17p is required for import of both peroxisome membrane and lumenal proteins and interacts with Pex19p and the peroxisomal targeting signal-receptor docking complex in *Pichia pastoris*. *Mol. Biol. Cell* **10,** 4005–4019.
7. Hinnen, A., Hicks, J. B., and Fink, G. R. (1978) Transformation of yeast. *Proc. Natl. Acad. Sci. USA* **75,** 1929–1934.
8. Becker, D. M. and Guarante, L. (1991) High-efficiency transformation of yeast by electroporation. *Methods Enzymol.* **194,** 182–187.
9. Lin-Cereghino, J., Wong, W. W., Xiong, S., Giang, W., Luong, L. T., Vu, J., Johnson, S. D., and Lin-Cereghino, G. P. (2005) Condensed protocol for competent cell preparation and transformation of the methylotrophic yeast *Pichia pastoris*. *BioTechniques* **38,** 44–48.
10. Klebe, R. J., Harriss, J. V., Sharp, Z. D., and Douglas, M. G. (1983) A general method for polyethylene-glycol-induced genetic transformation of bacteria and yeast. *Gene* **25,** 333–341.
11. Ito, H., Fukuda, Y., Murata, K., and Kimura, A. (1983) Transformation of intact yeast cells treated with alkali cations. *J. Bacteriol.* **153,** 163–168.
12. *Pichia* Expression Kit Instruction Manual, Version E, Invitrogen, San Diego CA, pp. 63.

13. Grey, M. and Brendel, M. (1995) Ten-minute electrotransformation of *Saccharomyces cerevisiae*, in *Methods in Molecular Biology Vol. 47: Electroporation Protocols for Microorganisms*, (Nickoloff, J. A. ed.), Humana Press, Totowa, NJ, pp. 269–272.

14. Stowers, L., Gautsch, J., Dana, R., and Hoekstra, M. F. (1995) Yeast transformation and preparation of frozen spheroplasts for electroporation, in *Methods in Molecular Biology Vol. 47: Electroporation Protocols for Microorganisms*, (Nickoloff, J. A. ed.), Humana Press, Totowa, NJ, pp. 261–267.

15. Lorow-Murray, D. and Jesse, J. (1991) High efficiency transformation of *Saccharomyces cerevisiae* by electroporation. *Focus* **13,** 65–68.

4

Rational Design and Optimization of Fed-Batch and Continuous Fermentations

Wenhui Zhang, Mehmet Inan, and Michael M. Meagher

Abstract

This chapter provides rational approaches to design and optimize fed-batch and continuous fermentations of both Mut$^+$ and Mut^s (methanol utilization plus and slow) *Pichia pastoris* strains. The methods are described in detail for glycerol batch, glycerol fed-batch, transition, and methanol fed-batch/mixed feed/ continuous stirred tank reactor (CSTR) phases of the process based on glycerol and methanol consumption models. Cell density, broth volume, substrate feed rate, and the length of each phase are rationally designed to conduct runs with selected parameters for optimizing a process. The optimization is anchored by the impact of a specific growth rate/dilution time (for CSTRs) on productivity. Equations for simulation of a process with optimal parameters are derived for an optimal process design. This protocol can be used as a practical manual for process development of a *P. pastoris* recombinant fermentation, and also as a reference for fermentation of other microorganisms.

Key Words: *Pichia pastoris*; fed-batch fermentation; CSTR; continuous fermentation; optimization; growth model; methanol; glycerol; modeling.

1. Introduction

Pichia pastoris has become a popular expression system for production of recombinant proteins. It has the advantages of the alcohol oxidase I (*AOX1*) promoter, one of the strongest, most regulated promoters known; the ability to grow to high cell density in a simple defined medium; stable integration of expression plasmids into the *P. pastoris* genome; and the availability of the expression system as a kit from Invitrogen Co. (San Diego, CA, USA) (*1*). It brings together the benefits of *Escherichia coli* and eukaryotic expression systems: high-level expression, easy scale-up, inexpensive growth, protein processing, folding, and other posttranslational modifications.

From: *Methods in Molecular Biology, vol. 389:* Pichia *Protocols, Second Edition*
Edited by: J. M. Cregg © Humana Press Inc., Totowa, NJ

In the genome of wild-type *P. pastoris*, there are two copies of the alcohol oxidase (*AOX*) gene, designated *AOX1* and *AOX2*. These genes enable the cells to assimilate methanol as the sole carbon and energy source. The *AOX1* promoter regulates 85% of alcohol oxidase production whereas the *AOX2* promoter is less active *(2)*. Recombinant *P. pastoris*, constructed by insertion of the "*AOX1* promoter-gene of interest" expression cassette into the genome, uses the *AOX1* promoter to drive heterologous protein expression with methanol as the inducer. Depending on whether the *AOX1* gene is functional, two different phenotypes of *P. pastoris* expression strains can be generated: either a methanol utilization plus (Mut⁺) strain or a methanol utilization slow (Mutˢ) strain *(2–4)*. The former contains both *AOX1* and *AOX2*, whereas the later only *AOX2*.

P. pastoris is typically grown in a fed-batch manner to a high cell density in order to obtain a high productivity. A fermentation guideline for both Mut⁺ and Mutˢ strains is available from Invitrogen Co. *(5)*, a company authorized by RCT (Research Corporation Technologies, Tucson, AZ, USA) to develop and sell the *P. pastoris* expression system. Stratton et al. also wrote a protocol for *P. pastoris* high cell-density fermentation *(6)*, in which detailed procedures are provided. However, the methanol feed rate profiles described by Stratton et al. are fixed and may be inapplicable for those strains whose ability to utilize methanol has changed as a result of the expression of heterologous genes. For a fed-batch fermentation process, the substrate feed rate usually needs to be optimized to maximize productivity. Various methods for such optimization have been reported in other production systems *(7–18)*. For *P. pastoris*, we presented the optimization of methanol feed rate based on the effect of specific growth rate on product formation *(19,20)*. In this chapter, we will present a detailed protocol for the reader to optimize substrate feed rates in a fed-batch or continuous fermentation process of Mut⁺ and Mutˢ *P. pastoris* strains. The protocol is for small-scale processes with a working volume less than 20 L. For larger-scale processes, some procedures in the protocol may need to be modified accordingly. Additional details for equipment set up and operation are provided in Stratton et al. *(6)*.

2. Materials

2.1. Media

2.1.1. Seed Flask Medium

Buffered glycerol-complex medium (BMGY) is commonly used for shake flask culture (*see* **Note 1**) and is composed of 1% yeast extract, 2% peptone, 100 m*M* potassium phosphate (pH 6.0), 1.34% yeast nitrogen base w/o amino acids (YNB), 4×10^{-5}% biotin, and 1% glycerol. To prepare 1 L dissolve 10 g yeast extract and 20 g peptone in 700 mL water and autoclave. Add 100 mL sterile 1M potassium phosphate buffer (pH 6.0) (dissolve 23 g K_2HPO_4 and 118.1 g KH_2PO_4 in 1 L water), 100 mL of a filter sterilized YNB solution (13.4 g/100 mL and 2 mL of a filter sterilized 20 mg/100 mL biotin solution), and 100 mL of a sterile 10 g/100 mL glycerol solution.

2.1.2. Fermentor Media

Basal salt medium (BSM) is the defined medium commonly used. The recipe is (per L): 26.7 mL 85% H_3PO_4, 0.93 g $CaSO_4$, 18.2 g K_2SO_4, 14.9 g $MgSO_4 \cdot 7H_2O$, 4.13 g KOH, and 40.0 g glycerol. After autoclaving, adjust the pH to 5.0 with 28% NH_3 (concentrated ammonium hydroxide). It is normal that the medium becomes cloudy at pH 5.0. The added NH_3H_2O serves as nitrogen source during fermentation (*see* **Note 2**).

FM22 is another medium we use which contains (per L): 42.9 g KH_2PO_4, 5 g $(NH_4)_2SO_4$, 1.0 g $CaSO_4 \cdot 2H_2O$, 14.3 g K_2SO_4, 11.7 g $MgSO_4 \cdot 7H_2O$, and 40 g glycerol. The pH is approx 4.5 after autoclaving (*see* **Note 3**).

2.2. Solutions

2.2.1. 5% (w/v) Antifoam Solution

Dissolve 25 g KFO™ 673 antifoam (food grade, KABO Chemicals, Inc., WY, USA) in 500 mL water. The solution is an emulsion with light cloudiness. Filter sterilize, add 2 mL/L to the starting medium, and a minimum amount needed to eliminate foam during the fermentation.

2.2.2. PTM1 Trace Salts Solution

Per liter: 6.0 g $CuSO_4 \cdot 5H_2O$, 0.08 g NaI, 3.0 g $MnSO_4 \cdot H_2O$, 0.2 g $Na_2MoO_4 \cdot 2H_2O$, 0.02 g H_3BO_3, 0.5 g $CoCl_2$, 20.0 g $ZnCl_2$, 65.0 g $FeSO_4 \cdot 7H_2O$, 0.2 g biotin and 5.0 mL H_2SO_4. Filter sterilize, and add 4.35 mL/L to the starting medium after the medium is autoclaved (*see* **Note 4**).

2.2.3. 63% (w/v) Glycerol Feed Solution

Dissolve 630 g glycerol in water and fill up to 1 L. After autoclaving, add 12 mL/L PTM1. It serves as glycerol feed solution in glycerol fed-batch and mixed feed phases (*see* **Note 5**).

2.2.4. 100% Methanol Feed Solution

100% methanol with addition of 12 mL/L PTM1. No sterilization needed. It serves as the methanol feed solution for the methanol fed-batch phase or CSTR fermentations (*see* **Notes 5** and **6**).

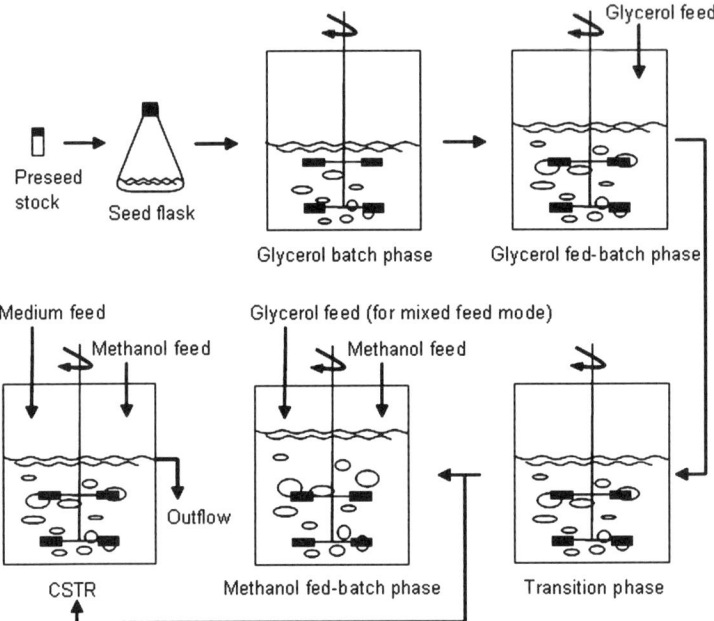

Fig. 1. Flow chart of the fed-batch or CSTR fermentation process.

2.2.5. Base and Acid for pH Control During Fermentation

Base: about 28% NH_3OH(concentrated ammonium hydroxide). Acid: about 10% HCl, or 30% H_2SO_4 (add 3 portions of concentrated acid to 7 portions water). No sterilization needed.

3. Methods

3.1. Flow Chart for the Fed-Batch and CSTR Fermentation Processes

Figure 1 shows the process flow chart. The preseed stock comes from a working cell bank, stored in the vapor phase of a liquid nitrogen dewar. It is used to inoculate the seed flask to prepare the inoculum for the fermentor. The glycerol batch and fed-batch phases are considered as growth phases in which cells only propagate and no recombinant protein production occurs. The transition phase is a 3-h period during which cells adapt to methanol before the methanol feed phase starts. The methanol fed-batch phase and CSTR phases are two different running modes for production.

3.2. Seed Flask Cultivation (for Both Mut⁺ and Mutˢ Strains)

Use BMGY as the medium. Set the volume needed $V_{seed} = V_0/10$. V_0 is the fermentation start volume. The medium volume in each shake flask should not be more than 20% of the total volume of the shake flask. Set the temperature to 30°C and an incubator speed of 250 to 300 rpm. Flask cultures are grown to an $OD_{600} = 2-10$. To accurately measure $OD_{600} > 1.0$, dilute the sample 10-fold before reading. The incubation time will be around 24 h (depending on the volume and cell density of the preseed inoculate).

3.3. Glycerol Batch Phase (for Both Mut⁺ and Mutˢ Strains)

Use BSM medium and set the volume (V_0) to approx one half of the bioreactor's maximum working volume (V_0 will be exactly determined after the process is optimized). After autoclaving, adjust the pH to 5.0. Take a sample and confirm the pH with an external pH meter (*see* **Note 7**). Add 12 mL/L PTM1 trace salts solution and 2 mL/L 5% antifoam to BSM. Calibrate the dissolved oxygen (DO) to 100% saturation, and then inoculate the fermentor with seed culture, the volume of which is approx 10% of the bioreactor initial volume. Control pH at 5.0, temperature at 30°C, and DO above 20%. The 5% antifoam solution is added when necessary to eliminate the foaming produced during fermentations (*see* **Note 8**).

The batch phase ends when the glycerol is exhausted, which is indicated by an abrupt increase in DO (so-called DO spike). The cell density at the end of the batch phase, X_{gb}, is around 105 g WCW/L (wet cell weight obtained by centrifugation at 2000g; 1 g WCW/L ≈ 0.27 g dry cells/L ≈ 1 OD_{600}). The specific growth rate, μ_{gb}, is 0.177/h⁻¹ in batch phase when there is no limiting factor on growth rate. By taking a sample at some point t_s during the batch phase and measuring the cell density X_s, the end time of batch phase, t_{gb}, can be predicted by **Eq. (1)**:

$$t_{gb} = t_s + \frac{1}{\mu_{gb}} \ln\left(\frac{X_{gb}}{X_s}\right) = t_s + \frac{1}{0.177} \ln\left(\frac{105}{X_s}\right) \tag{1}$$

3.4. Glycerol Fed-Batch Phase (for Both Mut⁺ and Mutˢ Strains)

Glycerol is fed a growth limiting rate in order to further increase the cell density and to prime methanol-utilization pathway enzymes before induction. Feed rate F_{gfb} and feed time t_{gfb} will depend on the desired cell density at the end of the glycerol fed batch phase, X_{gfb}, which may vary from 150 to 450 g WCW/L, and should be optimized based on the production time course (discussed in later sections on the production phase). We use 63% (*w/v*) glycerol as the feed solution (*see* **Note 9**) and control pH at 5.0, T 30°C, and DO above 20% (*see* **Note 8**). There are two feeding methods: one in which the fed rate is exponentially increased and the other in which the feed rate is held constant.

3.4.1. Exponentially Increasing Feed Rate

This feeding strategy results in exponential growth at a desired specific growth rate μ_g. The relationship between μ_g and the glycerol consumption rate v_g is:

$$v_g = 0.503 \, \mu_g + 0.0065 \tag{2}$$

where v_g is a unit of g glycerol/g WCW/h, and has a maximum value for $\mu_g = 0.177/h$. With **Eq. (2)**, we can design an exponentially increasing F_{gfb} that will result in the desired μ_g:

$$F_{gfb} = v_g (X_{gb} V_{gb}) e^{\mu_g t} = (0.503 \, \mu_g + 0.0065) \left(X_{gb} V_{gb} \right) e^{\mu_g t} \tag{3}$$

X_{gb} and V_{gb} can be measured by taking a sample at the end of glycerol batch phase. Within the bioreactor's capacity for heat transfer and oxygen supply, the desired μ_g can be set close to the maximum (0.177/h) in order to reach the desired X_{gfb} in a t_{gfb} as short as possible. Note in **Eq. (3)**, F_{gfb} has units of g glycerol/h. The total fed volume of 63% *(w/v)* glycerol at the end of the fed-batch phase, V_g, will be:

$$V_g = (0.503 \, \mu_g + 0.0065)\left(X_{gb} V_{gb} \right)\left(e^{\mu_g t_{gfb}} - 1 \right) \Big/ \left(630 \, \mu_g \right) \tag{4}$$

where 630 is the concentration of glycerol feed solution with a unit of g/L, and V_g has a unit of L. If we neglect the change in broth volume caused by factors other than glycerol feed, the broth volume at the end of the fed-batch phase, V_{gfb}, can be estimated as:

$$V_{gfb} = V_{gb} + V_g \tag{5}$$

To obtain the desired X_{gfb}, the feed time t_{gfb} will be:

$$t_{gfb} = \frac{1}{\mu_g} \ln \left(\frac{X_{gfb} V_{gfb}}{X_{gb} V_{gb}} \right) \tag{6}$$

Combining **Eq. (4–6)** yields:

$$t_{gfb} = \frac{1}{\mu_g} \ln \left(\frac{630 \, \mu_g X_{gfb} - X_{gb} X_{gfb} \left(0.503 \, \mu_g + 0.0065 \right)}{630 \, \mu_g X_{gb} - X_{gb} X_{gfb} \left(0.503 \, \mu_g + 0.0065 \right)} \right) \tag{7}$$

While X_{gb}, μ_g, and X_{gfb} are known, t_{gfb} can be calculated with **Eq. (7)**. For example, if $X_{gb} = 105$ g WCW/L, $\mu_g = 0.177/h$, and desired $X_{gfb} = 300$ g WCW/L, we obtain $t_{gfb} = 7.1$ h. With **Eq. (3)** we obtain a feed rate of $F_{gfb}/V_{gb} = 15.9 e^{0.177t}$ mL 63% *(w/v)* glycerol/L broth/h $= 17.59 e^{0.177t}$ g 63% *(w/v)* glycerol/L broth/h (density of 63% *(w/v)* glycerol is 1.1 g/mL), and with **Eq. (4)**

we obtain the total amount of glycerol fed, $V_g/V_{gb} = 0.225$ L 63% *(w/v)* glycerol/L broth. If the concentration of glycerol feed solution used is not 630 g/L, just replace the value of 630 in **Eqs. (4)** and **(7)** with the actual glycerol concentration used (*see* **Note 9**). If a particular *P. pastoris* expression strain has a glycerol consumption rate different from **Eq. (2)**, modify **Eqs. (3), (4),** and **(7)** with the actual rate to calculate t_{gfb}, F_{gfb}/V_{gb}, and V_g/V_{gb} (*see* **Notes 10** and **11**).

3.4.2. Constant Feed Rate

This feeding strategy is simple and easy to perform compared with the exponential feed which requires some modest level of programming. The shortcoming of the constant feed is that it will take longer to obtain the desired X_{gfb} as compared with the exponential feed. Whereas F_{gfb} is constant, μ_g is decreasing with time. If the feed rate is set to a value resulting in a μ_{g0} (≤ 0.177/h) at the initial point, F_{gfb} can be decided based on **Eq. (3)** with setting $t = 0$ and $\mu_g = \mu_{g0}$, that is:

$$F_{gfb} = (0.503\,\mu_{g0} + 0.0065)X_{gb}V_{gb} \tag{8}$$

where F_{gfb} has unit of g glycerol/h. And the V_g (L 63% *(w/v)* glycerol) is:

$$V_g = (0.503\,\mu_{g0} + 0.0065)X_{gb}V_{gb}t_{gfb}/630 \tag{9}$$

Note μ_g, v_g, X, and V are changing with time. With $v_g = F_{gfb}/XV$, **Eq. (2)** becomes:

$$F_{gfb}/(XV) = 0.503\,\mu_g + 0.0065 \tag{10}$$

μ_g is expressed as:

$$\mu_g = \frac{d(XV)}{(XV)dt} \tag{11}$$

Substituting **Eq. (10)** into **(11)**, by integration we obtain:

$$e^{-\frac{0.0065t_{gfb}}{0.503}} = \frac{F_{gfb} - 0.0065X_{gfb}V_{gfb}}{F_{gfb} - 0.0065X_{gb}V_{gb}} \tag{12}$$

Taking approximation by a first-order Taylor series expansion as:

$$e^{-\frac{0.0065t_{gfb}}{0.503}} \approx 1 - \frac{0.0065t_{gfb}}{0.503} \tag{13}$$

Combination of **Eqs. (5), (8), (9), (12),** and **(13)** yields:

$$t_{gfb} = \frac{630\left(X_{gfb} - X_{gb}\right)}{630X_{gb}\,\mu_{g0} - X_{gb}X_{gfb}\left(0.503\,\mu_{g0} + 0.0065\right)} \tag{14}$$

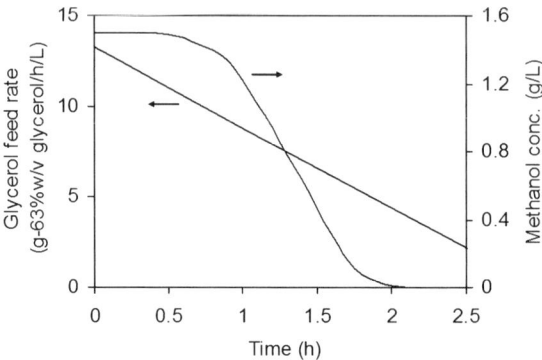

Fig. 2. A typical profile of transition phase.

Within the fermentor capacities of heat transfer and oxygen supply, μ_{g0} can be set close to the maximum μ_g (0.177/h) in order to reach the desired X_{gfb} in a t_{gfb} as short as possible. Whereas X_{gb}, μ_{g0}, and X_{gfb} are given, t_{gfb} can be calculated with **Eq. (14)**. For example, if X_{gb} = 105 g WCW/L, μ_{g0} = 0.177/h, and the desired X_{gfb} = 300 g WCW/L, we obtain t_{gfb} = 14.1 h, which is much longer than the exponential feed time (7.1 h) for reaching the same X_{gfb} as indicated previously. With **Eq. (8)** we obtain a feed rate F_{gfb}/V_{gb} = 15.9 mL 63% *(w/v)* glycerol/L broth/h = 17.5 g 63% *(w/v)* glycerol/L broth/h, and with **Eq. (9)** we obtain the total amount of glycerol fed, V_g/V_{gb} = 0.225 L 63% *(w/v)* glycerol/L broth, which is the same as the exponential feed described above. If the concentration of glycerol feed solution used is not 630 g/L, or the glycerol consumption rate is different from **Eq. (2)**, just modify **Eqs. (8–10)** and **(12–14)** accordingly for the calculation of t_{gfb}, F_{gfb}/V_{gb}, and V_g/V_{gb} (*see* **Notes 9–11**).

3.5. Transition Phase (for Mut⁺ Strains)

At the end of the glycerol fed-batch phase, 2 mL methanol/L broth is injected into the fermentor to start the induction. No consumption of methanol occurs until cells have synthesized some amount of the alcohol oxidase (AOX), which controls the first step of methanol catabolism. This period for cells to adapt to methanol is defined as the transition phase. We have observed that feeding growth limiting amounts of glycerol during transition phase may accelerate AOX synthesis and therefore shorten the length of the transition phase. **Figure 2** shows a typical profile for this transition protocol. At the starting point, 2 mL/L methanol is injected, and simultaneously a glycerol feed is initiated with a feed rate that is set to ramp down linearly from 13.3 g 63% *(w/v)* glycerol/L broth/h to 0 over a 3-h period. During the first 30 min, cells consume almost no methanol. After that period, the methanol concentration starts to decrease and reaches zero by approx 2 h, which indicates the cells have adapted to methanol and are ready for methanol feeding. A methanol sensor is useful for determining the end of the transition phase by direct monitoring of methanol concentration. If you do not have a methanol sensor, the transition phase end point can be estimated by DO response to methanol addition. If the DO decreases abruptly after addition of several drops of methanol to the bioreactor, the cells have adapted to methanol (*see* **Note 12**). During the transition phase the cell density will increase about 20 g WCW/L due to the glycerol feed.

During the production phase, if the pH and/or temperature are different from that in growth phase, adjust them to the new values before the end of the transition phase. A gradual multistep adjustment is typically better than a one-step change to protect cells from the pH and/or temperature shocks.

For Mut^s strains, cells utilize methanol very slowly, so the above transition protocol is not applicable. In this case, the production phase, in which a mixed glycerol-methanol feed is usually employed, follows immediately after the glycerol batch phase.

3.6. Methanol Fed-Batch Phase

Once cells are ready to consume methanol, the methanol feed can be initiated. For Mut^s and some Mut$^+$ strains a glycerol feed may be needed along with the methanol feed during the entire production phase with the goal of increasing the productivity of the fermentation, a process called a mixed feed method. Here we consider three factors affecting the production: methanol feed profile, pH and temperature, and also assume the former has no interaction with the later two.

3.6.1. Optimization of pH and Temperature (for Both Mut$^+$ and Mut^s Strains)

pH and temperature may be critical for protein expression. Before optimizing the methanol feed profile, one should first look for the optimal pH and T if they are unknown. According to our experience, levels of pH may be set to 3.5, 5.0, and 6.5; and temperature to 25°C and 30°C (*see* **Note 13**). A total of 6 runs are needed to cover all combinations for this initial optimization. For all the runs, methanol feed is controlled by methanol sensor to maintain the methanol concentration at 2 to 5 g/L (*see* **Note 14**). For Mut^s strains, an exponential glycerol feed with $\mu_g = 0.015$/h is simultaneously conducted (refer to **Eq. [3]** for the feed rate design, replace X_{gb} and V_{gb} with kX_{tr} and V_{tr}, the cell density and broth volume at the end of transition phase, respectively). Set $X_{gfb} = 200$ g WCW/L (refer to **Subheading 3.4.** for the design), and end the fermentation while methanol feed time $t_{mfb} = 72$ h or the cell density $X_{mfb} \geq 450$ g WCW/L, whichever comes first. Take samples every 8-h interval for cell density and protein measurement, also record the broth volume and total consumed methanol

Q_m (g) at each sampling point. Plot the time course of protein content in cells α (g/g WCW, for intracellular production) or the protein concentration in supernatant P (g/L, for extracellular production) to find the peak of α or P of each run. Compare the peaks of all runs and obtain the optimal pH and T from the run that has the highest α or P.

For Mut+ strains, calculate the specific growth rate μ_{max} and specific methanol consumption rate ν_{max} of the optimal run with **Eqs. (15)** and **(16)** *(19)*:

$$\ln \frac{XV}{X_2 V_2} = \mu_{max} t \tag{15}$$

$$Q_m = \nu_{max}(XV - X_2 V_2)/\mu_{max} \tag{16}$$

X_2 and V_2 are the X and V of the second sample (that is $t_{mfb} = 8$ h). Slope of plot $\ln[(XV)/(X_2 V_2)] \sim t$ and $Q_m \sim (XV - X_2 V_2)/\mu_{max}$ gives μ_{max} and ν_{max}, respectively. We observed that X_2 was lower than X_1 (cell density at $t_{mfb} = 0$), which indicates the wet cell weight is affected by the growth substrate. We observed that for the same number of cell counts the wet weight of cells growing on methanol (WCW_m) is less than that growing on glycerol (WCW_g), and $WCW_m/WCW_g = k \approx 0.86$. X_1 is the wet weight of cells growing on glycerol, so we used the time course data starting from the second sample for calculation of the μ_{max} and ν_{max}. For some of the strains we have worked with *(19)*, we obtained $\mu_{max} = 0.071/$h and $\nu_{max} = 0.068$ g methanol/g WCW/h at pH 5.0 and temperature at 30°C, which are close to that of the host strain GS115. On the other hand, some strains showed a much lower μ_{max} and ν_{max} because of the effect of recombinant protein production on methanol assimilation (*see* **Note 15**).

3.6.2. Optimization of Methanol Feed Profile

Once the optimal pH and T are determined, different methanol feed profiles are investigated to maximize production. For Mut+ strains, focus on the optimization of μ_m. For Mut^s, a mixed-feed method is typically conducted and one should optimize the μ_g contributed by the glycerol feeding while maintaining the methanol concentration at a constant level. We treat Mut+ strains as if they were Mut^s if their μ_{max} obtained from **Eq. (15)** are below 0.02/h as we observed the $\mu_{max} = 0.012/$h for the Mut^s host strain (KM71H).

3.6.2.1. METHANOL FEED FOR MUT+ STRAINS

A series of fermentations are performed at different exponential feed rates to generate the desired μ_m, which should be constant during the whole fed-batch phase of each run. The objective is to determine the effect of growth rate on recombinant protein production. The feed rate F_{mfb} is set to:

$$F_{mfb} = \mu_m \acute{\imath}_{max} k X_{tr} V_{tr} e^{\acute{\imath}_m t} / \mu_{max} \tag{17}$$

where $k = 0.86$, X_{tr} and V_{tr} are the cell density and broth volume at the end of the transition phase, respectively. v_{max} and μ_{max} have been obtained from a previous run with optimal pH and T (**Eqs. [15]** and **[16]**). Set $\mu_m = 0.2\mu_{max}$, $0.4 \, \mu_{max}$, $0.6 \, \mu_{max}$, and $0.8 \, \mu_{max}$, then conduct these 4 runs with feed rates as described in **Eq. (17)** (*see* **Note 16**). For all the runs, set $X_{gfb} = 200$ g WCW/L (refer to **Subheading 3.4.** for the design), and end the fermentation when the methanol feed time $t_{mfb} = 72$ h or the cell density $X_{mfb} \geq 450$ g WCW/L, whichever comes first. Take samples at 8-h intervals for cell density and protein measurement, also record the broth volume and total consumed methanol Q_m (g) at each sampling point. Calculate the actual μ_m and v_m of each run with **Eqs. (15)** and **(16)**. Plot $v_m \sim \mu_m$ (including v_{max} and μ_{max}) and obtain the linear fit equation:

$$v_m = a\mu_m + b \tag{18}$$

Eq. (18) describes the methanol consumption model of a strain. We observed $a = 0.86$ and $b = 0.0071$ for one of the strains we worked with *(19)*, which may be used as a reference.

Plot the time course of α or P to find the peak of α or P of each run. Compare the peaks of all runs (including the run with μ_{max}) and obtain the optimal growth rate $\mu_{m,c}$ from the run with the highest α or P. For one of our strains, we obtained $\mu_{m,c} = 0.0267/h$ *(19)*. The t_{mfb} at which the α or P reaches plateau in the optimal run is the optimal length, $t_{m,c}$, of the methanol fed-batch phase. $t_{m,c}$ depends on strains and intra/extracellular production. It could be as short as around 10 h *(19)* for intracellular production while much longer for extracellular production (*see* **Note 17**).

If we neglect the change of V caused by factors other than methanol feed, with $_{m,c}$ and $t_{m,c}$ known, we obtain an optimal process as follows:

$$F_{mfb} = \left(a\mu_{m,c} + b \right)\left(kX_{tr}V_{tr} \right)e^{\mu_{m,c}t} \tag{19}$$

$$V_m = \left(a\mu_{m,c} + b \right)\left(kX_{tr}V_{tr} \right)\left(e^{\mu_{m,c}t_{m,c}} - 1 \right) \Big/ \left(790\,\mu_{m,c} \right) \tag{20}$$

$$V_{mfb} = V_{tr} + V_m \tag{21}$$

$$t_{m,c} = \frac{1}{\mu_{m,c}} \ln \left(\frac{790\,\mu_{m,c}X_{mfb} - kX_{tr}X_{mfb}\left(a\mu_{m,c} + b \right)}{790\,\mu_{m,c}kX_{tr} - kX_{tr}X_{mfb}\left(a\mu_{m,c} + b \right)} \right) \tag{22}$$

where V_m is the volume of total consumed methanol at the end of the fed-batch phase, 790 is methanol density in g/L. Set X_{mfb} to $400 - 450$ g WCW/L (*see* **Note 18**), and V_{mfb} to the maximum working volume of fermentor. With X_{mfb} and V_{mfb} given, we will obtain X_{tr}, V_{tr} from **Eqs. (20–22)**. For example, if $X_{mfb} = 400$ g WCW/L, $V_{mfb} = 4$ L, $a = 0.086$, $b = 0.0071$, $\mu_{m,c} = 0.03$/h, and $t_{m,c} = 30$ h, we derive $X_{tr} = 267$ g WCW/L and $V_{tr} = 2.8$ L. Note that the actual obtained X_{mfb} and V_{mfb} may deviate from the desired values because of various reasons similar to that in the glycerol fed-batch phase (*see* **Note 10**). Recall that the increase of cell density in the transition phase is about 20 g WCW/L, so we may set $X_{gfb} = X_{tr} - 20$, also $V_{gfb} = V_{tr}$ if we ignore the change of V in transition phase. Knowing X_{gfb} and V_{gfb}, we can design the glycerol fed-batch phase with **Eqs. (3–7)**, refer to **Subheading 3.4**. To this point we obtain the complete design for the whole optimal process (*see* **Note 19**).

3.6.2.2. MIXED GLYCEROL-METHANOL FEED (FOR MUT[S] AND MUT[+] STRAINS)

According to our observation, the total growth rate μ_{mg} on a mixed feed can be approximately represented as:

$$\mu_{mg} = \mu_m + \mu_g \tag{23}$$

where μ_m and μ_g are the growth rates contributed by methanol and glycerol feeds, respectively. μ_m is constant while maintaining methanol concentration at a constant level (2–5 g/L). We may set $\mu_g = 0, 0.015, 0.03, 0.045, 0.06$/h to conduct a series of runs with different μ_{mg} (*see* **Note 20**). Note that we conducted the run with $\mu_g = 0.015$/h while optimizing pH and temperature (*see* **Subheading 3.6.1.**). From the run with $\mu_g = 0$ (no glycerol feed) we will obtain the μ_{max} and v_{max} of the Mut[s] strain with **Eqs. (15)** and **(16)**. For one of the Mut[s] strains we worked with, we observed $\mu_{max} = 0.008$/ h and $v_{max} = 0.014$ g/g WCW/h. The glycerol feed rate corresponding to the different desired μ_g is set as described in **Eq. (3)** (replace X_{gb} and V_{gb} with kX_{tr} and V_{tr}, respectively). Using methods similar to the "methanol only feed," conduct all runs and obtain the optima $\mu_{g,c}$ and $t_{g,c}$ from the run yielding the highest α or P. μ_{max} is usually much less than $\mu_{g,c}$, so we can neglect the growth contributed by the methanol and have the optimal process design similar to the glycerol fed-batch phase, refer to **Eqs. (4–7)** (replace μ_g, t_{gfb}, X_{gb}, X_{gfb}, V_{gb}, V_{gfb} with $\mu_{g,c}$, $t_{g,c}$, kX_{tr}, X_{mfb}, V_{tr}, V_{mfb}, respectively) (*see* **Note 21**).

For some Mut[+] strains, a mixed feed may be more favorable than methanol feed alone for production, especially when expression of the heterologous protein is toxic to the host strain. To optimize the μ_g, we may set μ_m to $\mu_{m,c}$ obtained from "methanol feed alone," and set $\mu_g = 0.5 \mu_{m,c}$, $\mu_{m,c}$, $1.5 \mu_{m,c}$, and $2 \mu_{m,c}$ to conduct a series of runs with selected μ_{mg} (*see* **Note 20**). The feed rates for methanol (F_m) and glycerol (F_g) are:

$$F_m = \left(a\mu_{m,c} + b\right)\left(kX_{tr}V_{tr}\right)e^{\mu_{mg}t} \tag{24}$$

$$F_g = \left(0.503\,\mu_g + 0.0065\right)\left(kX_{tr}V_{tr}\right)e^{\mu_{mg}t} \tag{25}$$

where $\mu_{mg} = \mu_{m,c} + \mu_g$ according to **Eq. (23)**. Complete all the runs with the methods described previously and obtain the optima $\mu_{g,c}$ and $t_{mg,c}$, then we derive the optimal process design as:

$$V_g = \left(0.503\,\mu_{g,c} + 0.0065\right)\left(kX_{tr}V_{tr}\right)\left(e^{\mu_{mg,c}\,t_{m,c}} - 1\right)\Big/\left(630\,\mu_{mg,c}\right) \tag{26}$$

$$V_m = \left(a\mu_{m,c} + b\right)\left(kX_{tr}V_{tr}\right)\left(e^{\mu_{mg,c}\,t_{m,c}} - 1\right)\Big/\left(790\,\mu_{mg,c}\right) \tag{27}$$

$$V_{mg} = V_{tr} + V_m + V_g \tag{28}$$

$$t_{mg,c} = \frac{1}{\mu_{mg,c}}\ln\left(\frac{X_{mg} - kX_{tr}X_{mg}\delta}{kX_{tr} - kX_{tr}X_{mg}\delta}\right) \tag{29}$$

where X_{mg} and V_{mg} are the cell density and broth volume at the end of the mixed fed-batch phase, $\mu_{mg,c}$ and δ are represented as:

$$\mu_{mg,c} = \mu_{m,c} + \mu_{g,c} \tag{30}$$

$$\delta = \frac{\left(a\mu_{m,c} + b\right)}{790\,\mu_{mg,c}} + \frac{\left(0.503\,\mu_{g,c} + 0.0065\right)}{630\,\mu_{mg,c}} \tag{31}$$

While the constraints X_{mg} and V_{mg} are given, X_{tr} and V_{tr} can be derived with **Eqs. (26–31)** for completing the whole optimal process design, as described previously.

3.7. Continuous Fermentation

Figure 3 shows a schematic of a CSTR. Before starting the CSTR, the glycerol batch, fed-batch, and transition phases are conducted with the same protocols designed as previously described to obtain a desired cell density of X. pH and temperature are also optimized as in **Subheading 3.6.1.**. FM22 without glycerol is used as medium feed in CSTR and its pH adjusted to the optimum. At chemostat, we have:

$$\mu = D = F_{in}/V \tag{32}$$

where D is the dilution rate (h^{-1}) and F_{in} with unit L/h. For Mut$^+$ strains with methanol feed alone, if the growth is limited by methanol, methanol concentration $S \approx 0$, and X can be approximated as (*see* **Note 22**):

$$X = \frac{790\,F_m\,\mu_{max}}{F_{in}\,v_{max}} \tag{33}$$

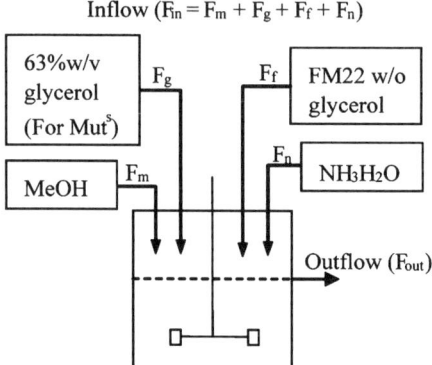

Fig. 3. CSTR schematic diagram.

where μ_{max} and v_{max} are obtained from **Eqs. (15)** and **(16)**. F_m is methanol flow rate with unit L/h. There will be not much NH_3H_2O consumed because the pH of the FM22 has been adjusted to the working pH. We may neglect F_n and have $F_{in} = F_m + F_f$, and with **Eqs. (32)** and **(33)**, obtain the F_m and F_f for the desired D and X. With this design, we can run the CSTR with different D and X and obtain their optima.

For Mut^s strain (also for Mut$^-$) (*see* **Note 21**) with mixed feed, we may add 5 g/L methanol in the FM22 feed solution. Because very little methanol is consumed, some level of methanol will be maintained at chemostat, and no separate methanol feed is needed ($F_m = 0$). The cell growth is mainly supported by the glycerol feed. While $S \approx 0$, with **Eq. (2)**, X can be approximated as:

$$X = \frac{630 \, F_g \, D}{F_{in}\left(0.503 \, D + 0.0065\right)} \tag{34}$$

where F_g is the feed rate of 63% *(w/v)* glycerol with unit L/h. Neglecting F_n, we have $F_{in} = F_g + F_f$. With **Eqs. (32)** and **(34)**, we obtain the F_g and F_f to run the CSTR with desired D and X for their optimization.

For Mut$^+$ strains with mixed feed, cell growth is supported by both methanol and glycerol. While $S \approx 0$, we have (*see* **Note 22**):

$$X = X_m + X_g \tag{35}$$

$$X_m = \frac{790 \, F_m \, \mu_{max}}{F_{in} \, v_{max}} \tag{36}$$

$$X_g = \frac{630 \, F_g \, \mu_g}{F_{in}\left(0.503 \, \mu_g + 0.0065\right)} \tag{37}$$

$$\frac{\mu_m}{\mu_g} = \frac{X_m}{X_g} \tag{38}$$

$$D = \mu_m + \mu_g \tag{39}$$

$$F_{in} = F_m + F_g + F_f \tag{40}$$

Combining **Eqs. (32)** and **(35–40)**, we can derive F_m, F_g, and F_f to conduct the CSTR runs with desired X, D, and μ_m/μ_g for their optimization.

4. Notes

1. In some cases, the use of products from animal sources, such as peptone, is strongly discouraged by the FDA. To meet this requirement, BMGY medium can be replaced by MD or MG medium, which consists of: 1.34% YNB (w/o amino acids), $4 \times 10^{-5}\%$ biotin, and either 1% dextrose (for MD medium) or 1% glycerol (for MG medium). The medium is filter sterilized rather than autoclaved.

2. At pH higher than 6.0, BSM medium will form precipitates resulting from the insolubility of magnesium and calcium phosphates. In one of processes that required pH 7.0 for expression, we modified the BSM medium by reducing the amount of 85% H_3PO_4 from 26.7 mL/L to 4.25 mL/L, and adding 5 g/L $(NH_4)_2SO_4$. This modification, named WZ1 medium, diminished the precipitation greatly and worked well. Keep in mind that BSM is not an optimized medium though it works well for most recombinant fermentations. You may modify it to fit your applications. Hellwig et al. *(21,22)* reported a salt-reduced medium containing approx one-eighth the salt concentrations of BSM could also support a growth with high cell density up to 450 g WCW/L. Their medium consists of (per L): 4.25 mL 85% H_3PO_4, 0.17 g $CaSO_4$, 2.86 g K_2SO_4, 2.32 g $MgSO_4 \cdot 7H_2O$, 0.64 g KOH, and 50.0 g glycerol.

3. FM22 has about the same salt concentrations as BSM. The difference is that the pH of FM22 is close to 5.0 before pH adjusting while BSM around 1.5. FM22 can serve as a medium feed in CSTR fermentations. FM22 was originally reported by Laroche et al. in their isotopic labeling studies *(23)*.

   For some protein production, a rich medium may be more favorable than a defined medium such as BSM or FM22 because of the protein stability issues (e.g., protease activity, aggregation, folding, hydrophobicity/philicity) Here is a typical recipe of a rich medium we have used: (per L) 40 g glycerol, 10 g cassamino acids, 10 g yeast extract, 20 g peptone or soytone, 13.4 g YNB without amino acids, and 2 mL 5% antiform.

4. When PTM1 is added to the medium, precipitates are produced as a result of the insolubility of cupric, manganese, cobaltous, zinc, and ferrous phosphates. No effect of these precipitates on growth was observed. If precipitation is a concern, an alternative trace salts solution, PTM4 may be used, which is the same as PTM1 except it only has one-third the amount of copper sulfate, ferrous sulfate, zinc chloride, and sulfuric acid.

5. Brierley *(24)* reported that no addition of PTM1 in glycerol and methanol feed solutions was needed, and there was only one single addition of 2 mL PTM1/L of starting medium in his production of human insulin-like growth factor I. On the other hand, Siegel et al. *(25)* found that trace mineral deficiencies could occur at high cell density and that may decrease cell yield and limit protein expression. This indicates that the PTM1 requirement may vary for different processes and proteins. It is recommended that excess PTM1 be used if one is not sure how a low level will impact cell growth or protein expression.

   Also, if there is a significant amount of foam produced during the glycerol or methanol fed-batch phase, add 10 mL of the 5% antifoam per L of the glycerol or methanol feed solution to avoid the foaming contamination.

6. For secreted production of some protein that is sensitive to serine protease, which typically are produced by *Pichia* strains according to our observation, the protease inhibitor PMSF (phenylmethylsulfonyl fluoride) may be added to the methanol and fed together to protect the protein from degradation. The workable concentration of PMSF in methanol may range from 100 to 400 m*M* depending on the methanol consumption rate of the strain. Higher consumption rate needs lower concentration. Because PMSF is very unstable in aqueous solution, the methanol employed should be of absolute grade, and PTM1 should be added to fermentor separately rather than mixed with the methanol feed solution.

7. The response of the pH probe can shift after autoclave. We found it is necessary to confirm the pH reading of the fermentor is correct by taking a sample and measuring the pH with an external pH meter. If the pH value of the fermentor (say 5.0) is significantly different from the read value of the external pH meter (say 4.7), the difference should be taken into account for setting the pH set-point (that will be 5.3 instead of 5.0).

8. Glycerol batch and fed-batch phases are considered as growth phase, and methanol fed-batch phase as production phase. Usually pH is set to 5.0 and temperature to 30°C in growth phase for cells to grow well, whereas different pH and temperature may be applied in the production phase depending on the properties of expressed proteins. Nonetheless, we found pH 3.5 in growth phase resulted in less secretion of impurities than pH 5.0 in one process, which shows a lower pH in growth phase may be better for some strains.

9. The concentration of glycerol feed solution should be no more than 70% *(w/v)* considering the high viscosity of glycerol. We mostly use 63% *(w/v)* glycerol. Some protocols recommend 50% *(v/v)* glycerol *(5,6)*.

10. The actual X_{gfb} obtained could deviate from the desired X_{gfb}. The deviation could attribute to inaccuracies from several aspects such as: glycerol consumption model **Eq. (2)**; measurement of V_{gb}; neglect of broth volume change caused by factors other than glycerol feeding, such as evaporation, ammonium hydroxide addition; and performance of the glycerol feed pump.

11. For *Pichia* strains that use glycerolaldehyde-3-phosphate dehydrogenase gene (GAP) promoter instead of the *AOX*1 promoter *(26)*, the glycerol fed-batch phase is the production phase because the GAP promoter provides constitutive expression

of the recombinant protein on a variety of carbon source including glycerol. Based on the feeding strategies described here, different μ_g (for exponential feed) or F_{gfb} (for constant feed) can be designed and conducted to obtain their optimal values for achieving the maximum productivities.

12. The length of transition phase is usually 1 to 3 h. It depends on properties of the expressed proteins. For some strains, expression of the protein may change the cells capability of assimilating methanol, which results in a long adaptation period and slow methanol consumption rate.

13. The levels of pH and temperature set for optimization could vary for proteins. Modify these levels as necessary. For example, one protein we worked with required pH above 6.3 for activity, so we simply set the pH to 7.0 for its production. If the protein is very sensitive to pH and temperature, the Response Surface Method (RSM) may be needed for the experimental design to find the optimal set point.

14. We advise that the control is performed with PI or PID (proportional, integral, and derivative) control mode to maintain a stable methanol concentration *(27)*. If there is no methanol sensor available, alternative methods should be used to maintain the methanol concentration. For Mut^s strains, manual addition of methanol may be needed. The maximum methanol consumption rate we observed is around 0.014 g/g WCW/h, which can be used as a reference to guide the manual addition of methanol. For Mut$^+$, methanol feed may be controlled with DO-stat method, which is, based on the response of DO to methanol, using methanol feed to maintain DO at some set-point around 20 to 40% where aeration and agitation rates are fixed.

15. For those Mut$^+$ strains in which the protein expression significantly affects the methanol utilization, plot $\ln[(XV)/(X_2V_2)] \sim t$ may not fit a straight line, that is, the growth is not exponential. The μ_m is changing during the whole fed-batch phase and could be far below the μ_{max} of the host strain. In this case, **Eqs. (15)** and **(16)** are inapplicable, and the strain will be treated as Mut^s for the optimization.

16. μ_m can be set differently. For strains with μ_{max} smaller than 0.04/h, two settings such as 0.4 μ_{max} and 0.7 μ_{max} may be good enough.

17. For extracellular production, an advanced or more strict optimization to obtain the $\mu_{m,c}$ and $t_{m,c}$ will need modeling of relationship between specific production rate ρ and μ_m as well as employment of the Pontryagin's Maximum Principle *(13,14,16)*. While the model $\rho = \alpha r(\mu_m)$ is obtained, the objective function J, that is, the total amount of the product produced in t_{mfb} will be:

$$J = \int_0^{t_{mfb}} \rho XV dt \tag{41}$$

Define $Z = XV$, and consider Z and J as the state variables, μ_m as the control variable, we obtain the Hamiltonian of the system equation:

$$H = \left(\lambda\mu_m - \rho\right)Z \tag{42}$$

where λ is an adjoint variable expressed as:

$$\frac{d\lambda}{dt} = \rho - \lambda\mu_m \tag{43}$$

By the Maximum Principle, the optimal trajectory of μ_m that makes H minimum is the optimal $\mu_{m,c}$ that maximizes J. From **Eqs. (42)** and **(43)**, whereas the process constrains are given, the problem to obtain the $\mu_{m,c}$ and $t_{m,c}$ can be solved with an iterative algebraic manipulation or some computer program *(28)*.

18. If the high cell density has no inhibition to the production, X_{mfb} may be set to around 450 g WCW/L that is the maximum the BSM/FM22 medium can yield, otherwise a lower X_{mfb} should be set for the process design. The optimal X_{mfb} can be determined from the runs to optimize μ_m. In the case the fermentor capacities for oxygen supply and/or heat transfer can not support a high X_{mfg} growing at $\mu_{m,c}$, X_{mfb} will also need to be lowered. Or, instead of lowering X_{mfg}, we may lower μ_m to meet the fermentor capacities and run a suboptimal process.

19. A simple methanol fed-batch phase with an arbitrary constant feed rate may need to be run in some cases just for testing the production or not focusing on the μ_m optimization. The design for such an non-optimal process will be:

$$F_{mf\,b} = \left(a\mu_{m0} + b\right)kX_{tr}V_{tr} \tag{44}$$

$$V_m = \left(a\mu_{m0} + b\right)kX_{tr}V_{tr}\,t_{mf\,b}/890 \tag{45}$$

$$t_{mf\,b} = \frac{890\left(X_{mf\,b} - kX_{tr}\right)}{890\,kX_{tr}\,\mu_{m0} - kX_{tr}X_{mf\,b}\left(a\mu_{m0} + b\right)} \tag{46}$$

where μ_{m0} is a value below μ_{max} to determine the F_{mfb}. $t_{mfb}(X_{mfb})$ can be calculated while μ_{m0}, $X_{mfb}(t_{mfb})$, X_{tr}, and V_{tr} are given.

20. Settings of μ_g may vary for proteins. Note excess glycerol may repress the *AOX1* promoter. Conduct the runs in the order from low to high μ_g to observe the production and modify the settings accordingly.

21. Methods described here for Mut[s] strains is also applicable for Mut[-] strains, which contain neither *AOX*1 or *AOX*2 genes and are unable to utilize any methanol. For Mut[-], methanol just functions as an inducer.

22. If a and b in **Eq. (18)** have been obtained before CSTR runs, **Eqs. (33)** and **(36)** should accordingly change to:

$$X = \frac{790\,F_m D}{F_{in}\left(aD + b\right)} \tag{47}$$

$$X_m = \frac{790\,F_m\,\mu_m}{F_{in}\left(a\mu_m + b\right)} \tag{48}$$

References

1. Cereghino, G. P. and Cregg, J. M. (1999) Applications of yeast in biotechnology: protein production and genetic analysis. *Curr. Opin. Biotechnol.* **10,** 422–427.
2. Cregg, J. M., Madden, K. R., Barringer, K. J., Thill, G. P., and Stillman, C. A. (1989) Functional characterization of the two alcohol oxidase genes from the yeast *Pichia pastoris. Mol. Cell. Biol.* **9,** 1316–1323.

3. Romanos, M. A., Scorer, C. A., and Clare, J. J. (1992) Foreign gene expression in yeast: a review. *Yeast* **8,** 423–488.
4. Cregg, J. M., Cereghino, J. L., Shi, J., and Higgins, D. R. (2000) Recombinant protein expression in *Pichia pastoris*. *Mol. Biotechnol.* **16,** 23–52.
5. Invitrogen Co. (2002) *Pichia* Fermentation Process Guidelines, *www.invitrogen. com*. Invitrogen Co., San Diego, CA.
6. Stratton, J., Chiruvolu, V., and Meagher, M. M. (1998) High cell-density fermentation, in *Pichia Protocols*, (Higgins, D. R. and Cregg, J. M., eds.), Humana, Totowa, NJ, pp. 107–120.
7. Shimizu, H., Kozaki, Y., Kodama, H., and Shioya, S. (1999) Maximum production strategy for biodegradable copolymer P(HB-co-HV) in fed-batch culture of *Alcaligenes eutrophus*. *Biotechnol. Bioeng.* **62,** 518–525.
8. Chim-Anage, P., Shioya, S., and Suga, K. (1991) Maximum histidine production by fed-batch culture of Brevibacterium flavum. *J. Ferment. Bioeng.* **71,** 186–190.
9. Lim, H. C., Tayeb, Y. J., Modak, J. M., and Bonte, P. (1986) Computational algorithms for optimal feed rates for a class of fed-batch fermentation: Numerical results for penicillin and cell mass production. *Biotechnol. Bioeng.* **28,** 1408–1420.
10. Modak, J. M., Lim, H. C., and Tayeb, Y. J. (1986) General characteristics of optimal feed rate profiles for various fed-batch fermentation processes. *Biotechnol. Bioeng.* **28,** 1396–1407.
11. Parulekar, S. J. and Lim, H. C. (1985) Modeling, optimization and control of semibatch bioreactors, in *Advances in biochemical engineering/biotechnology,* vol. 32 (Fiechter, A., ed.), Springer, Berlin, pp. 207–258.
12. O'connor, G. M., Sanchez-Riera, F., and Cooney, C. L. (1992) Design and evaluation of control strategies for high cell density fermentations. *Biotechnol. Bioeng.* **39,** 293–304.
13. Yamane, T. and Shimizu, S. (1984) Fed-batch techniques in microbial processes in *Advances in biochemical engineering biotechnology,* vol. 30, (Fiechter, A., ed.), Springer, Berlin, pp. 147–194.
14. Shioya, S. (1992) Optimization and control in fed-batch bioreactors, in *Advances in biochemical engineering biotechnology,* vol. 46, (Fiechter, A., ed.), Springer, Berlin, pp. 111–142.
15. Fishman, V. M. and Biryukov, V. V. (1974) Kinetic model of secondary metabolite production and its use in computation of optimal conditions. *Biotechnol. Bioeng. Symp.* **4,** Pt. 2, 647–662.
16. Yamane, T., Kume, T., and Sada, E. (1977) A simple optimization technique for fed-batch culture. *J. Ferment. Technol.* **55,** 587–598.
17. Ohno, H. and Nakanishi, E. (1976) Optimal control of a semibatch fermentation. *Biotechnol. Bioeng.* **18,** 847–864.
18. Yamane, T. and Tsukano, M. (1977) Effect of several substrate-feeding modes on production of extracellular beta-amylase by fed-batch culture of *Bacillus megaterium*. *J. Ferment. Technol.* **55,** 233–242.
19. Zhang, W., Bevins, M. A., Plantz, B. A., Smith, L. A., and Meagher, M. M. (2000) Modeling *Pichia pastoris* growth on methanol and optimizing the production of a

recombinant protein, the heavy-chain fragment C of botulinum neurotoxin, serotype A. *Biotechnol. Bioeng.* **70,** 1–8.

20. Zhang, W., Inan, M., and Meagher, M. M. (2000) Fermentation strategies for recombinant protein expression in the methylotrophic yeast *Pichia pastoris*. *Biotechnol. Bioprocess Eng.* **5,** 275–287.
21. Hellwig, S., Robin, F., Drossard, J., Raven, N. P. G., Vaquero-Martin, C., Shively, J. E., and Fischer, R. (1999) Production of carcinoembryonic antigen (CEA) N-A3 domain in *Pichia pastoris* by fermentation. *Biotechnol. Appl. Biochem.* **30,** 267–275.
22. Hellwig, S., Emde, F., Raven, N. P. G., Henke, M., Van der Logt, P., and Fischer, R. (2001) Analysis of single-chain antibody production in *Pichia pastoris* using on-line methanol control in fed-batch and mixed-feed fermentations. *Biotechnol. Bioeng.* **74,** 344–352.
23. Laroche, Y., Storme, V., De Meutter, J., Messens, J., and Lauwereys, M. (1994) High-level secretion and very efficient isotopic labeling of tick anticoagulant peptide (TAP) expressed in the methylotrophic yeast, *Pichia pastoris*. *Biotechnology (N Y)* **12,** 1119–1124.
24. Brierley, R. A. (1998) Secretion of recombinant human insulin-like growth factor I (IGF-I), in *Pichia Protocols*, vol. 103, (Cregg, J. M., ed.), Humana Press, Totowa, New Jersey, pp. 149–177.
25. Siegel, R. S., and Brierley, R. A. (1989) Methylotrophic yeast *Pichia pastoris* produced in high-cell-density fermentations with high cell yields as vehicle for recombinant protein production. *Biotechnol. Bioeng.* **34,** 403–404.
26. Waterham, H. R., Digan, M. E., Koutz, P. J., Lair, S. V., and Cregg, J. M. (1997) Isolation of the Pichia pastoris glyceraldehyde-3-phosphate dehydrogenase gene and regulation and use of its promoter. *Gene* **186,** 37–44.
27. Zhang, W., Smith, L. A., Plantz, B. A., and Meagher, M. M. (2002) Design of Methanol Feed Control in *Pichia pastoris* Fermentations Based upon a Growth Model. *Biotechnol Prog.* **18,** 1392–1399.
28. Zhang, W., Sinha, J., Smith, L. A., Inan, M., and Meagher, M. M. (2005) Maximization of production of secreted recombinant proteins in *Pichia pastoris* fed-batch fermentation. *Biotechnol. Prog.* **21,** 386–393.

Appendix

a	coefficient in **Eq. (18)**, g methanol/g WCW
b	coefficient in **Eq. (18)**, g methanol/g WCW/h
D	dilution rate (h^{-1})
F	feed rate (g/h)
k	wet weight ratio of cells growing on methanol to that growing on glycerol for the same number of cell counts, $k \approx 0.86$
P	product concentration in fermentor supernatant (g/L)
Q_m	total consumed methanol at sampling point (g)
t	process time (h)

$t_{g,c}$	optimal length of mixed feed for Mut^s strains (h)
$t_{m,c}$	optimal length of methanol fed-batch phase (h)
$t_{mg,c}$	optimal length of mixed feed for Mut$^+$ strains (h)
V	broth volume (L)
X	cell density, g WCW/L (wet cell weight obtained by centrifuge with 2000g)

Greek Symbols

α	product content in cells (g/g WCW)
μ	specific growth rate (h^{-1})
$\alpha m_{g,c}$	optimal μ_g (h^{-1})
$\alpha m_{m,c}$	optimal μ_m (h^{-1})
ν	Specific consumption rate (g/gWCW/h)
ρ	specific production rate (g/g WCW/h)

Subscripts

c	optimal
g	glycerol
gb	glycerol batch phase
gfb	glycerol fed-batch phase
m	methanol
max	maximum
mfb	methanol fed-batch phase
mg	methanol and glycerol
s	sampling
S	methanol or glycerol concentration in CSTR (g/L)
tr	transition phase
0	start point

5

Saturation of the Secretory Pathway by Overexpression of a Hookworm *(Necator americanus)* Protein (Na-ASP1)

Mehmet Inan, Sarah A. Fanders, Wenhui Zhang, Peter J. Hotez, Bin Zhan, and Michael M. Meagher

Abstract

Human hookworm infection is one of the most significant parasitic infections, and a leading global cause of anemia and malnutrition of adults and children in rural areas of the tropics and subtropics. *Necator americanus* secretory protein (Na-ASP1), which is a potential vaccine candidate against hookworm infections, has been expressed in *Pichia pastoris*. *Na*-ASP1 protein was expressed extracellulary by employing the leader sequence of the α-mating factor of *Saccharomyces cerevisiae*. Most of the protein produced by single copy clones was secreted outside the cell. The *Na-ASP1* steady state mRNA levels of the clones were correlated to their *Na-ASP1* gene copy number. However, increasing gene copy number of Na-ASP1 protein in *P. pastoris* saturated secretory capacity and therefore, decreased the amount of secreted protein in clones harboring multiple copies of *Na-ASP1* gene.

Key Words: *Pichia pastoris*; heterologous protein secretion; *Necator americanus* secretory protein (Na-ASP1).

1. Introduction

Pichia pastoris is an industrially important organism because of its capacity to produce recombinant proteins at high yields at the industrial scale. A unique feature of the expression system is the promoter used to drive heterologous gene expression. The system capitalizes on the characteristic of *P. pastoris* to produce peroxisomes during growth on methanol. The alcohol oxidase enzyme comprises a large fraction of the protein within the peroxisomes and the corresponding enzyme is expressed from two different *AOX* genes, namely *AOX1* and *AOX2* *(1,2)*. The *AOX1* promoter is derived from the methanol regulated

From: *Methods in Molecular Biology, vol. 389:* Pichia *Protocols, Second Edition*
Edited by: J. M. Cregg © Humana Press Inc., Totowa, NJ

alcohol oxidase I gene (*AOX1*), one of the most efficient and tightly regulated promoters known *(3)*. *AOX2* encodes a protein that shares 97% homology, and has about the same specific activity as that of *AOX1* *(4)*. However, *AOX2* is responsible for only a small percentage of the total alcohol oxidase activity in methanol-grown cell cultures *(1)*.

Hookworm infection is one of the most significant parasitic infections of humans, affecting 740 million people in rural areas of the tropics and subtropics *(5)*. Despite the inability of the human host to develop naturally acquired immune responses to hookworm, there is evidence for the feasibility of developing a vaccine based on the success of immunizing laboratory animals with either attenuated larval vaccines or antigens *(6)*. The vaccine development for hookworm infection has focused primarily on antigens secreted by third stage infective hookworm larvae (L_3) in vitro under host-like stimulatory conditions *(7)*. A useful source of infective hookworm larvae (L_3) is from the dog hookworm *Ancylostoma caninum,* the most common of all species. Among the proteins released from L_3 under activating conditions, are a metalloprotease *(8)* and two proteins of unknown function that belong to a recently discovered class of proteins known as the *Ancylostoma* secreted proteins (ASPs). The ASPs are the most abundant proteins released by activated L_3 *(7,9,10)*. *A. caninum-ASP-1* (ASP-1) cDNA encodes a 45-kD polypeptide with sequence similarity to the major antigens found in *Hymenoptera* venoms, as well as a large number of cysteine rich secretory proteins (CRISP) *(11)*.

We have employed the *P. pastoris* expression system to express a *Necator americanus* secretory protein (Na-ASP1) for development as a vaccine candidate.

2. Materials

2.1. Host Strains, Vectors, and Reagents

1. Easy Select *Pichia pastoris* expression system (Invitrogen, Carlsbad, CA).
2. *Escherichia coli* TOP10 strains (Invitrogen, Carlsbad, CA).
3. *Na-ASP1* cDNA in pBlueScript/Na-ASP1.
4. Rabbit anti-ASP1 serum.
5. Goat anti-Rabbit IgG (KPL, Gaithersburg, MD).
6. Oligonucleotide primers (Eurogentec North America Inc, San Diego, CA).
7. Restriction enzymes (New England Biolabs, Beverly, MA).
8. MasterPure Yeast DNA Purification Kit (Epicentre Inc., Madison, WI).
9. DIG Labeling and Detection Kit (Roche, Indianapolis, IN).
10. *Pfu* DNA polymerase (Stratagene, La Jolla, CA).
11. Agarose gel electrophoresis system.
12. 50-mL baffled shake flasks (Bellco Glass Inc., Vineland, NJ).
13. Sodium dodecyl sulfate-polyacrylamide gel electrophoresis (SDS-PAGE) system and reagents (Invitrogen, Carlsbad, CA).

14. Centrifuge.
15. Zeocin (Invitrogen, Carlsbad, CA).
16. Nylon membrane, Zeta-Probe GT Genomic Tested and PVDF (BioRad, Hercules, CA).
17. Western reagents (Amersham Bioscience, Piscataway, NJ).
18. Zirconia/Silica Beads (0.5mm) (BioSpec Products, Bartlesville, OK).
19. Bead Beater (BioSpec Products, Bartlesville, OK).
20. Hybri-Dot blot apparatus (Biometra, Göttingen, Germany).
21. TRI reagent (Molecular Research Center Inc., Cincinnati, OH).
22. Gene Pulser electroporation system (BioRad, Hercules, CA).
23. Electroporation cuvets (2 mm gap) (BioRad, Hercules, CA).

2.2. Culture Media

1. YPD: 1% yeast extract, 2% peptone, and 2% dextrose.
2. Buffered glycerol-complex medium (BMGY): 1% yeast extract, 2% peptone, 100 mM potassium phosphate (pH 6.0), 1.34% yeast nitrogen base w/o amino acids (YNB), 4×10^{-5} % biotin, and 1% glycerol.
3. Buffered methanol-complex medium (BMMY): 1% yeast extract, 2% peptone, 100 mM potassium phosphate (pH 6.0), 1.34% YNB, 4×10^{-5} % biotin, and 1% methanol.
4. Luria Bertani (LB) Lennox: 10% pancreatic digest of casein, 5% yeast extract, and 5% sodium chloride.

3. Methods

3.1. Expression Vectors

The construction of expression plasmids for native and codon optimized gene sequences are described in **Subheadings 3.1.1.** and **3.1.2.** Native and codon optimized *Na-ASP1* genes were fused to the DNA sequence encoding the prepro region of the α-mating factor of *S. cerevisiae* in pPICZαA plasmid for extracellular production.

3.1.1. Construction of Native Na-ASP1 Gene Expression Vector

DNA manipulations were performed following standard recombinant DNA methods (*12*) and are not described here in details because of space limitations.

1. Use pBluescript/Na-ASP1 containing cDNA encoding native Na-ASP1 as a template for construction of pPICZαA/Na-ASP1 plasmid in a PCR reaction. Amplify the *Na-ASP1* gene using the following primers:
   ASP15F - 5′-TCT**CTCGAG**AAGAGATCTCCAGCAAGAGACAGCTTC-3′
   ASP13R - 5′-GG**GGTACC**TTAAGGAGCGCTGCACAAGCC-3′.
   The primers introduce *Xho*I and *Kpn*I sites at the 5′- and 3′-ends of the gene, respectively. Because the *Xho*I upstream of the Kex2 cleavage site in the expression vector is used to clone the *Na-ASP1* gene, the sequences from *Xho*I to the

arginine codon are included in the forward primer. This primer design results in Na-ASP1 protein with a native N-terminus and flushed with Kex2 cleavage site.

2. Digest the PCR product with *Xho*I and *Kpn*I, and then ethanol precipitate to remove excess enzymes and buffers. Ligate 1200-bp fragment into vector pPICZαA, which is partially digested with *Xho*I and *Kpn*I (*see* **Note 1**).

3. Introduce ligation mixture into chemically competent *E. coli* TOP10 cells by heat shock. Plate the cells on LB Lennox plates containing zeocin (25 µg/mL) and incubate overnight at 37°C. Select single colonies and grow overnight in LB Lennox broth with zeocin. Isolate the plasmid DNA and check for the presence of the insert and for the correct orientation using restriction enzyme digestion and DNA sequencing.

3.1.2. Construction of a Codon Optimized Na-ASP1 Expression Vector

Initial expression of native Na-ASP1 protein in *P. pastoris* X-33 resulted in low-expression levels (*see* **Note 2**). Upon analyzing the *Na-ASP1* gene for codon usage, we found many suboptimal codons for *P. pastoris*. Therefore, the *Na-ASP1* codons were optimized according to *P. pastoris* codon preferences (*see* **Note 3**). The codon optimized *Na-ASP1* gene (Na-ASP1O) was cloned into the pPICZαA vector at its *Xho*I and *Xba*I sites.

3.2. Transformation and Selection

We used electrocompetent *P. pastoris* X-33 cells, which were prepared as described by Cregg (*see* Chapter 3).

1. Digest 10 µg of pPICZαA/Na-ASP1 and pPICZαA/Na-ASP1O with *Pme*I restriction enzyme (which is located within the *AOX1* promoter) to linearize the plasmids.

2. Ethanol precipitate, wash, dry and elute linear plasmid DNA in 10 µL of sterile water to prepare for electroporation.

3. Add up to 5 µg DNA into a tube containing 80 µL of frozen or fresh competent cells and transfer to a 2-mm gap electroporation cuvet held on ice.

4. Pulse the cells in Gene Pulser II system using following parameters: 1.5kV, 200 Ω resistance, 25 µF capacitance. Add 1 mL of 1 *M* sorbitol immediately.

5. Plate cells on YPD plates containing selected amounts of zeocin (100, 500, 1000, and 2000 mg/L) and incubate at 28°C for 2 d (*see* **Note 4**).

6. To isolate single colonies, select between 10 and 100 colonies and streak on YPD agar plates. Incubate at 28°C for 2 to 3 d.

3.3. Southern Blot Analysis and Copy Number Determination

1. Grow single colonies in 5 mL of YPD broth, at 28°C and 200 rpm, overnight.

2. Isolate genomic DNA from 2 mL of culture (~10 OD$_{600}$) using MasterPure Yeast DNA Purification Kit (Epicentre).

3. Digest 1 µg of genomic DNA with *Eco*RI and separate on a 0.8 % TAE agarose gel.

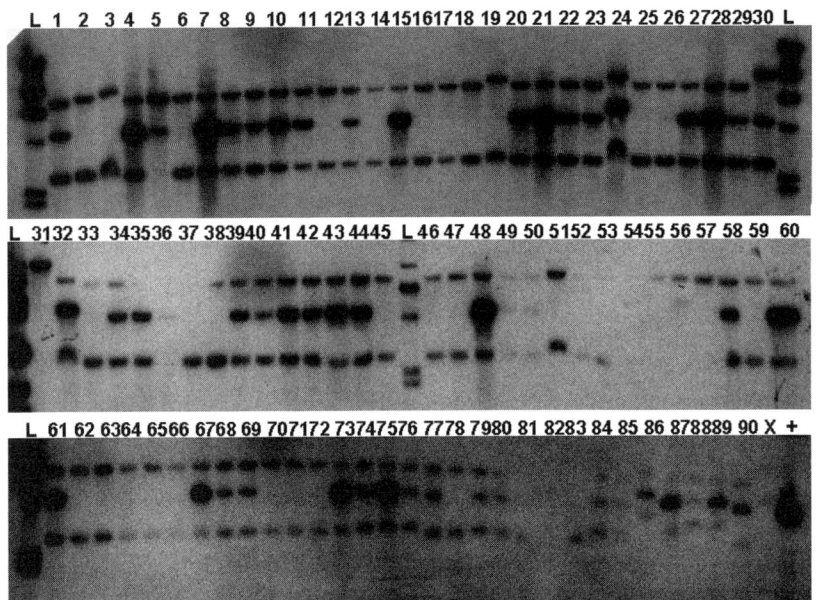

Fig. 1. Southern blot of codon optimized (X-33/Na-ASP1O) clones. Lane numbers correspond to clone numbers. L, DIG labeled DNA molecular weight Marker II (Roche) (23 kb, 9.4 kb, 6.5 kb, 4.4 kb, 2.3 kb, and 2 kb); X, untransformed X-33; +, plasmid pPICZαA/Na-ASP1O.

4. Transfer DNA to a Zeta-Probe GT Genomic Tested Blotting membrane (BioRad), using the method described by Southern *(13)*.
5. Bind DNA fragments to the membrane using an ultraviolet (UV)-crosslinker.
6. Prehybridize membrane for 2 h at 42°C in a hybridization solution from the DIG High Prime DNA Labeling and Detection Starter Kit II (Roche).
7. Hybridize membrane with the DIG labeled *AOX1* promoter, in the same hybridization buffer, for 16 h at 42°C.
8. Complete the washing and detection protocol as described in the kit manual.

Results of the Southern blot analysis of optimized *Na-ASP1* gene clones are shown in **Fig. 1**. The restriction enzyme *Eco*RI cuts the expression plasmid once within the *Na-ASP1* gene. Lane X contains *Eco*RI digested chromosomal DNA from untransformed *P. pastoris* X-33. The DIG labeled *AOX1* promoter hybridized to a ~5.5-kb chromosomal fragment of this host strain. The single plasmid copy clones (e.g., nos. 2, 3) resulted in two bands (a ~2.9-kb chromosomal copy and a ~7.3-kb plasmid copy), resulting from the presence of the plasmid at the *AOX1* locus. Two or higher copy clones resulted in three bands. Two of the three bands were the same as observed in the single-copy clones. The third band with a size of 4.7 kb is the size of the expression plasmid. The intensity of

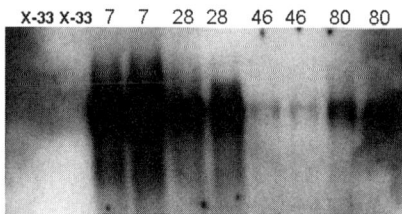

Fig. 2. Northern blot of selected copy number clones of X-33/Na-ASP1O strains. Each sample was run in duplicate. *Lane X-33* is untransformed host strain; *lane 7*:four-copy clone; *lane 28*: three-copy clone; *lane 46*: one-copy clone; *lane 80*: two-copy clone.

the 4.7-kb band increased with the additional copies of the plasmid. To calculate copy number, the intensity of the 4.7-kb band was compared with that of either the 2.9-kb or 7.3-kb bands. For example, the intensity of each of the three bands in clone no. 1 was the same. Therefore there were two copies of the expression plasmid. In *lane 7*, the ratio of the 4.7-kb band to 2.9-kb band was three, indicating a 4-copy clone. The ratio of the 4.7-kb band to either band plus one provided the copy number.

3.4. Northern Blot Analysis

We measured the *Na-ASP1* steady state mRNA levels of selected copy number clones to confirm that the transcription of the *Na-ASP1* gene was not limiting by observing that an increasing gene copy number yielded a higher mRNA level.

1. Grow inoculum for selected clones (nos. 46, 80, 28, and 7 representing 1, 2, 3, and 4-copy clones, respectively) in YPD broth at 28°C and 220 rpm.
2. Inoculate 15 mL BMGY medium and grow at 28°C and 220 rpm for 18 h to between 10 and 15 OD_{600}.
3. Harvest cells at 3000g at room temperature. Resuspend in 15 mL BMMY medium containing 1% methanol, and induce for 12 h at 28°C and 220 rpm.
4. Extract total RNA from the induced cells using the TRI reagent (MRC, Inc.) by the zirconia/silica beads (0.5 mm) disruption method.
5. Run RNA on a 0.9% formaldehyde denaturing gel, and transfer to a nylon membrane.
6. The remaining steps for Northern blot analysis are the same as the Southern analysis except use of a DIG labeled *Na-ASP1O* gene as a probe for hybridization, instead of the *AOX1* probe.

The results of the Northern blot analysis are shown in **Fig. 2**. A 1.5-kb band was detected in all samples except the untransformed X-33 strain because the *Na-ASP1O* gene was used as the probe and no *Na-ASP1O* gene sequences were expected in this control. There was a good correlation between copy number and *Na-ASP1O* specific mRNA levels.

3.5. Shake Flask Induction

We evaluated extracellular Na-ASP1 expression at small scale by semiquantitative dot-blot western analysis. This method enabled us to screen fermentation broth of as many as 90 clones in 2 to 3 d.

1. Set up shake flask experiment using 50-mL baffled shake flasks (Bellco) containing 15 mL of BMGY.
2. Inoculate with 100 μL of YPD broth grown cultures.
3. Grow at 28°C for 18 to 20 h or until the cells reached an OD_{600} of between 10 and 15.
4. Harvest the cells at 3000*g* at room temperature.
5. Induce the cells in 15 mL BMMY medium, containing 1% methanol, for 12 h at 28°C.
6. After induction, collect the supernatant by centrifugation at 5000*g* for dot blot western analysis (*see* **Note 5**).

3.5.1. Semi Quantitative Dot Blots

1. Apply 30 μL of supernatant to each well of the Hybri-Dot blot apparatus containing a presoaked (in methanol) PVDF membrane (BioRad).
2. Using a vacuum, draw the samples through the apparatus.
3. Block the membrane overnight with 5% Blotto (Dry Milk).
4. Treat the membrane with a 1:8000 dilution of Rabbit anti-ASP-1 serum (2 hours) and a 1:5000 dilution of Goat anti-rabbit serum (1 h).
5. Develop the X-ray films following Western Chemiluminescent detection kit.

The results of the dot-blot protein analysis are shown in **Fig. 3**. There were variations in extracellular Na-ASP1 production in selected clones. Some clones did not produce any Na-ASP1 extracellularly. The best Na-ASP1 producing clones (e.g., nos. 2, 3, and 46) were single-copy clones. The X-33 host strain did not produce any Na-ASP1. Consistently, a higher number of copies of the *Na-ASP1* gene resulted in lower amounts of extracellular Na-ASP1 protein. The clone no. 7 (4 copies) produced the lowest amount of Na-ASP1 extracellularly. These results were contradictory to Northern results, in which a gene copy increase correlated with higher *Na-ASP1* mRNA production (*see* **Fig. 2**).

3.5.2. Western Blot Analysis of Cell Extract and Fermentation Broth

Because we have observed a wide range of Na-ASP1 extracellular production in 90 clones, and high-copy clones did not produce Na-ASP1 extracellularly, we selected 10 clones for further evaluation of both intra and extracellular Na-ASP1 production. Five clones with a high-copy-number (nos. 7, 15, 28, 48, and 67) and 5 clones with a low-copy-number (nos. 16, 18, 46, 63, and 80) were induced for Na-ASP1 production.

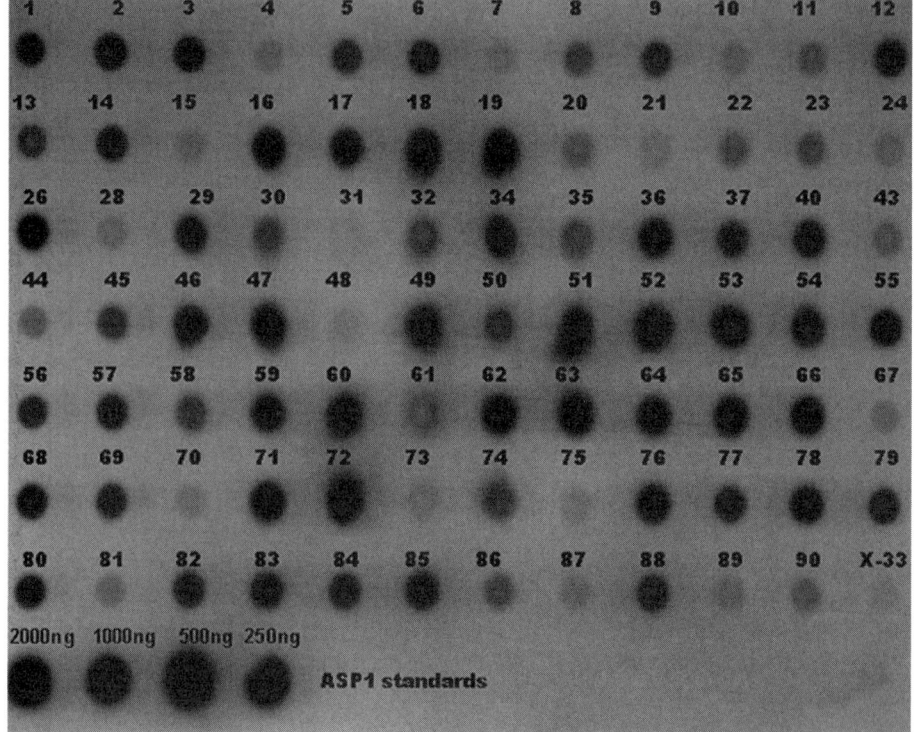

Fig. 3. Semiquantitative dot-western blot. Sample number reflects clone number.

1. Inoculate 15 mL of BMGY medium with 200 µL of cells grown in YPD broth.
2. Place shake flask cultures in a 28°C incubator/shaker until cells reach an OD_{600} of between 10 and 15.
3. Harvest cells by centrifugation ($3000g$) and resuspend in 15 mL BMMY medium.
4. Collect supernatants after a 12-h induction with methanol.
5. Run 20 µL of the fermentation broth on 4 to 20% Tris-Glycine gels.
6. Examine intracellular accumulation of Na-ASP1, by disrupting the cells using a Bead Beater.
7. Perform a total protein assay and separate 20 µg of total protein on the 4 to 20% Tris-Glycine SDS-PAGE gel for each clone.
8. Transfer the proteins to a PVDF membrane for Western blotting (*see* **Fig. 4**).
9. Treat the membranes with a 1:8000 dilution of Rabbit anti-ASP-1 serum (2 h) and a 1:5000 dilution of Goat anti-rabbit serum (1 h).
10. Develop the X-ray films following Western Chemiluminescent detection kit.

The results of the western blot analysis of cell lysates and fermentation broth is shown in **Fig. 4**. The higher copy number clones were not able to secrete

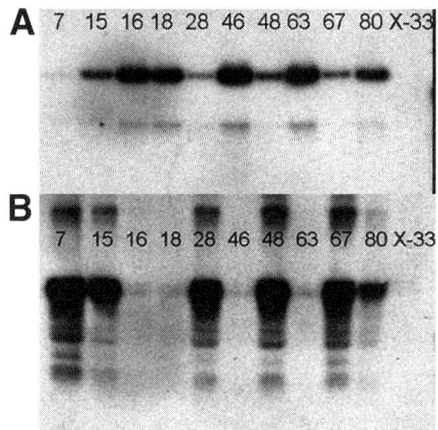

Fig. 4. Western blot of selected clones. (**A**) Extracellular or (**B**) intracellular Na-ASP1 protein. Na-ASP1 production in 12 h induced selected clones. Lane number indicates clone number and X-33 is induced untransformed host sample. *Lane 7:* Four-copy clone; *lanes 15, 28, 48, 67*: three-copy clones; *lane 80:* two-copy clone; and *lanes 16, 18, 46, 63:* single-copy clones.

Na-ASP1 and most of the protein appears to be trapped inside the cell (*see* **Fig. 4B**). The 4-copy clone (no. 7) had higher Na-ASP1 production but almost all of the Na-ASP1 produced accumulated inside the cells. The intracellular Na-ASP1 also had a molecular weight that was higher than expected, indicating that the bottleneck could be in the secretory pathway before the Golgi where α-mating factor is cleaved by Kex2, a dibasic endopeptidase. The unprocessed proregion of α-mating factor would account for a 9-kDA increase in protein size. This problem may in part be caused by the presence of 20 cysteine residues in the Na-ASP1 protein sequence. The large number of cysteines would require a large number of disulfide bonds. If these did not form correctly, the protein may aggregate and remain misfolded in the secretory pathway.

There was a negative correlation with extracellular Na-ASP1 production vs copy number of the *Na-ASP1* gene (*see* **Fig. 5**). However, intracellular Na-ASP1 production did increase with copy number (*see* **Note 6**). A quantitative method was not available at the time of these experiments. Therefore, a qualitative method was used to compare Na-ASP1 production rates. The band intensities on developed films were quantified with an Alpha Innotech 8800 gel imaging system using Alpha EaseFC Software (Alpha Innotech, San Leandro, CA). These results indicate that copy-number increase yields higher Na-ASP1 production intracellularly but the secretory system did not appear to be able to process the protein and successfully secrete it.

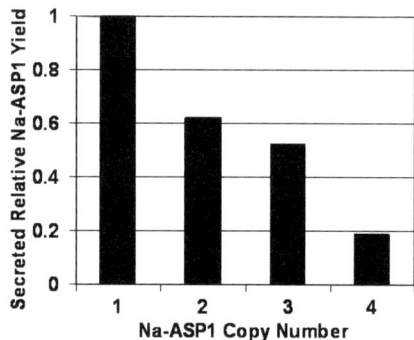

Fig. 5. Effect of copy number on Na-ASP1 secretion levels. Extracellular amount of Na-ASP1 levels were calculated from band intensities of **Fig. 4A** and reported relative to extracellular amount of Na-ASP1 produced from single *Na-ASP1* copy clone number 46.

4. Notes

1. We fused the *Na-ASP1* gene with α-mating factor at the *Xho*I site of pPICZαA. However pPICZαA has another *Xho*I site downstream of *Kpn*I at the multiple cloning site. We were able to clone *Xho*I/*Kpn*I double digested *Na-ASP1* PCR product by only partially digesting pPICZαA vector with *Xho*I and *Kpn*I. The construct was confirmed by sequencing the plasmid.

2. In the first round of transformations, we screened 10 single *Na-ASP1* copy clones by inducing in a shake flask. We observed Na-ASP1 secretion of these clones but expression levels were low (10 mg/L).

3. *P. pastoris* codon usage table was obtained from http://www.kazusa.or.jp/codon/ and compared with the *Na-ASP1* cDNA codon table. There were many suboptimal codons. Therefore we constructed a *Na-ASP1* gene synthetically using a commercial source.

4. We observed varying activity of zeocin depending on the YPD plate lot to lot preparation. Although a 100 mg/L Zeocin concentration gave usually single-copy integrants, we obtained 3 to 4 copy integrants with the same concentrations of zeocin plates.

5. *P. pastoris* clones usually produce enough protein after 12 h of induction for clone screening purposes. The first 12 h induction gives an indication of the rate of protein production rather than total protein production. However, if you induce for longer than 12 h, the methanol concentration should be measured externally by a GC method or with a methanol sensor to make sure the culture does not run out of methanol.

6. We usually see an increase in intracellular recombinant protein production with increased copy number up to four copies. With this particular protein (Na-ASP1), increased copy number had a negative impact on secreted Na-ASP1 level. We suggest the selection of a range of copy number containing clones rather than just obtaining a few high-copy-number clones.

References

1. Cregg, J. M., Madden, K. R., Barringer, K. J., Thill, G. P., and Stillman, C. A. (1989) Functional characterization of the two alcohol oxidase genes from the yeast *Pichia pastoris. Mol. Cell Biol.* **9,** 1316–1323.

2. Cregg, J. M. and Madden, K. R. (1987) Development of yeast transformation systems and construction of methanol-utilization-defective mutants of *Pichia pastoris* by gene disruption. *Biological Research on Industrial Yeast* II, 1–18.

3. Cregg, J. M., Vedvick, T. S., and Raschke, W. C. (1993) Recent advances in the expression of foreign genes in *Pichia pastoris. Biotechnology (N Y)* **11,** 905–910.

4. Koutz, P., Davis, G. R., Stillman, C., Barringer, K., Cregg, J., and Thill, G. (1989) Structural comparison of the *Pichia pastoris* alcohol oxidase genes. *Yeast* **5,** 167–177.

5. de Silva, N. R., Brooker, S., Hotez, P. J., Montresor, A., Engels, D., and Savioli, L. (2003) Soil-transmitted helminth infection: updating the global picture. *Trends Parasitol.* **19,** 547–551.

6. Hotez, P. J., Zhan, B., Bethony, J. M., et al. (2003) Progress in the development of a recombinant vaccine for human hookworm disease: the Human Hookworm Vaccine Initiative. *Int. J. Parasitol.* **33,** 1245–1258.

7. Hawdon, J. M. and Hotez, P. J. (1996) Hookworm: developmental biology of the infectious process. *Curr. Opin. Genet. Dev.* **6,** 618–623.

8. Zhan, B. H., Hotez, P. J., Wang, Y., and Hawdon, J. M. (2002) A developmentally regulated metalloprotease secreted by host-stimulated *Ancylostoma caninum* third stage infective larvae is a member of astacin family of proteases. *Mol. Biochem. Parasitol.* **120,** 291–296.

9. Hawdon, J. M., Jones, B. F., Hoffman, D. R., and Hotez, P. J. (1996) Cloning and characterization of *Ancylostoma*-secreted protein; A novel protein associated with the transition to parasitism by infective hookworm larvae. *J. Biol. Chem.* **271,** 6672–6678.

10. Hawdon, J. M., Narasimhan, S., and Hotez, P. J. (1999) *Ancylostoma* secreted protein 2: cloning and characterization of a second member of a family of nematode secreted proteins from *Ancylostoma caninum. Mol. Biochem. Parasitol.* **99,** 149–165.

11. Zhan, B., Hawdon, J., Shan, Q., et al. (1999) *Ancylostoma* secreted protein 1 (ASP-1) homologues in human hookworms. *Mol. Biochem. Parasitol.* **98,** 143–149.

12. Sambrook, J. R., DW. (2001) *Molecular Cloning; A Laboratory Manual*, Third ed. Cold Spring Harbor, New York, Cold Spring Harbor Laboratory Press.

13. Southern, E. (1975) Detection of specific sequences among DNA fragments separated by gel electrophoresis. *J. Mol. Biol.* **98,** 503–517.

6

Purification of the N- and C-Terminal Subdomains of Recombinant Heavy Chain Fragment C of Botulinum Neurotoxin Serotype C

Jicai Huang, Rick Barent, Mehmet Inan, Mark Gouthro, Virginia P. Roxas, Leonard A. Smith, and Michael M. Meagher

Abstract

The N-terminal and C-terminal portions of the heavy chain fragment C from botulinum neu-rotoxin serotype C [rBoNT(H$_C$)] were expressed in *Pichia pastoris* and purified by ion-exchange chromotography (IEC). The N-terminal fragment, rBoNTC(H$_C$)-N, was purified in three IEC steps: a Q Sepharose Fast Flow (FF) capture step followed by a negative SP Sepharose FF step, and finally, Q Sepharose FF as a polishing step. The purification process resulted in greater than 90% pure rBoNTC(H$_C$)-N based on SDS-PAGE, and yielded up to 1.02 g of rBoNTC(H$_C$)-N/kg of cells. Alternately, the C-terminal fragment, rBoNTC(H$_C$)-C, was purified by using a SP Sepharose FF capture step followed by a second SP Sepharose FF step, and finally a Q Sepharose FF as a polishing step. This purification process resulted in greater than 95% pure rBoNTC(H$_C$)-C based on SDS-PAGE, and yielded up to 0.2 g of rBoNTC(H$_C$)-C/kg cells. The final protein yield is a function of protein expression level during fermentation and the purification methods, and usually final protein yield between 0.1 and 2 mg/g cells is acceptable. Another concern is protein degradation. Especially with *Pichia*, protease activity during cell lysis and purification is always an issue. The importance of N-terminal degradation depends on product and its function. N-terminal sequencing revealed that the purified rBoNTC(H$_C$)-N is missing the first eight amino acids of the N-terminus of the protein, whereas the purified rBoNTC(H$_C$)-C protein is intact. After a mouse bioassay test, both the intact rBoNTC(H$_C$)-C and the rBoNTC(H$_C$)-N missing the first eight amino acids of the N-terminus have vaccine potency; consequently, partial degradation did not have an impact on these protein's utility.

Key Words: *Pichia pastoris;* botulinum neurotoxin serotype C [rBoNTC(H$_C$)] protein purifi-cation; SP sepharose fast flow; Q sepharose fast flow; anion exchange chromatography; cation exchange chromatography; SDS-PAGE; isoelectric point.

From: *Methods in Molecular Biology, vol. 389:* Pichia *Protocols, Second Edition*
Edited by: J. M. Cregg © Humana Press Inc., Totowa, NJ

1. Introduction

The bacterium *Clostridium botulinum* produces seven serologically distinct toxins (A–G), each of which is extremely toxic to man *(1)*. Exposure to these toxins results in the disease botulism, which leads to fatal paralysis of the respiratory muscles *(2)*. Active botulinum neurotoxins are produced by post-translational modification of a 150-kDa precursor. The modification results in 2 peptides connected by a disulfide bond consisting of a 100-kDa heavy chain and a 50-kDa light chain *(3,4)*. The C-terminal portion of the heavy chain (H_c) binds to specific receptors on the cholinergic nerve cells *(5–10)*. The N-terminus of the heavy chain plays a role in translocation of the light chain, a zinc-endopeptidase, which targets nerve cells, across the phospholipid membrane *(11–14)*. The H_c fragments of the botulinum neurotoxins were found to provide increased levels of protective immunity in animal models *(15)*. Recombinant vaccine candidates that are stable, safe, and highly efficacious have been produced for serotypes A, B, C, E and F *(16–24)*.

The heavy chain fragment C from botulinum neurotoxin serotype C [rBoNT(H_c)] has been successfully produced and purified in *Pichia* and determined to be an effective vaccine in a mouse model. That work has been submitted for publication by this group. The objective of this work was to determine the effectiveness of the N-terminal and C-terminal halves of the BoNTC(H_c) as vaccine candidates. Though the N- and C-terminal portions were found to be ineffective as vaccines, this work illustrates the purification of recombinant proteins expressed intracellularly in *Pichia pastoris* that we have conducted at the University of Nebraska-Lincoln Biological Process Development Facility over the last 12 yr.

Here we express the N-t and C-terminal portions of the heavy chain fragment C from rBoNT(H_c), then use ion exchange chromotography (IEC) to purify them from soluble cell lysate. IEC is a method commonly used for protein purification. IEC works in three steps. First, a soluble cell lysate containing the target protein is mixed with a binding buffer and passed through a matrix (resin); here the target protein should be electrostatically adsorbed to the resin. Second, washing buffers are passed through the column to remove contaminating proteins. Third, an elution buffer is passed through to elute the target protein. The ability for the target protein to bind depends on several factors such as the target protein's isoelectric point (pI), the pH of the buffers, the column working volume, ionic strength (salt content and concentration) of the washing buffers. By altering the pH and salt concentration of subsequent buffers, we can first elute contaminating proteins, then the target protein. (For greater detail, IEC is more thoroughly described by Scope and colleagues in *[27]*.)

Good IEC depends on selection of the proper resin. There are two kinds of IEC resins: anionic and cationic. When the pH of a buffer solution for dissolving the protein is higher than the protein's pI, the protein will have a net negative charge and will bind to an anionic exchange resin. When the pH of the buffer is lower than the protein's pI point, the protein will have a net positive charge and will bind to a cationic exchange resin. A number of ion exchange resins are commercially available with different market brand names such as Q sepharose, Poros Q, and Source Q series for strong anionic exchange resins, and DEAE and PI series for weak anion exchange resins, SP sepharose, Poros SP and Source S series for strong cationic exchange resins, CM series for weak cationic exchange resins (*see* catalogs of GE Healthcare, TosoHaas, PerSeptive Biosystems BioRad, and EM Science). The terms strong and weak refer to the extent that the ionization state of the functional groups varies with pH. For strong ion exchangers, they are fully charged at all unstable pH values, usually pH 1 to 10 *(27)*. For additional reading on selection and use of ion exchange resins, GE Lifesciences (formerly Amersham Biosciences) offers the following guide *Ion Exchange Chromatography & Chromatofocusing, Principles and Methods* (cat. no. 11-0004-21).

In order to purify the N- and C-termini of the rBoNTC(H_c) protein, we used the following resins based on theoretical pI calculations. First, the theoretical pI of rBoNTC(H_c)-N protein is 4.47. Thus we start with a Q Sepharose FF strong anionic exchange column at pH 7.5 for the capture step, followed by a negative SP Sepharose FF strong cationic exchange column at pH 5.5 for the purification step, and a final Q Sepharose FF strong anionic exchange column at pH 5.5 was used for polishing step. In the second case, the theoretical pI point of rBoNT C(H_c)-C protein is 8.71, a SP Sepharose FF strong cationic exchange column at pH 6.0 was used for the capture step, a SP Sepharose FF strong cationic exchange column at pH 7.0 was used for purification step and a final negative Q Sepharose FF strong anionic exchange column at pH 7.0 was used for polishing step.

2. Materials

2.1. Fermentation Media and Equipment

1. Minimal Dextrose medium (MD): 1.34% YNB (w/o amino acids), $4 \times 10^{-5}\%$ biotin, and 2% glucose.
2. YPD agar: 1% yeast extract, 2% peptone, 2% dextrose, and 2% agar.
3. Geneticin (G418 Sulfate) (cat. no. 10131-035; Invitrogen, Carlsbad, CA).
4. Buffered minimal glycerol complex medium (BMGY): 1% yeast extract, 2% peptone, 1% glycerol, 1.34% YNB (w/o amino acids), $4 \times 10^{-5}\%$ biotin and 100 *mM* potassium phosphate (pH 6.0).
5. 1-L baffled shake flask.

6. PTM1 trace mineral salts: 6.0 g/L cupric sulfate pentahydrate, 0.08 g/L sodium iodide, 3.0 g/L manganese sulfate monohydrate, 0.20 g/L sodium molybdate dehydrate, 0.020 g/L boric acid, 0.50 g/L cobalt chloride, 20.0g/L zinc chloride, 65 g/L ferrous sulfate hepahydrate, 0.2 g/L biotin, and 5 mL/L sulfuric acid.

7. Glycerol feed: 63% *(w/v)* glycerol with 12 mL/L PTM1.

8. Basal salts medium: 26.7 mL/L phosphoric acid, 85%, 0.93 g/L calcium sulfate-dihydrate, 18.2 g/L potassium sulfate, 14.9 g/L magnesium sulfate-heptahydrate, 4.13 g/L potassium hydroxide, and 40 g/L glycerol.

9. Concentrated ammonium hydroxide solution (30% *[w/v]*).

10. Methanol (100%).

11. MC-168 Methanol Monitor and Controller (PTI Instruments, Kathleen, GA) (*see* **Note 1**).

12. Innova 4000 shaker (New Brunswick Scientific, Edison, NJ).

13. Bioflo3000 5 L fermenters (New Brunswick Scientific).

2.2. Protein Purification, Chromotography, and Analysis

2.2.1. Chromotography Equipment

1. ABI BioCad or Vision Chromatography System (Applied Biosystems) or equivalent chromatography system.

2. 10-mm diameter × 100-mm length chromatography columns.

3. Bead blender homogenizer and zirconia/silica beads (0.5 mm) (BioSpec, Bartlesville, OK).

4. Q Sepharose Fast Flow (cat. no. 17-0510-10; GE Healthcare Life Sciences, Piscataway, NJ).

5. SP Sepharose Fast Flow (cat. no. 17-0729-10; GE Healthcare Life Sciences).

6. Centrifuge and centrifuge bottles.

7. 0.2 μm Fluorodyne II filters and bottle top 0.2 μm filters (Pall Corporation, East Hills, NY).

8. Omega membrane with 10 KDa molecular weight cut-off (Pall Corportation).

9. Spectra/Por dialysis tubing, pore size 6-8 kDA (Spectrum Laboratories, Dominguez Hills, CA), or Slide-a-Lyzer 3.5 kDA cassette (Pierce, Rockford, IL).

10. Amicon stirred cell (cat. no. 5124; Millipore, Billerica, MA).

11. Bicinchoninic acid assay (BCA) (cat. no. 23225; Pierce, Rockford, IL).

12. SDS-PAGE equipment and supplies.

13. Western immunoblotting equipment and supplies.

14. Primary antibody: rabbit anti-rBoNTC(Hc), kindly provided by United States Army Medical Research Institute of Infectious Diseases (USAMRIID), Integrated Toxicology Division, Fort Detrick, MD.

2.2.2. Buffers for Purifying rBoNTC(H$_c$)-N

1. Phosphate buffered saline (PBS) buffer: 10 mM Na phosphate, 120 mM NaCl, and 2.7 mM KCl (pH 7.4).

2. Cell lysis buffer for rBoNTC(H$_c$)-N: 50 mM Na phosphate, 2 mM ehtylene diamine tetraacetic acid (EDTA), and 2 mM phenylmethylsulphonylfluoride (PMSF) (pH 7.5).
3. Equilibration and wash buffer for Q Sepharose FF Capture step of rBoNTC(H$_c$)-N: 50 mM Na phosphate, and 2 mM EDTA (pH 7.5).
4. Elution buffer for Q Sepharose FF capture step for rBoNTC(H$_c$)-N: 50 mM Na phosphate, 1M NaCl, and 2 mM EDTA (pH 7.5).
5. Dialysis buffer for Q Sepharose FF capture step for rBoNTC(H$_c$)-N: 50 mM Na-phosphate, and 2 mM EDTA (pH 5.5).
6. Equilibration and wash buffer for Negative SP Sepharose FF second step and Final Q Sepharose Fast Flow polishing step for rBoNTC(H$_c$)-N: 50 mM Na phosphate, and 2 mM EDTA (pH 5.5).
7. Elution buffer for Negative SP Sepharose FF second step and Final Q Sepharose Fast Flow polishing step for rBoNTC(H$_c$)-N: 50 mM Na phosphate, 2 mM EDTA, and 1M NaCl (pH 5.5).

2.2.3. Buffers for Purifying of rBoNTC(H$_c$)-C

1. PBS buffer: 10 mM Na phosphate, 120 mM NaCl, and 2.7 mM KCl (pH 7.4).
2. Cell lysis buffer for rBoNTC(H$_c$)-C: 50 mM Na phosphate, 2 mM EDTA, and 2 mM PMSF (pH 6.0).
3. Equilibration and wash buffer for SP Sepharose FF capture step for rBoNTC(H$_c$)-C: 50 mM Na phosphate, and 2 mM EDTA (pH 6.0).
4. Elution buffer for SP Sepharose FF capture step for rBoNTC(H$_c$)-C: 50 mM Na phosphate, 2 mM EDTA, and 1M NaCl (pH 6.0).
5. Dialysis buffer for SP Sepharose FF second step and negative Q Separose Fast Flow step for rBoNTC(H$_c$)-C: 50 mM phosphate-Na, and 2 mM EDTA (pH 7.0).
6. Equilibration and wash buffer for SP Sepharose FF second step and Negative Q Sepharose Fast Flow polishing step for rBoNTC(H$_c$)-C: 50 mM Na phosphate, and 2 mM EDTA (pH 7.0).
   a. Elution buffer for SP Sepharose FF second step and Negative Q Sepharose FF polishing step for rBoNTC(H$_c$)-C: 50 mM Na phosphate, 2 mM EDTA, and 1M NaCl, (pH 7.0).

3. Methods

The methods below outline: (1) the cloning and expression of the rBoNTC(H$_c$)-N and rBoNTC(H$_c$)-C fragments, (2) the preparation of *P. pastoris* cells for extraction of soluble proteins, (3) the purification of proteins, and (4) the analysis of the final products.

3.1. Cloning and Fermentation

The N-terminal and C-terminal subdomains were cloned from an engineered gene encoding the nontoxic 50 kDa carboxy terminal fragment Hc of BoNTC. rBoNTC(H$_c$)-N was created by incorporating stop codons (TAA and TAG) at position 252 in the protein sequence by polymerase chain reaction (PCR).

For rBoNTC(H$_c$)-C, the 5′ PCR primer was designed to introduce an ATG codon before amino acid position 253. PCR fragments were ligated into a vector (pCR2.1 from Invitrogen, Carlsbad, CA) and then subcloned into the *Pichia pastoris* expression vector pHILD4. The expression cassette (*see* **Fig. 1**) was integrated into the chromosomal alcohol oxidase of *P. pastoris* GS115 strain by electroporation.

The transformed cells were plated onto MD plates lacking histidine. After 2 to 3 d, cells were pooled and spread onto YPD agar containing selected amounts (0.25–2.0 mg/mL) of Geneticin (G418 sulfate). Yeast transformants expressing the selectable markers were isolated and characterized. Positive clones were selected for growth in a 5-L scale benchtop fermentor under the control of methanol induction.

For both proteins, fermentation was performed as follows (for more complete details about optimization of *P. pastoris* fermentation, please *see* **Chapter 4** and *[25,26]*) strains of *P. pastoris* transformed with derivatives of plasmid pHILD4 containing either the BoNTC(H$_c$)-N or BoNTC(H$_c$)-C were grown as seed cultures by innoculating a 1-L baffled shake flask containing 200 mL BMGY with 1 mL of the frozen stock. The seed cultures were incubated in a shaker at 28°C and 200 rpm until A$_{600}$ between 4 and 6 was reached. The entire seed culture was then used to inoculate a 5-L fermentor containing 2 L basal salts medium plus 4.35 mL/L PTM1 trace mineral salts. Note, prior to inoculation, the pH in the fermentor was adjusted to 5.0 using concentrated ammonium hydroxide solution. The fermentation was run initially in batch mode at 30°C with a pH 5.0 and dissolved oxygen (DO) of 30% saturation. DO was maintained by adjusting agitation rate and pure oxygen supply. pH was maintained by further addition of ammonium hydroxide. When glycerol in the batch medium was exhausted (seen as a DO spike), we switch to a fed-batch mode using a regulated glycerol feed initiated at 13.3 g/L/h for 1 h, then linearly ramped down from 13.3 g/L/h to 0 over the next 3 h. (Note, a cell density of about 100 g/L [wet cell weight] was achieved by the end of batch phase.)

After 1 h of glycerol feed, 2 mL/L methanol was added to the fermentor. Once the initial methanol was consumed, as indicated by the DO spike or by a methanol sensor (*see* **Note 1**), a methanol feed was initiated using an exponential feed rate protocol to maintain a growth rate of μ = 0.0267/h. The feed solution was 100% methanol with 12 mL/L PTM1 salts. The total induction time on methanol was 27 h. Cells were harvested by centrifugation and stored frozen at –80°C.

3.2. Purification of the N-Terminus Fragment of the rBoNTC(H$_c$) Protein

Described below are the three IEC steps used to purify the N-terminal fragment of the rBoNTC(H$_c$) protein. Chromatographic separations were performed on an ABI BioCad or ABI Vision chromatography workstation (Applied Biosciences, Foster City, CA). First, a Q Sepharose Fast Flow (FF) resin

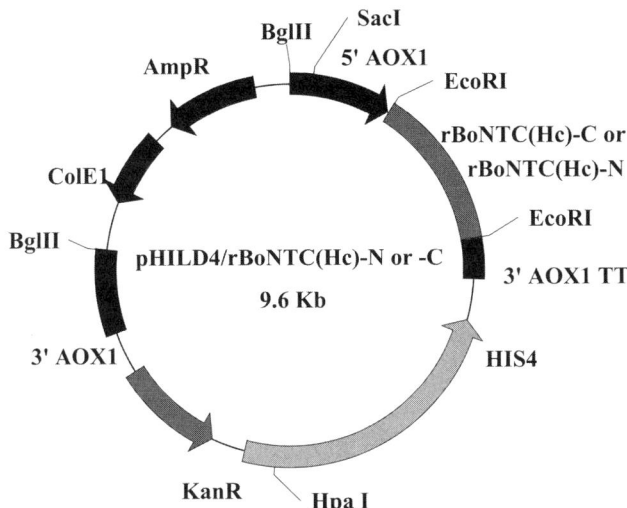

Fig. 1. Diagram of plasmids pHILD4/rBoNTC(H$_c$)-N and pHILD4/rBoNTC(H$_c$)-C.

(*see* **Note 2**) was used to capture the rBoNTC(H$_c$)-N protein from soluble cell lysates. Next, a negative SP Sepharose FF (*see* **Note 2**) was used for further purification. Finally, a Q Sepharose FF column was used for final polishing to further increase the product purity. As the theoretical PI for rBoNTC(H$_c$)-N protein is 4.47, while the theoretical PI for rBoNTC(H$_c$)-C protein is 8.71, a strong anion exchange column was used for capture the rBoNTC(H$_c$)-N protein and a strong cation exchange column was selected for capture the rBoNTC(H$_c$)-C protein (*see* **Note 2**).

3.2.1. Cell Disruption

1. Wash *P. pastoris* cell mass once with cell lysis buffer (*see* **Subheading 2.2.2.**, **item 2**).
2. Suspend the washed cell mass at a ratio of 5 g cell/25 mL in cell lysis buffer (*see* **Note 3**). Mix 25 mL of 0.5 mm zirconia/silica beads and 25-mL cell suspension solution into the bottle of a bead blender homogenizer. Disrupt the cells by 5 cycles of 30 s on then 60 s off in an ice-water bath. Separate the beads from cellular materials by a sintered glass filter with 0.1-mm pore size and wash the beads 3× using 100-mL lysis buffer. To collect the most product possible, the washes and filtered solution were mixed together to maximize yield prior to centrifugation.
3. Centrifuge the cell lysate at 11,300*g* for 30 min at 4°C to remove cell debris. Filter the supernatant with a 0.2 μm fluorodyne II membrane filter and store in ice bucket or 4°C cold room.

3.2.2. Q Sepharose FF Capture Step

1. Pack a chromotography column with Q Sepharose FF and equilibrate with equilibration buffer (*see* **Subheading 2.2.2.**, **item 3**) at flow rate of 200 cm/h for

5 column volumes (CVs) (the volume of resin). Flow rate is usually measured in mL/min. But when comparing results between different size columns or when scaling-up, it is useful to use linear flow rate, cm/hr. The formula to convert between volumetric flow and linear flow rate is: linear flow rate (cm/hr) = volumetric flow rate (mL/min) × 60/column cross section area (cm²).

2. Load the cell lysate at a flow rate of 150 cm/h. After loading, wash the column with wash buffer (*see* **Subheading 2.2.2.**, **item 3**) at a flow rate of 200 cm/h for 5 CVs or until the absorbance at 280 nm returns to that of the buffer or near to the buffer baseline.

3. Elute the product using a 20 CVs linear gradient from wash buffer to elution buffer (*see* **Subheading 2.2.2.**, **item 4**) (*see* **Note 4**). Pool the protein peaks that contain the product and analyze by SDS-PAGE, western blot and BCA protein assay (*see* **Fig. 2**). Fractions can be collected using the Vision fraction collector (set volume) or manually. Peaks containing the protein are identified by SDS-PAGE and Western immunoblot. Because this is a capture step, the objective is to collect as much of the product as possible. As a general rule of thumb, if a fraction contains at least 15 to 20% of the product based on the western immunoblot and SDS-PAGE, the fraction is pooled.

4. Dialyze the rBoNTC(H$_c$)-N protein using a 6- to 8-kDa Spectra/Por dialysis tubing at 4°C overnight against at least a 50-fold excess volume of dialysis buffer (*see* **Subheading 2.2.2.**, **item 5**) (*see* **Note 5**).

3.2.3. Negative SP Sepharose FF Second Step

1. Pack a chromotography column with SP Sepharose FF and equilibrate with equilibration buffer at a flow rate of 150 cm/h for 5 CVs.

2. Load the dialyzed sample at a flow rate of 70 cm/h. Collect the rBoNTC(H$_c$)-N protein, which is found in the flowthrough. After loading, wash the column with wash buffer (*see* **Subheading 2.2.2.**, **item 6**) at a flow rate of 150 cm/h for 5 CVs or until the absorbance at 280 nm returns to that of the buffer.

3. Elute protein using 5 CVs of a linear gradient from wash buffer to elution buffer (*see* **Subheading 2.2.2.**, **item 7**). Collect the elution fractions. Analyze the flowthrough and elution fractions by SDS-PAGE, western blotting and BCA protein assay (*see* **Fig. 3**).

3.2.4. Final Q Sepharose FF Polishing Step

1. Pack a chromotography column with Q Sepharose FF and equilibrate with equilibration buffer (*see* **Subheading 2.2.2.**, **item 6**) at a flow rate of 200 cm/h for 5 column volumes (CVs).

2. Load the flowthrough from the negative SP Sepharose FF column at a flow rate of 150 cm/h. After loading, wash the column with wash buffer (*see* **Subheading 2.2.2.**,

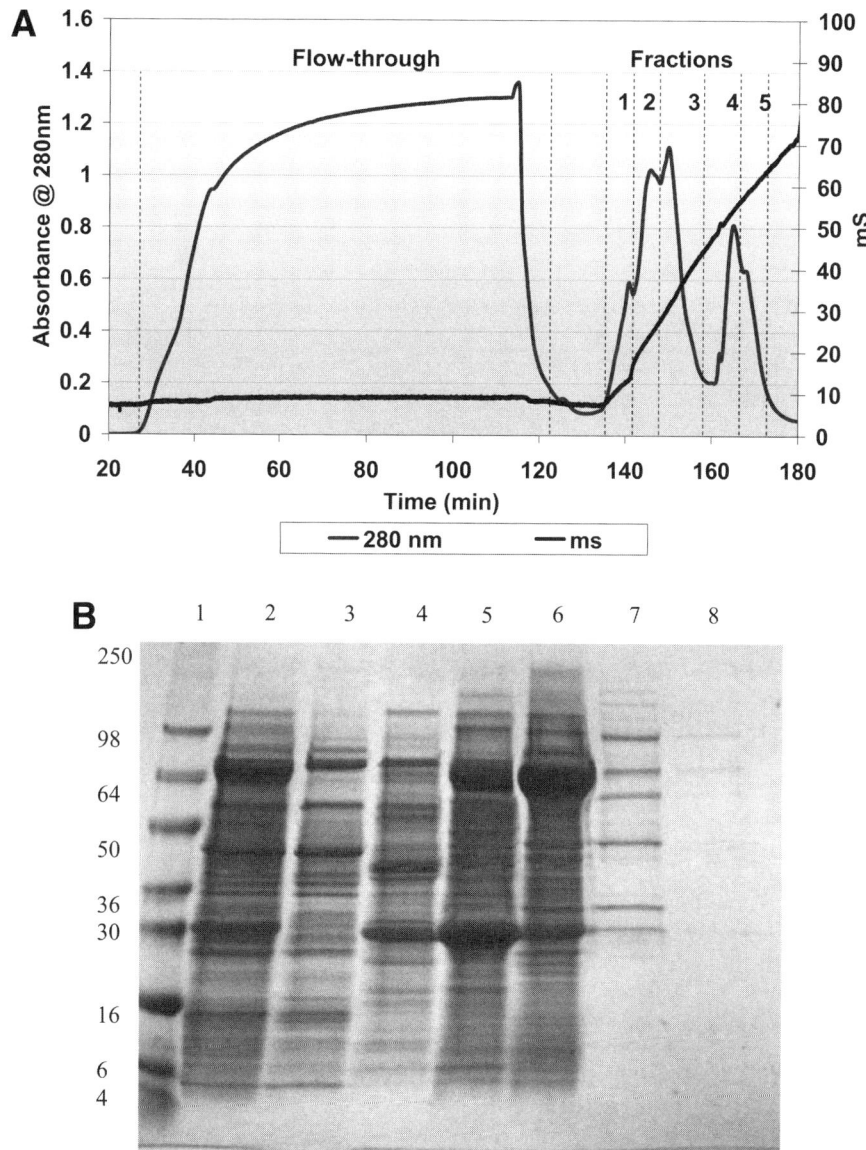

Fig. 2. (**A**) Chromatogram of Q Sepharose Fast Flow capture step for rBoNTC(H$_c$)-N purification. (**B**) SDS-PAGE of Q Sepharose Fast Flow capture step for rBoNTC(H$_c$)-N. Fractions were collected and protein separated on a 4 to 20% Tris–Glycine gel (Novex). The gel was stained with Commassie Brilliant Blue. *Lane 1:* Novex See Blue Marker; *lane 2:* preload sample; *lane 3:* flow through; *lanes 4–8:* fractions 1–5.

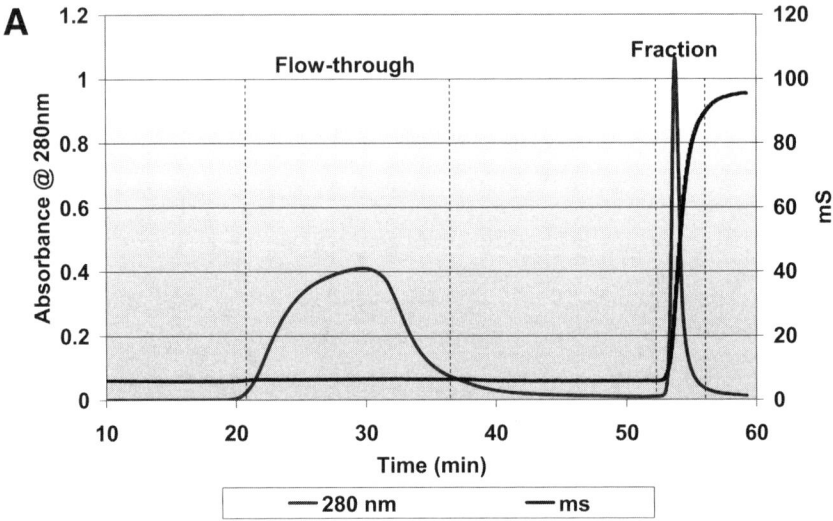

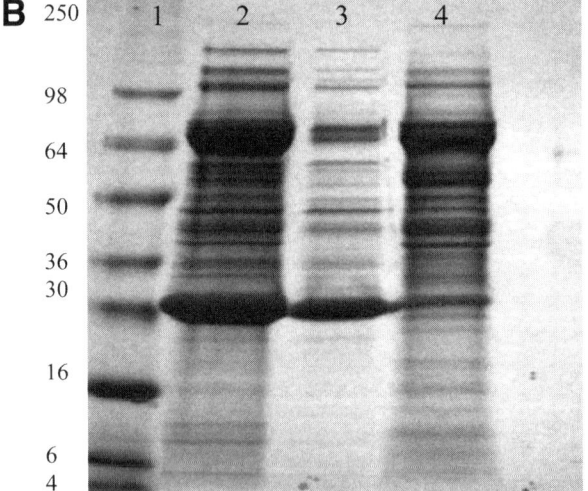

Fig. 3. **(A)** Chromatogram of Negative SP Sepharose Fast Flow second step for rBoNTC(H$_c$)-N purification. **(B)** SDS-PAGE of SP Sepharose Fast Flow second step for rBoNTC(H$_c$)-N. Fractions were collected and protein separated on a 4 to 20% Tris-Glycine gel from Novex. The gel was stained with Commasie Brilliant Blue. *Lane 1:* Novex See Blue Marker; *lane 2:* preload sample; *lane 3:* flow through; *lane 4:* eluted fraction.

    item 6) at a flow rate of 200 cm/h for 5 CVs or until the absorbance at 280 nm returns to that of the buffer.
3. Elute the product using a 20 CVs linear gradient from wash buffer to elution buffer (*see* **Subheading 2.2.2.**, **item 7**). Pool the protein peaks and analyze by SDS-PAGE, western blotting and BCA protein assay (*see* **Fig. 4**).

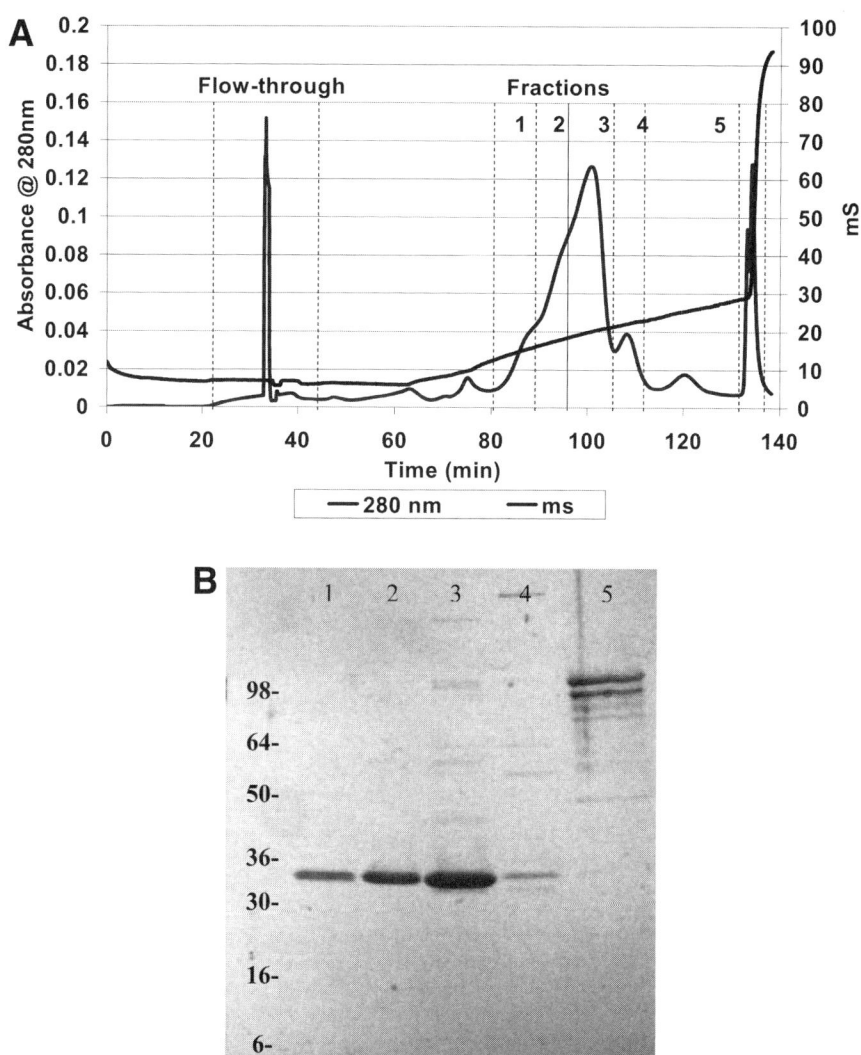

Fig. 4. (**A**) Chromatogram of final Q Sepharose Fast Flow polishing step for rBoNTC(H$_c$)-N purification. (**B**) SDS-PAGE of the final Q Sepharose Fast Flow polishing step for rBoNTC(H$_c$)-N. Fractions were collected and protein separated on a 4 to 20% Tris-Glycine gel (Novex). The gel was stained with Commasie Brilliant Blue. *Lanes 1–5:* fractions 1–5.

4. Dialyze the peak fraction containing rBoNTC(H$_c$)-N protein using a 3.5-kDa Slide-A-Lyzer cassette at 4°C overnight against at least a 50-fold excess volume of PBS buffer (*see* **Subheading 2.2.2.**, **item 1**). Final product analysis was performed using SDS-PAGE (*see* **Fig. 5**), BCA protein assay and N-terminal amino acid sequencing.

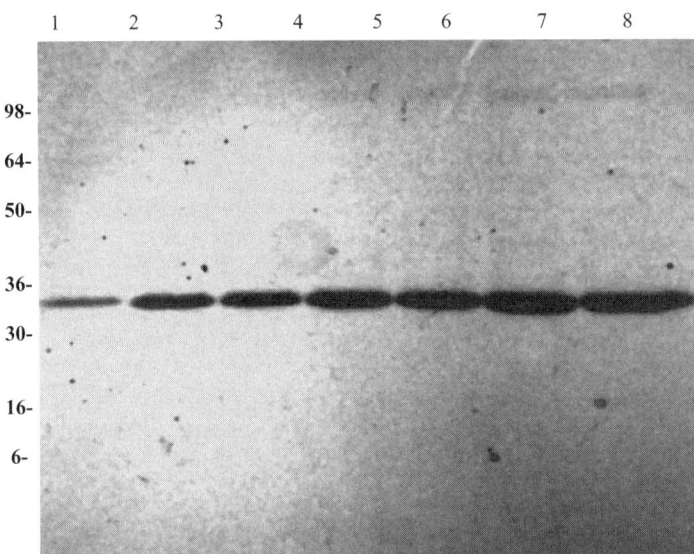

Fig. 5. SDS-PAGE of rBoNTC(H_c)-N protein. The dialyzed product from fraction 3 of the Q Sepharose Fast Flow final step was separated on a 4 to 20% TrisGlycine Novex gel. The gel was stained with commassie blue. Selected concentrations of protein were loaded to check purity. *Lanes 1, 2:* 1 μg rBoNTC(H_c)-N protein sample; *lanes 3, 4:* 3 μg sample; *lanes 5, 6:* 5 μg sample; *lanes 7, 8:* 7.25 μg sample. The protein concentration was determined by Pierce BCA kit using BSA as standard.

3.3. Purification of the C-Terminus Fragment of rBoNTC(H_c)

Described below are the three IEC steps used to purify the C-terminal fragment of the rBoNTC(H_c) protein. First, a SP Sepharose FF column was used to capture the rBoNTC(H_c)-C protein from soluble cell lysates. A second SP Sepharose FF column was used for a second chromatographic step. Finally, a final negative Q Sepharose FF column was used for final polishing to further increase the product purity.

3.3.1. Cell Disruption

1. Wash *P. pastoris* cell mass once using cell lysis buffer (*see* **Subheading 2.2.3., item 2**).
2. Suspend the washed cell mass at a ratio of 5 g cell/25 mL in cell lysis buffer (*see* **Note 3**). Mix 25 mL of 0.5 mm zirconia/silica beads and 25-mL cell suspension solution into the bottle of a bead blender homogenizer. Disrupt the cells by 5 cycles of 30 s on then 60 s off in an ice-water bath. Separate the beads from cellular materials by a sintered glass filter with 0.1-mm pore size and wash the beads 3× using 100-mL lysis buffer. The wash buffer and filtered solution were mixed together to maximize yield prior to centrifugation.

3. Centrifuge the cell lysate at 11,300*g* for 30 min at 4°C to remove cell debris. Filter the supernatant with a 0.2-μm fluorodyne II membrane filter and store on ice.

3.3.2. SP Sepharose FF Capture Step

1. Pack a chromotography column with SP Sepharose FF column and equilibrate with equilibration buffer (*see* **Subheading 2.2.3., item 3**) at a flow rate of 200 cm/h for 5 CVs.
2. Load the cell lysate at a flow rate of 150 cm/h. After loading, wash the column with wash buffer (*see* **Subheading 2.2.3., item 3**) at a flow rate of 200 cm/hr for 5 CVs or until the absorbance at 280 nm returns to that of the buffer or near to the buffer baseline.
3. Elute the product using a 30 CVs linear gradient from wash buffer to elution buffer (*see* **Subheading 2.2.3., item 4**) (*see* **Note 4**). By increasing the length of the gradient we were able to improve separation and the purity of the product. Pool the protein peaks and analyze by SDS-PAGE, western blotting and BCA protein assay (*see* **Fig. 6**).
4. Dialyze the rBoNT C(H$_c$)-C protein using a 6- to 8-kDa Spectra/Por dialysis tubing at 4°C overnight against at least a 50-fold excess volume of dialysis buffer (*see* **Subheading 2.2.3., item 5**).

3.3.3. SP Sepharose FF Second Step

1. Pack a second chromotography column with SP Sepharose FF and equilibrate with equilibration buffer (*see* **Subheading 2.2.3., item 6**) at flow rate of 150 cm/h for 5 CVs. (Note, it is not uncommon to use the same resin for a subsequent step as chromatographic behavior of proteins can vary as a function of the type and amount of contaminating materials present from the same starting material.)
2. Load the dialyzed sample of first SP Sepharose column at a flow rate of 70 cm/h. After loading, wash the column with wash buffer (*see* **Subheading 2.2.3., item 6**) at a flow rate of 150 cm/h for 5 CVs or until the absorbance at 280nm returns to that of the buffer.
3. Elute the product using a 20 CVs linear gradient from wash buffer to elution buffer (*see* **Subheading 2.2.3., item 7**). Collect the elution fractions. Analyze the fractions (protein peaks) by SDS-PAGE, western blotting and BCA protein assay to identify where the product elutes (*see* **Fig. 7**).
4. Dialyze the rBoNT C(H$_c$)-C protein using a 6–8 kDA Spectra/Por dialysis tubing at 4°C overnight against at least a 50-fold excess volume of dialysis buffer (*see* **Subheading 2.2.3., item 5**) (*see* **Note 5**).

3.3.4. Negative Q Sepharose FF Polishing Step

1. Pack a chromotography column with Q Sepharose FF and equilibrated with equilibration buffer (*see* **Subheading 2.2.3., item 6**) at a flow rate of 150 cm/h for 5 CVs.
2. Load the dialyzed sample at a flow rate of 100 cm/h. Collect the flowthrough. After loading, wash the column with wash buffer (*see* **Subheading 2.2.3., item 6**) at a flow rate of 150 cm/h for 5 CVs or until the absorbance at 280 nm returns to that of the buffer.

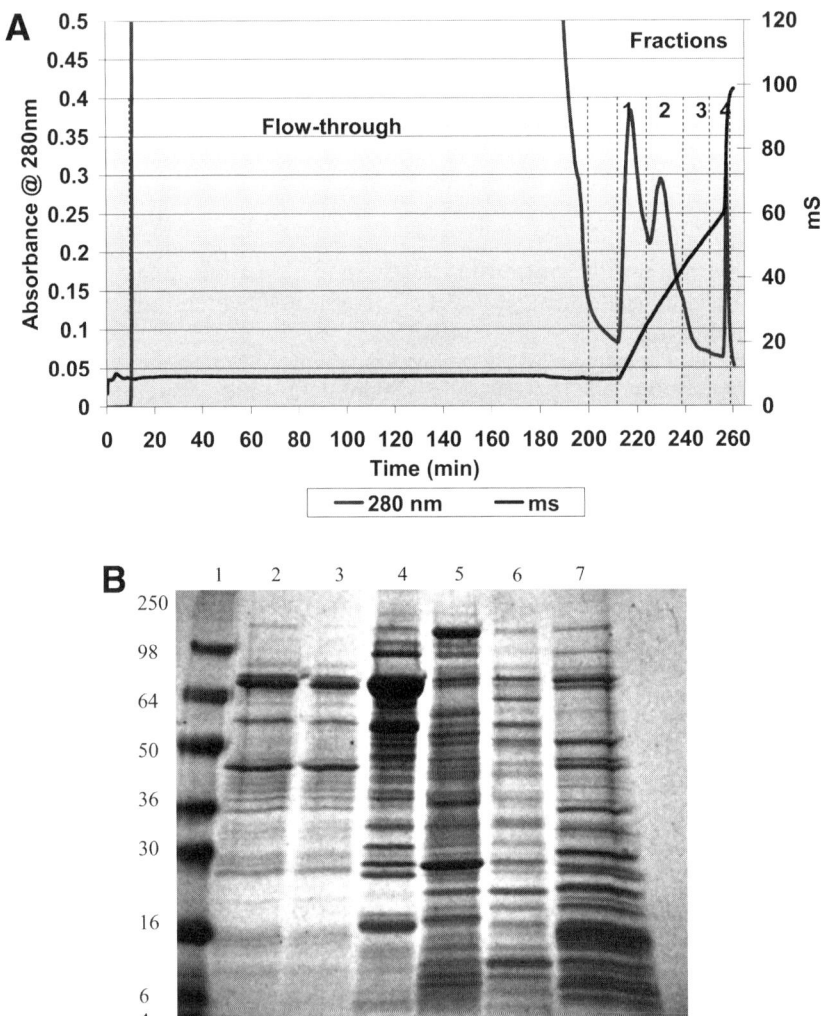

Fig. 6. (**A**) Chromatogram of SP Sepharose Fast Flow capture step for rBoNTC(H$_c$)-C purification. (**B**) SDS-PAGE of SP Sepharose Fast Flow capture step for rBoNTC(H$_c$)-C. Fractions were collected and protein separated on a 4 to 20% Tris–Glycine gel from Novex. The gel was stained with Commasie Brilliant Blue. *Lane 1:* Novex See Blue Marker; *lane 2:* preload sample; *lane 3:* flowthrough; *lanes 4–7:* fractions 1–4.

3. Elute protein using a 5 CVs linear gradient from wash buffer to elution buffer (*see* **Subheading 2.2.3., item 7**). Collect the elution fractions. Analyze the flowthrough and elution fractions by SDS-PAGE, western blotting and BCA protein assay (*see* **Fig. 8**).

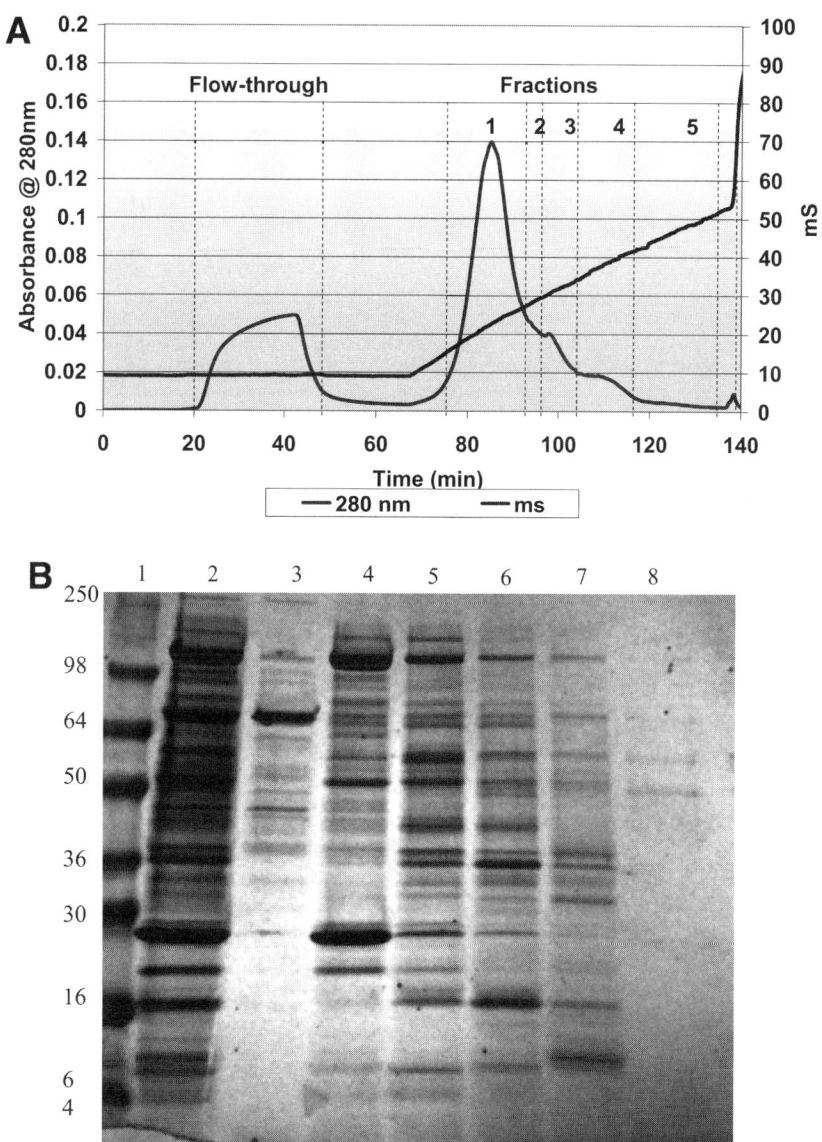

Fig. 7. (**A**) Chromatogram of SP Sepharose Fast Flow second step for rBoNTC(H$_c$)-C purification. (**B**) SDS-PAGE of the SP Sepharose Fast Flow second step for rBoNTC(H$_c$)-C. Fractions were collected and protein separated on a 4 to 20% Tris–Glycine gel from Novex. The gel was stained with Commasie Brilliant Blue. *Lane 1:* Novex See Blue Marker; *lane 2:* preload sample; *lane 3:* flowthrough; *lanes 4–8:* fractions 1–5.

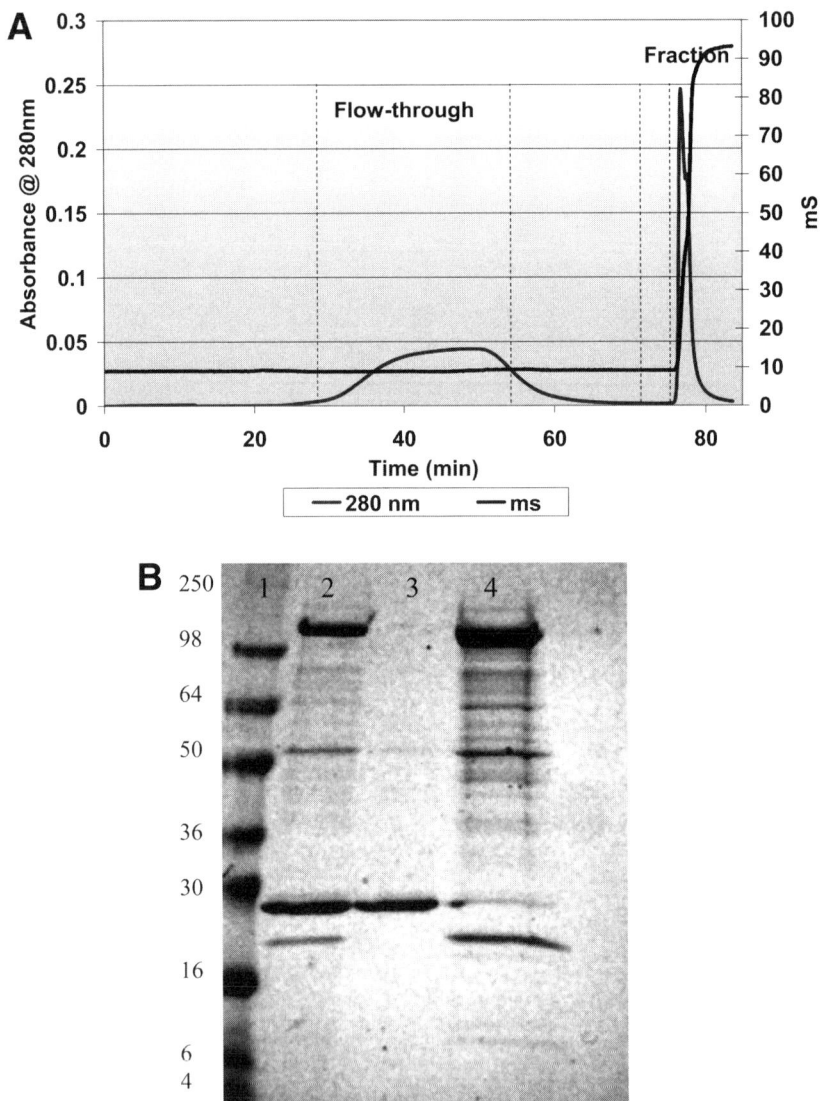

Fig. 8. (**A**) Chromatogram of Negative Q Sepharose Fast Flow polishing step for rBoNTC(H$_c$)-C purification. (**B**) SDS-PAGE of the Negative Q Sepharose Fast Flow polishing step for rBoNTC(H$_c$)-C. Fractions were collected and protein separated on a 4 to 20% Tris–Glycine gel from Novex. The gel was stained with Commasie Brilliant Blue. *Lane 1:* Novex See Blue Marker; *lane 2:* preload sample; *lane 3:* flowthrough; *lane 4:* eluted fraction.

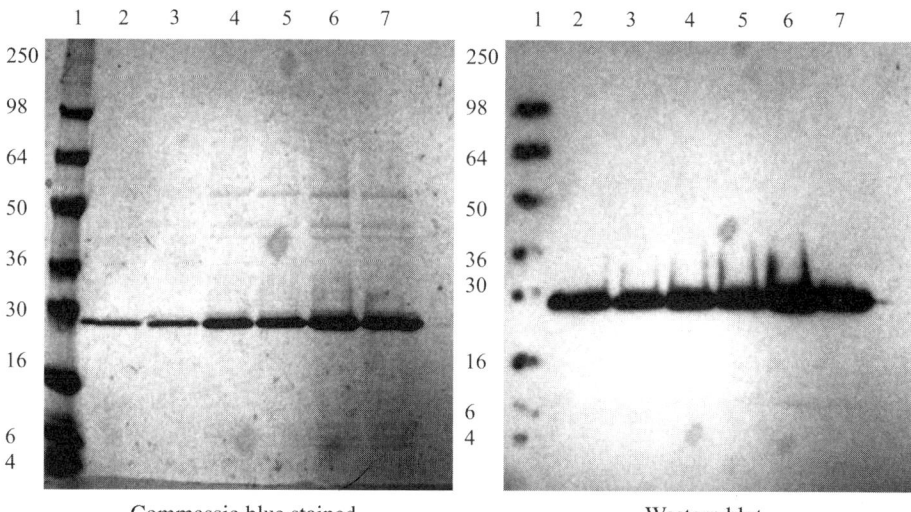

Fig. 9. SDS-PAGE of rBoNTC(H$_c$)-C protein. (**A**) The dialyzed product from concentrated flowthrough of the negative Q Sepharose Fast Flow polishing step was run on a 4 to 20% Tris–glycine gel (Novex). The gel was stained with Commassie Brilliant Blue. (**B**) Western immunoblot. Rabbit anti-rBoNTC(H$_c$) was used as the primary antibody. Affinity-purified peroxidase-labeled Goat anti-Rabbit IgG (H+L) (Kirkegaard [r2] and Perry Laboratory) was used as secondary antibody. ECL plus western blotting detection system (GE Healthcare Lifesciences) was used. Selected concentrations of protein were loaded to check purity. *Lane 1:* Novex See Blue Marker; *lanes 2, 3:* 1 µg rBoNTC(H$_c$)-C protein; *lanes 4, 5:* 3 µg rBoNTC(H$_c$)-C protein; *lanes 6, 7:* 6 µg rBoNTC(H$_c$)-C protein. Protein concentration was determined by Pierce BCA kit using BSA as a standard.

4. Concentrate the flowthrough which contained rBoNT C(H$_c$)-C protein on an Amicon stirred ultrafiltration cell with a 10-kDa Omega membrane (*see* **Note 6**) and dialyze using a 3.5-kDA Slide-A-Lyzer cassette at 4°C overnight against at least a 50-fold excess volume of PBS buffer (*see* **Subheading 2.2.3.**, **item 1**). Final product analysis was performed using SDS-PAGE, western blot (*see* **Fig. 9**), BCA protein assay and N-terminal amino acid sequencing.

4. Notes

1. Protein purification from *P. pastoris* can be improved by optimizing fermentation conditions to increase production and minimize degradation. Methanol is both the inducer and the carbon source during *P. pastoris* fermentations, thus adaptation of the cells to methanol and control of the feed rate can be critical for production (*25*). When switching from glycerol to methanol, it is important to verify that

adaptation to methanol metabolism has occurred. Optimizing the growth rate on methanol is key for good protein production, and the optimum growth rate will vary with each clone. Equations for calculating methanol feed rates and determining the methanol consumption rates are documented *(26)*. Using an exponential feed assumes that the clone has no restrictions upon methanol consumption. If the clone has difficulty utilizing methanol, a methanol sensor can be used to control feed to prevent methanol accumulation. This is determined by measuring methanol concentration and making sure it is being utilized and not building up in the fermentor. Whereas we have used a direct methanol sensor, PTI Instruments, Inc., no longer, manufactures this device. Instead, an off-gas analyzer for detecting methanol can be used; these can be purchased from Bioengineering AG (Wald, Switzerland) or RavenBiotech (Vancouver, Canada). Choosing when to harvest is also important for maximizing production and limiting degradation of your protein. Harvesting early may limit degradation and the release of proteases, but you also reduce the cell mass and product from the fermentation. This can be compensated for by increasing cell mass with glycerol prior to induction.

2. Ion exchange (IE) columns which are packed with IE resins are commonly used for protein purification. The choice of resin to use depends upon the target protein's isoelectric point (pI), pH of the buffer solutions used for dissolving the protein, column working volume, ionic strength of the buffer and the resin matrix *(27)*. Based on the functional group that binds to the resin matrix, there are two kinds of IE resins: anionic and cationic. When the pH of a buffer solution for dissolving the protein is higher than the protein's pI, the protein will have a net negative charge and will bind to an anionic exchange resin. When the pH of the buffer is lower than the protein's pI point, the protein will have a net positive charge and will bind to a cationic exchange resin. A number of ion exchange resins are commercially available with different market brand names such as Q sepharose, Poros Q, and Source Q series for strong anionic exchange resins, and DEAE and PI series for weak anion exchange resins, SP sepharose, Poros SP and Source S series for strong cationic exchange resins, CM series for weak cationic exchange resins. As the theoretical pI of rBoNTC(H_c)-N protein is 4.47, a Q Sepharose FF strong anionic exchange column at pH 7.5 was used for the capture step, a negative SP Sepharose FF strong cationic exchange column at pH 5.5 was used for purification step and a final Q Sepharose FF strong anionic exchange column at pH 5.5 was used for polishing step. As the theoretical pI point of rBoNT C(H_c)-C protein is 8.71, a SP Sepharose FF strong cationic exchange column at pH 6.0 was used for the capture step, a SP Sepharose FF strong cationic exchange column at pH 7.0 was used for purification step and a final negative Q Sepharose FF strong anionic exchange column at pH 7.0 was used for polishing step.

3. Degradation of proteins by native proteases can be a problem when purifying proteins. During cell disruption, protease inhibitors are often used to prevent

degradation of the protein of interest. An individual protease inhibitors (e.g., PMSF, Pepstatin A, Bestatin, or E64) or a protease inhibitor cocktail may be used. In addition, fermentation conditions can be adjusted to decrease the effects of proteases (e.g., adding casamino acids or decreasing fermentation time). Also, work has been done on creating *Pichia pastoris* strains with various protease genes knocked out. (For a more thorough discussion of dealing with proteases, *see* **ref. 28**).

4. The column flow rate mainly depends on the type of exchanger resin used, the protein binding capacity and the resolution of peaks *(29)*. Here we are describing the flow rate as *linear flow rate* (cm/h), which is independent of cross sectional area of the column (i.e. column diameter). When scaling up a chromatography step, the linear flow rate is kept constant, but that actual flow rate will increase with the column size, assuming the column height is constant. Although a fast flow rate can reduce the time for chromatography, usually it is better to start with a moderate flow rate (100–300 cm/h) in order to be sure that the protein solution has an adequate opportunity to equilibrate and adsorb to the ion exchanger. Faster chromatography flow rates also can reduce the resolution of peaks. A gradient elution generally gives a more complete separation of peaks. Total volume for a gradient elution should be approx 5 to 10 column volumes (CVs). High gradient volumes (20–30 CVs) will give better resolution of the peaks.

5. Dialysis techniques are commonly used in protein purification for buffer exchange, removing unwanted low-molecular-weight solutes, and desalting *(27)*. A semipermeable membrane with different molecular-weight cut-off ranging from 1000 to 20,000 Da can be used in dialysis processes. The standard procedure is to chose a membrane pore size that is one-tenth that of the protein to be retained. A large volume of buffer is preferable because fewer buffer changes are needed; usually the buffer volume used is more than 50× that of the dialysis bag. A well-mixed system using large volumes with multiple buffer changes can reach equilibration in 2 to 3 h. Alternately, dialysis just using a single large volume of buffer can be carried out overnight with equilibrium reached by morning; however, the possibility of proteolytic degradation may make overnight dialysis undesirable.

6. The most commonly employed system for concentrating dilute protein solutions is ultrafiltration, in which water is forced through a membrane with a selected molecular-weight cut-off, leaving a more concentrated protein solution behind *(27)*. The membrane selected mainly depends on the protein molecular weight, membrane material (low-protein binding) and permeate flow rate needed. The molecular weight of rBoNT C(H_c)-C protein is 23,470 Da. Hence, we used a low-protein binding Omega membrane with 10 kDa molecular-weight cut-off. Other membranes are also available (e.g., Pall Corporation, Amicon, Millipore, Koch and NCSRT).

Acknowledgments

Research was supported by The Medical Research and Materiel Command Contract No.: DAMD17-02-C-0107 and National Institute of Allergy and Infectious Disease Contract No.:1U01 AI 056514-01. We would like to thank the BPDF's Purification Development Laboratory staff for process development support throughout this project. N-terminal sequencing was performed by the University of Nebraska Medical Center, Protein Structure Core Facility.

References

1. Lamanna, C. and Hart, E. R. (1968) Relationship of lethal toxic dose to body weight of the mouse, *Toxicol. Appl. Pharmacol.* **13,** 307–315.
2. Anderson, J. H. and Lewis, G. E., Jr. (1981) Clinical evaluation of botulinum toxoids, in *Biomedical Aspects of Botulism,* (Lewis, G.E., ed.), Academic Press, New York, NY, pp. 233–246.
3. Simpson, L. L. (1986) Molecular pharmacology botulinum toxin and tetanus toxin, *Annu. Rev. Pharmacol. Toxicol.* **26,** 427–453.
4. DasGupta, B. R. (1989) The structure of botulinum neurotoxins, in *Botulinum Neurotoxin and Tetanus Toxin*, (L.L. Simpson,L.L., ed.), Academic Press, New York, NY, pp. 53–67.
5. Simpson. L. L. (1981) The Origin, Structure, and Pharmacological Activity of Botulinum Toxin. *Pharmacol. Rev.* **33,** 155–188.
6. Poulain, B., Weller, U., Binz, T., et al. (1993) Functional Roles of the Domains of Clostridial Neurotoxins: The Contribution from Studies on *Aplysia*, in *Botulinum and Tenanus Neurontoxins: Neurotransmission and Biomedical Aspects* (DasGupta, B. R., ed.), Plenum Press, New York, NY, pp 34–360.
7. Black, J. D. and Dolly, J. O. (1986) Interaction of ^{125}I-labeled Botulinum Neurotoxin with Nerve Terminals. I. Ultrastructual Autoradiographic Localization and Quantification of Distinct Membrane Acceptors for Types A and B on Motor Nerves. *J. Cell Biol.* **103,** 521–534.
8. Nishiki, T. -I., Kamata, Y., Nemoto, Y. et al. (1994) Identification of Protein Receptor for *Clostridium botulinum* Type B Neurotoxin in Rat Brain Synaptosomes. *J. Biol. Chem.* **269,** 10,498–10,503.
9. Shone, C. C., Hambleton, P., and Melling, J. (1985) Inactivation of *Clostridium botulinum* Type A Neurotoxin by Trypsin and Purification of Two Tryptic Fragments. Proteolytic Cleavage Near the COOH-terminus of the Heavy Subunit Destroys Toxin-Binding Activity. *Eur. J. Biochem.* **151,** 75–82.
10. Schiavo, G., Rossetto, O., Catsicas, S., et al. (1993) Identification of the Nerve Terminal Targets of Botulinum Neurotoxin Serotypes A, D and E. *E. J. Biol. Chem.* **268,** 23,784–23,787.
11. Black, J. D. and Dolly, J. O. (1986) Interaction of ^{125}I-labeled Botulinum Neurotoxin with Nerve Terminals. II. Autoradiographic Evidence for Its Uptake into Motor Nerves by Acceptor-mediated Endocytosis. *J. Cell Biol.* **103,** 535–544.
12. Shone, C. C., Hambleton, P., and Melling, J. (1987) A 50-kDa Fragment from the NH_2-terminus of the Heavy Subunit of *Clostridium botulinum* Type A Neurotoxin Forms Channels in Lipid Vesicles. *Eur. J. Biochem.* **167,** 175–180.
13. Blaustein, R. O., Germann, W. J., Finkelstein, A., and Dasgupta, B. R. (1987) The N-terminal Half of the Heacy Chain of Botulinum Type A Neurotoxin Forms Channels in Planar Phospholipid Bilayers. *FEBS Letters* **226,** 115–120.

14. Montal, M. S., Blewitt, R. Tomich, J. M., and Montal, M. (1992) Identification of an Ion Channel-forming Motif in the Primary Structure of Tetanus and Botulinum Neurotoxins. *FEBS Letters* **313**, 12–18.

15. Smith, L. A., Jensen M. J., Montgomery, V. A., Brown, D. R., Ahmed, S. A., and Smith, T. J. (2004) Roads from Vaccines to Therapies. *Mov. Disorders* **19** Suppl. 8., S48–S52.

16. Potter, K. J., Bevins, M. A., Vassilieva, E. V., et al. (1998) Production and Purification of the Heavy-Chain Fragment C of Botulinum Neurotoxin, Serotype B, Expressed in the Methylotrophic Yeast *Pichia pastoris*. *Prot. Exp. Purif.* **13**, 357–365.

17. Potter, K. J., Zhang, W., Smith, L. A., and Meagher, M. M. (2000) Production and Purification of the Heavy Chain Fragment C of Botulinum Neurotoxin, Serotype A, Expressed in the Methylotrophic Yeast *Pichia pastoris*. *Prot. Exp. Purif.* **19**, 393–402.

18. Fairweather, N. F., Lyness, V. A., and Maskell, D. J. (1987) Immunization of Mice Against Tetanus Toxin with Fragments of Tetanus Toxin Synthesized in *Escherichia coli*. *Infect. Immun.* **55**, 2541–2545.

19. Makoff, A. J., Ballantine, S. P., Smallwood, A. E., and Fairweather, N. F. (1989) Expression of Tetanus Toxin Fragment C in *E. coli*: Its Purification and Potenial Use as a Vaccine. *Biotechnology* **7**, 1043–1046.

20. Clayton, M. A., Clayton, J. M., Brown, D. R., and Middlebrook, J. L. (1995) Protective Vaccination with a Recombinant Fragment of *Clostridium botulinum* Neurotoxin Serotype A Expressed from a Synthetic Gene in *Escherichia coli*. *Infect. Immun.* **63**, 2738–2742.

21. Byrne, M. P., Smith, T. J., Montgomery, V. A., and Smith, L. A. (1998) Purification potency, and efficacy of the botilinum neurotoxin type A binding domain from Pichia pastoris as a recombinant vaccine candidate. *Infect. Immun.* **66**, 4817–4822.

22. Byrne, M. P., Titball, R. W., Holley, J., and Smith, L.A. (2000) Fermentation, Purification, and efficacy of a recombinant vaccine candidate against botulinum neurotoxin type F from Pichia pastoris. *Protein Expr. Purif.* **18**, 327–337.

23. Smith, L. A. and Byrne, M. P. (2002) Vaccines for preventing botulism, in *Scientific and therapeutic aspects of botulinum toxin*, (Brin M. F., Jankovic, J., Hallet, M., eds.), Lippincott, Williams and Wilkins, Philadelphia, PA, pp. 427–440.

24. Smith, L. A. (1998) Development of Recombinant Vaccines for Botulinum Neurotoxin. *Toxicon.* **36**, 1539–1548.

25. Zhang, W., Inan, M., and Meagher, M. M. (2000) Fermentation strategies for recombinant protein expression in methylotrophic yeast *Pichia pastoris*. *Biotechnol. Bioprocess. Eng.* **5**, 275–287.

26. Zhang, W., Bevins, M. A., Plantz, B. A., Smith, L. A., and Meagher, M. M. (2000). Modeling *Pichia pastoris* growth on methanol and optimizing the production of a recombinant protein, the heavy-chain fragment C of botulinum neurotoxin, serotype A. *Biotechnol. Bioeng.* **70**, 1–8.

27. Scope, R. K. (ed.) (1994) *Protein Purification – Principles and Practice* (Third Edition). Springer-Verlag, New York, NY.

28. Sinha, J., Plantz, B. A., Inan, M., and Meagher, M. M. (2005) Causes of Proteolytic Degradation of Secreted recombinant Proteins Produced in Methyotrophic Yeast Pichia pastoris: case Study with Recombinatn Ovine Interferon . *Biotechnol. Bioeng.* **89,** 102–112.
29. Bollag, D. M., Rozycki, M. D., and Edelstein, S. J. (1996) *Protein Methods,* Second Edition. Wiley-Liss, New York, NY.

7

Rapid Screening of Chromatography Resins for the Purification of Proteins

Sandra E. Ríos, Erin M. Giaccone, and Tillman U. Gerngross

Abstract

With an ever increasing number of proteins being expressed in the *Pichia* system, there is a growing need to rapidly develop scalable and robust purification schemes. This chapter describes a high-throughput method to screen for the optimal chromatography conditions and resin to capture and release a protein secreted by *Pichia pastoris*. The method involves a chromatography matrix involving four resins (Q-Sepharose, DEAE-Sepharose, SP-Sepharose, and CM-Sepharose), 4 pHs from 5.0 to 8.0, and 3 NaCl concentrations. The method was tested on three proteins and found to be reproducible and easily scalable.

Key Words: *Pichia pastoris*; protein purification; high-throughput screening; yeast; ion exchange chromotography resins.

1. Introduction

Generating a recombinant strain of *P. pastoris* that is able to secrete a protein of interest is becoming a routine procedure that has been successfully demonstrated on a large number of proteins (*see* http://faculty.kgi.edu/cregg/index.htm and Chapter 2) *(1)*. Following the generation of a strain that is able to produce and secrete a recombinant protein, purification typically becomes the next step in obtaining an enriched, stable and homogenous form of the protein of interest *(2,3)*. Because *P. pastoris* grows on chemically defined media and does not secrete high amounts of endogenous protein, secretion often becomes the first enrichment or "purification" step. Proteins secreted into the culture medium are relatively easy to purify once a clear supernatant is obtained either by cross-flow filtration or centrifugation. We routinely submit the clear supernatant to ion exchange chromatography to capture the protein of interest and to remove undesired contaminants. In order to quickly determine a feasible purification

From: *Methods in Molecular Biology, vol. 389:* Pichia *Protocols, Second Edition*
Edited by: J. M. Cregg © Humana Press Inc., Totowa, NJ

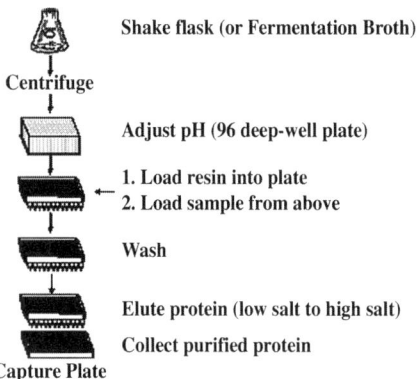

Fig. 1. Schematic diagram for setting up a protein purification screen in a 96-well format.

resin, we have developed a rapid screening protocol to test a series of chromatography resins and elution conditions in a single automated procedure. Once a proper resin is identified, scale-up can be readily accomplished by choosing a larger chromatography bed volume. We routinely scale from the 0.2-mL scale (screening) to 0.5 L (preparative scale), with a single ion exchange step often yielding protein that is 90 to 95% pure as judged by sodium dodecyl sulfate-polyacrylamide gel electrophoresis (SDS-PAGE).

Consequently, this protocol describes a fast approach to determine the capture/primary purification step of proteins expressed in *P. pastoris*.

2. Materials

2.1. Buffers, Resins, and Equipment

1. Resins: Q-Sepharose, DEAE-Sepharose, SP-Sepharose, and CM-Sepharose. (Amersham Biosciences, Piscataway, NJ).
2. Equilibration buffers:
   a. 50 mM NaAc (pH 5.0).
   b. 50 mM MES (pH 6.0).
   c. 50 mM Tris-HCl (pH 7.0).
   d. 50 mM HEPES (pH 8.0).
3. Elution buffers:
   a. 50 mM NaAc (pH 5.0) plus 100 mM, 250 mM, 500 mM, and 1 M NaCl, respectively.
   b. 50 mM MES (pH 6.0) plus 100 mM, 250 mM, 500 mM, and 1 M NaCl, respectively.
   c. 50 mM Tris-HCl (pH 7.0) plus 100 mM, 250 mM, 500 mM, and 1 M NaCl, respectively.
   d. 50 mM HEPES (pH 8.0) plus 100 mM, 250 mM, 500 mM, and 1 M NaCl, respectively.

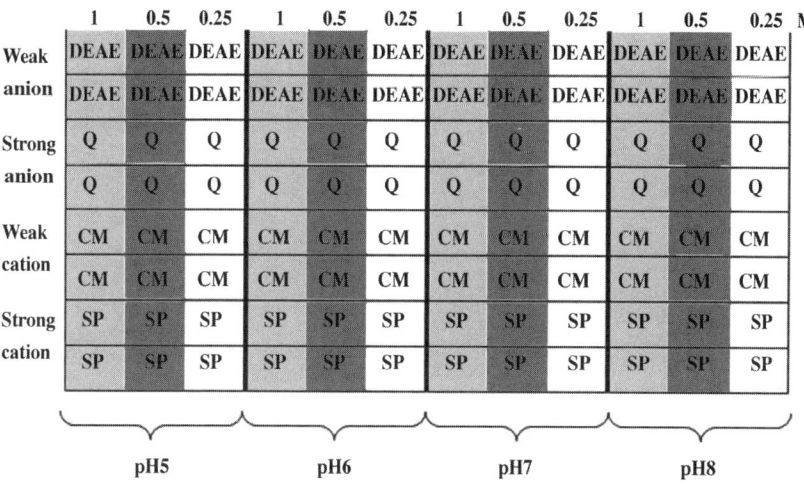

Fig. 2. Layout of the anionic (DEAE-Sepharose, Q-Sepharose) and cationic resins (CM-Sepharose, SP-Sepharose) with respective pH and NaCl concentrations utilized in the elution steps.

4. 96-well lysate clearing plate (Wizard SV96, Promega Corp, Madison, WI) and a Beckman 96-well plate manifold (any manifold can be used if the procedure is not automated).
5. Beckman BioMek 2000 laboratory robot (Beckman Instruments, Fullerton CA) (optional).

3. Methods

3.1. Protein Purification on a 96-Well Plate Format

The culture broth (at least 30 mL) from a *P. pastoris* strain secreting a protein of interest is cleared by centrifugation (2500g for 10 min) and the supernatant is divided into five equal fractions of 2.4 mL. The pH of each aliquot is adjusted by diluting the supernatant 1:5 with the equilibration buffers listed above (*see* **Note 1**). This results in 5 samples of 14.4 mL with respective pH values of 5.0, 6.0, 7.0, and 8.0. A purification plate is prepared with four resins (Q-, DEAE-, SP-, and CM-Sepharose), which are suspended in distilled water in a ratio of 1:3.5 (volume of settled resin:water) (*see* **Note 2**). Once the purification plate is prepared, the purification procedure (*see* **Fig. 1**) is performed using a 96-well filtration plate on a Beckman BioMek 2000 laboratory robot.

A robotic protocol was set up based on the following:

1. Manually prepare plate by transferring 700 µL of the fully suspended resins to each well and allow to drain. This will result in about 200 µL of settled resin. Follow the resin pattern described in **Fig. 2** (i.e., load the first row with DEAE).

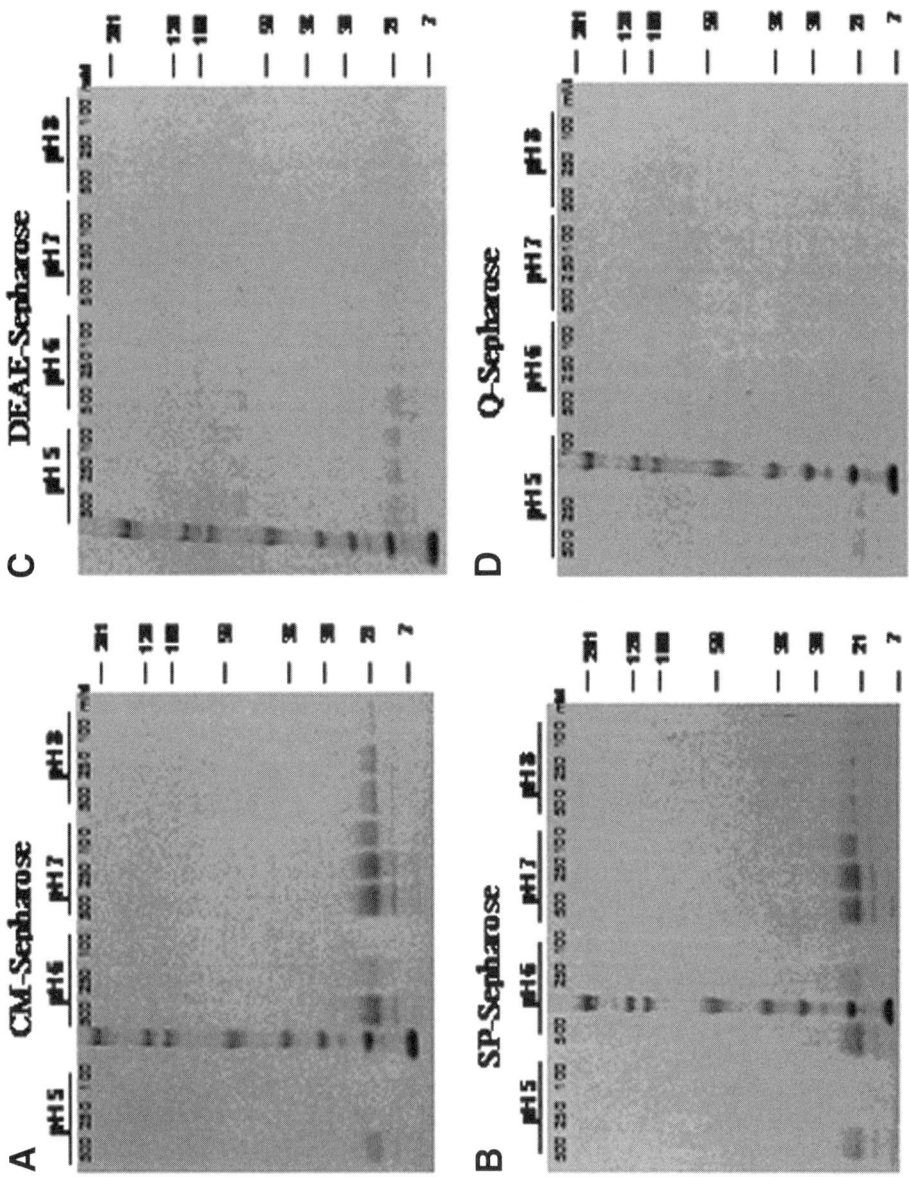

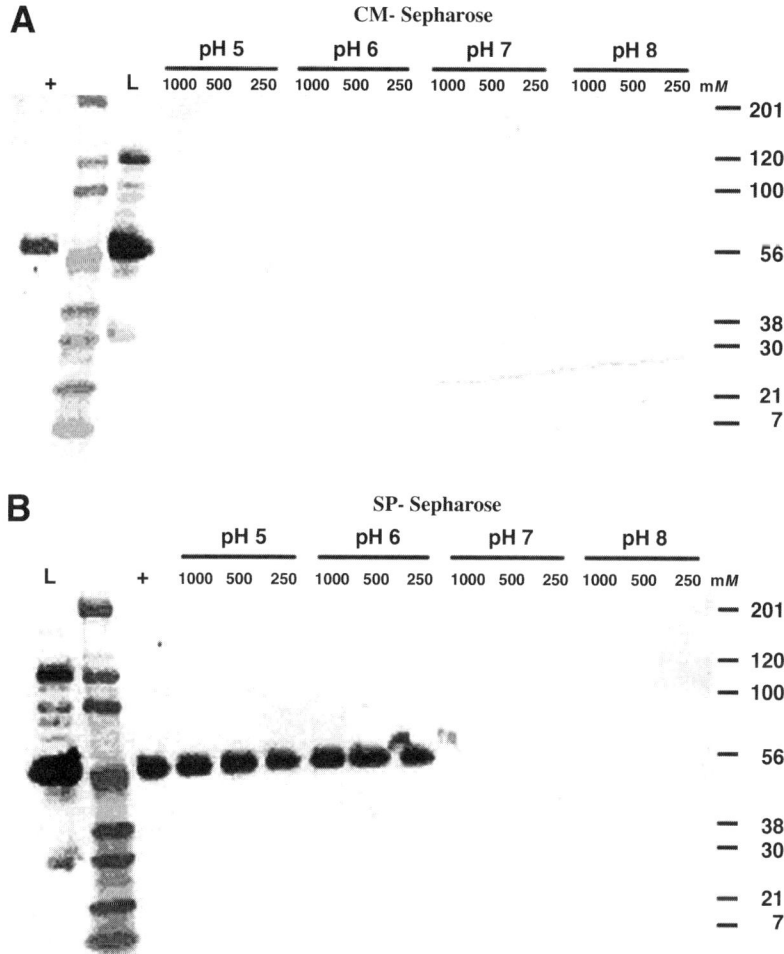

Fig. 4. *(Continued)*

2. Add 10X 0.2 mL of equilibration buffer to each column and allow to drain between additions. (This may be reduced to 2X 1 mL if performed manually).
3. Load samples (2 × 250 µL of pH adjusted sample from above) to each column; match pH of column with pH of sample. Follow pattern in **Fig. 2**. Sample volume may be increased if expression levels are low and detection methods are not sufficiently sensitive to detect the protein of interest.

Fig. 3. *(Opposite page)* Purification of sample Protein I. Supernatants were separated by the 96-well purification protocol described above. An aliquot of each well was analyzed by SDS-PAGE and visualized by staining with Coomasie Brilliant Blue. (**A**) CM-Sepharose, (**B**) SP-Sepharose, (**C**) DEAE-Sepharose, and (**D**) Q-Sepharose were used respectively.

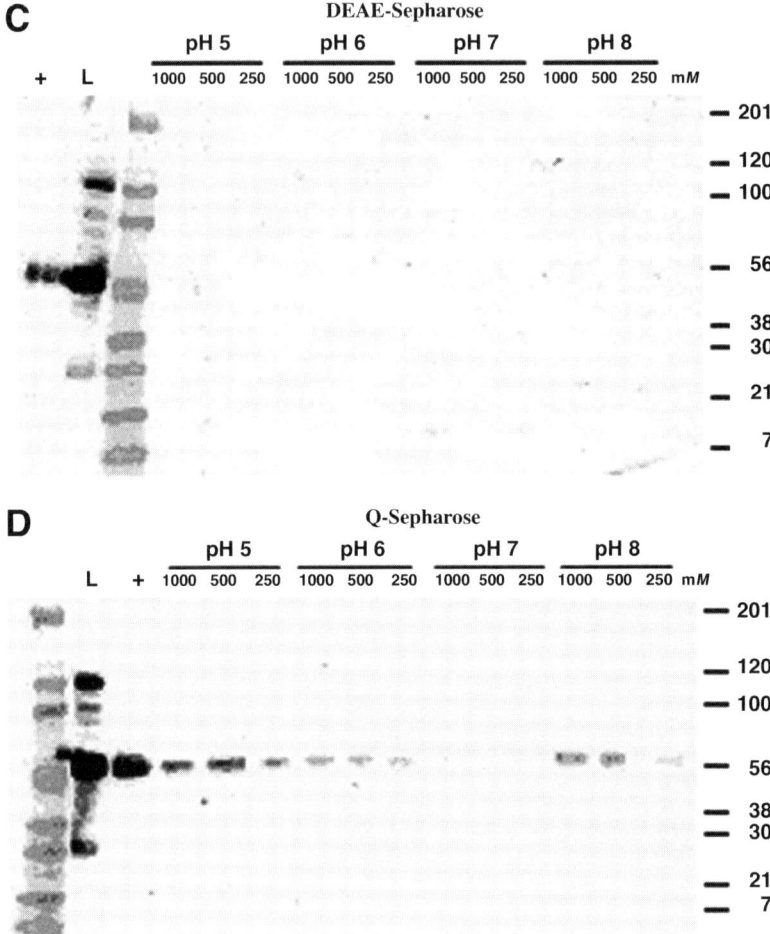

Fig. 4. Purification of sample Protein II. Western blot analysis of sample Protein II following purification on a 96-well chromatography plate as described above. (**A**, **B**, **C** and **D**) figures are the elution samples from the (**A**) CM-Sepharose, (**B**) SP-Sepharose, (**C**) DEAE-Sepharose, and (**D**) Q-Sepharose respectively. The + sign correspond to a positive control in each of the figures. The letter L corresponds to the sample loaded into the column.

4. Wash each well with 4X 0.2 mL of equilibration buffer (collection plate optional).
5. Elute protein with 300 µL of elution buffer. Each well receives elution buffer that contains one of three possible NaCl concentrations (i.e., 250 mM, 500 mM and 1 M) (*see* **Note 3**). Follow the elution pattern described in **Fig. 2**. Only the captured and then released proteins are collected. Proteins that do not bind ends up in the wash, and are not collected.

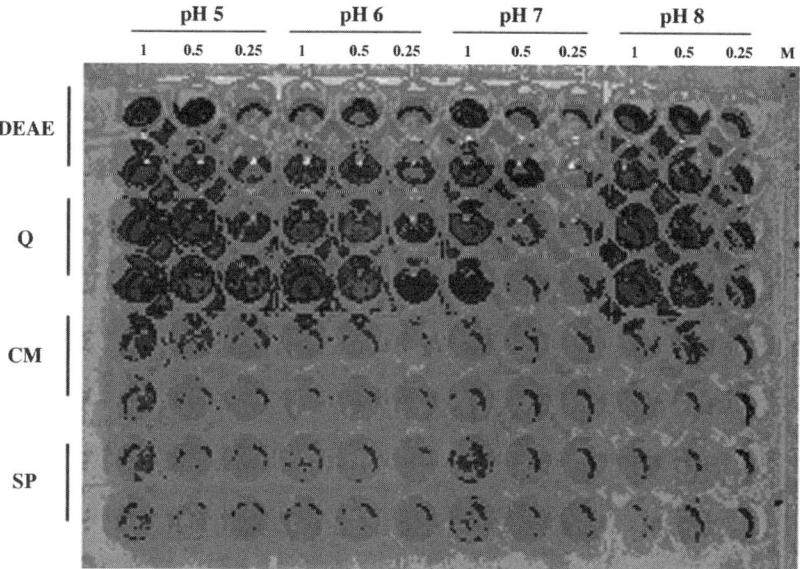

Fig. 5. ELISA of sample Protein III micro-scale purification. The figure shows the resin, pH, and NaCl concentration utilized in the micro-scale protein purification.

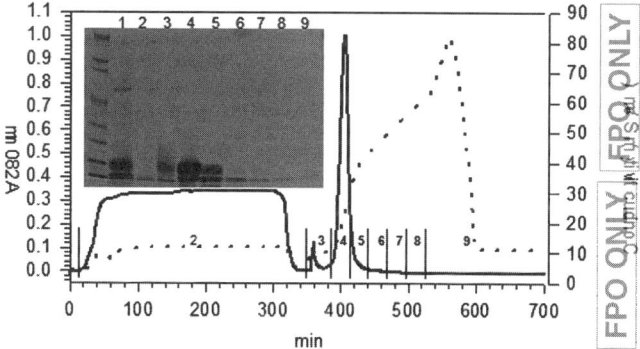

Fig. 6. Scale-up of sample protein I from a 200 μL column in a 96-well plate to a 500-mL column. The figure shows the chromatographic profile of the loading and elution of Protein I. The solid line corresponds to the absorbance at 280 nm and the doted line to the conductivity during the purification. *Insert:* SDS-PAGE of samples taken during the purification. *Lane 1:* pH adjusted supernatant (load); *lanes 2–9:* fractions 2–9 obtained throughout the purification run.

3.2. Protein Detection

Three sample proteins were submitted to the screening protocol described above (Protein I, II, and III) to illustrate the use of different detection methods following separation. The final collection plates containing 300 μL of eluent/well, in the respective elution buffer, were analyzed using the following techniques: SDS-PAGE for sample protein I (*see* **Fig. 3**), Western blot for sample Protein II (*see* **Fig. 4**), and enzyme-linked immunosorbent assay (ELISA) for sample protein III (*see* **Fig 5**).

SDS-PAGE analysis of the collection plate of sample Protein I showed that the protein is efficiently captured and released by the cation resins SP-Sepharose and CM-Sepharose at pH 7 (*see* **Fig. 3A,B**).

Sample Protein II binds only to SP-Sepharose and Q-Sepharose resins under the conditions we have tested. Western blots show that SP-Sepharose at pH 5 is the best condition to capture and release this protein (*see* **Fig. 4B,D**).

Sample Protein III was detected using an ELISA assay which revealed that this particular protein binds only to DEAE- and Q- resins, preferably at high and low pH (*see* **Fig. 5**).

3.3. Scale-Up

To exemplify the ease of scale-up from the milliliter to the liter scale, Protein I was purified on a 400 mL CM-Sepharose column. **Figure 6** shows the elution profile of a CM-Sepharose run with 50 mM Tris-HCl (pH 7.0), and a NaCl gradient from 0 to 1 M. The resin and the conditions utilized to run the column were taken directly from the information obtained from the micro-scale purification (*see* **Fig. 3**).

4. Notes

1. We have noticed that in some fermentations the final salt concentration in the culture supernatant can be very high resulting in a conductivity in excess of 30 mS/cm. It is important to ensure that the fivefold dilution in equilibration buffer, used to adjust the pH, results in conductivity below 4 mS/cm.
2. Resins are typically shipped in 20% ethanol; after allowing resin to settle by gravity, supernatant is decanted and replaced with water to give a 1 : 3.5 ratio of resin to water, and then loaded onto the filter plate.
3. We have also successfully used NaCl concentrations of 100 mM, 250 mM, and 500 mM (*see* sample Protein I). The choice of NaCl concentration is influenced by previous knowledge of salting-out effects that may preclude the use of high salt elution buffers with certain proteins

References

1. Cregg, J. M. and Madden, K. R. (1988) Development of the methylotrophic yeast Pichia pastoris as a host for the production of foreign proteins. *Dev. Ind. Microbiol.* **29**, 33–41.
2. Higgins, D. R. and Cregg, J. M. (1998) Introduction to *Pichia pastoris*, in *Pichia Protocols*, (Higgins, D. R. and Cregg, J. M., eds.), Humana, Totowa, NJ, pp. 1–15.
3. Sreekrishna, K., Brankamp, R. G., Kropp, K. E., et al. (1997) Strategies for optimal synthesis and secretion of heterologous proteins in the methylotrophic yeast Pichia pastoris. *Gene* **190**, 55–62.

8

Characterization of O-Linked Saccharides on Glycoproteins

Roger K. Bretthauer

Abstract

Recombinant and native proteins of *Pichia pastoris* can be O-mannosylated on serine and threonine residues, allowing further elongation reactions to generate short O-linked oligosaccharides of mannose. Methods for release from the protein with alkaline β-elimination with or without reduction of the released saccharides, and for subsequent chromatographic and enzymatic characterization of these saccharides are described.

Key Words: O-Mannosylation; β-elimination; reductive labeling; 2-aminobenzamide; oligosaccharide; mannosidase; high-pressure liquid chromatography (HPLC); size-exclusion chromatography.

1. Introduction

Pichia pastoris is commonly utilized to generate recombinant proteins because of high yields of secreted products, economical growth and expression media (methanol), and availability of numerous genetic markers and constructs for generating recombinant protein products *(1)*. Many recombinant proteins are secreted in a glycosylated form, and to properly evaluate the structure and biological function of a secreted glycoprotein, the structure and amino acid-site of the saccharide(s) must be known. Yeast in general will carry out glycosylation on asparagine residues (N-glycosylation) and on threonine and serine residues (O-glycosylation) of proteins *(2)*. Details of these glycosylation pathways have been derived predominantly from studies with *Saccharomyces cerevisiae*, and these will be referred to in the following discussion.

The N-type glycosylation is well-established as occurring on the side chain amide nitrogen of asparagine residues that are located in an Asn-X-Thr/Ser sequence with X being any amino acid except proline. The N-linked

From: *Methods in Molecular Biology, vol. 389:* Pichia *Protocols, Second Edition*
Edited by: J. M. Cregg © Humana Press Inc., Totowa, NJ

oligosaccharide is assembled in a precursor form on the lipid carrier dolichol phosphate in the endoplasmic reticulum, transferred to the nascent protein in a cotranslational event, and further modified to final structure as the glycoprotein proceeds through the Golgi apparatus. In most yeast, oligosaccharides of this type will contain only mannose residues, up to hundreds, on the protein-linked core $GlcNAc_2Man_8$ structure derived from the dolichol-linked precursor. Recent advances in genetic alteration of the N-glycosylation pathway in *Pichia* and other yeast have resulted in the ability of the modified cells to synthesize N-linked saccharides that lack this extended "yeast-like" mannose oligosaccharide chain, but can add N-acetylglucosamine residues to allow further addition of "human-like" saccharides such as galactose and sialic acid (reviewed in *[3]*). The potential ability of these engineered cells to synthesize and secrete, in large yields, recombinant glycoproteins with "human-like" N-linked saccharides is of obvious importance to the pharmaceutical industry. It is therefore also of importance to characterize any possible O-linked saccharides on these recombinant products that have potential human use.

Protein O-mannosylation in yeast has features distinct from N-glycosylation. The process is initiated in the endoplasmic reticulum with a mannose residue from GDP-mannose being activated in a precursor form as dolichylphosphoryl β-D-mannosylpyranoside. This mannose residue is transferred to appropriate serine or threonine residues of the protein, as catalyzed by a protein O-mannosyltransferase (Pmt) to form the mannosyl-α-*O*-ser/thr-protein product. Seven Pmts are found in *S. cerevisiae*. Although unique functions of these seven proteins, such as involvement in glycosylation of serine or threonine residues at different amino acid sequences or glycosylation of specific proteins, remain to be fully elucidated, early studies with gene mutation and deletion demonstrated the essential requirement of some for cell growth and vitality *(4)*. Also, some differences in protein substrate specificity were observed for Pmt1-4 *(5)*. The difficulty in evaluating the specific functions of these proteins is demonstrated by the observed formation of complexes between certain Pmts that are considered to be physiologically relevant *(6)*. That O-glycosylation could control N-glycosylation of a protein was demonstrated by the observed N-glycosylaton at a specific site in a cell wall protein (Ccw-5) when Pmt4 was mutated; wild-type cells did not N-glycosylate because the Thr residue in the N-glycosylation sequence Asn-Ala-Thr was O-mannosylated *(7)*. After the Pmt reaction, additional mannose residues are added directly from GDP-mannose as catalyzed by other mannosyl transferases. These elongation reactions occur as the protein proceeds through the Golgi apparatus to form the final unbranched oligosaccharide structure having usually up to only five mannose residues *(8)*. Phosphorylation of O-linked saccharides can also occur by transfer of mannopyranosyl-α1-phosphate from GDP-mannose to the carbon 6 hydroxyl group of, usually, the second α1,2-linked mannose residue in the oligosaccharide *(9)*. The resulting mannopyranosyl-α1-phosphoryl-6-mannopyranosyl diester group can then be converted to the phosphomonoester structure (6-phosphomannopyranosyl) by loss of the α-1-linked mannose residue.

The ability of *P. pastoris* to carry out O-glycosylation of expressed recombinant proteins has been documented in several studies *(10–17)*. Associated structural studies have revealed the disaccharide Manα1,2Man as a common O-linked product, but extensions up to the pentasaccharide and higher of α1,2 and α1,3 linked mannose residues have been reported. A phosphorylated O-linked saccharide from recombinant human bile salt-stimulated lipase has also been characterized *(16)*, documenting that this type of modification reaction occurs on O-linked saccharides as well as on N-linked saccharides of expressed proteins. Only in limited cases have specific amino acid sites of O-glycosylation been identified. The N-lobe of expressed human transferrin was shown to be glycosylated with a single hexose (mannose) residue on serine-32 located in the protein amino acid sequence Pro-Ser-Asp *(12)*. Recombinant human antithrombin III was shown to be glycosylated at three sites: with a disaccharide of mannose on threonine-386 in the protein sequence Ser-Thr-Ala; with the monosaccharide on serine-3 in the sequence Gly-Ser-Pro; and with the monosaccharide on threonine-9 in the sequence Cys-Thr-Ala *(13)*. The presence of proline residues near the O-glycosylation sites has been observed also in several proteins expressed in *S. cerevisiae*, but identification of specific amino acid sequences required by any of the PMT enzymes remains elusive. It is considered that, as initial O-glycosylation with PMT occurs in the endoplasmic reticulum, the glycosylation occurs on exposed serine and threonine residues before folding of the protein, or possibly in folded proteins with exposed serine and threonine residues in surface loops of the protein.

2. Materials

2.1. Alkaline β-Elimination

1. 1 *M* Sodium hydroxide in water.
2. 1 *M* Sodium borohydride in water.
3. Glacial acetic acid.
4. Dowex 50 (H$^+$) resin in water.
5. Amberlite MB-3 resin in water.
6. Methanol.

2.2. Reductive Labeling

1. Sodium [^{3}H]borohydride.
2. 2-Aminobenzamide (ProZyme; Sigma-Aldrich).

2.3. Size Analysis

1. BioGel P-2 or P-4 (Bio-Rad).
2. 0.01 *M* Ammonium acetate in water.

3. Propanol.
4. 0.05 *M* Hydrochloric acid.
5. Phosphomonoesterase (alkaline phosphatase) (*see* **Note 1**).

2.4. HPLC Analysis

1. High-performance anion exchange chromatography (HPAEC) system with pulsed amperometric detector (PAD) and PA1 columns (Dionex Corporation).
2. Standard saccharides of mannose (ProZyme; Sigma-Aldrich).

2.5. Glycosidase Analysis

1. Jack bean α-mannosidase (*see* **Note 1**).
2. *Aspergillus saitoi* α-1,2-mannosidase (*see* **Note 1**).
3. *Xanthomonas manihotis* α-1,2/3-mannosidase (*see* **Note 1**).

3. Methods

Following are methods of releasing and characterizing O-linked saccharides on glycoproteins that have been expressed in *P. pastoris*. These methods have been utilized in several laboratories to characterize the O-linked saccharides that are common to yeast, in particular to *S. cerevisiae* and in some cases to *P. pastoris* (*see* **Note 2**).

3.1. Alkaline β-Elimination of Saccharides With or Without Reduction

The secreted recombinant glycoprotein should be ascertained to be pure and free of other extracellular carbohydrate-containing material (*see* **Note 3**).

3.1.1. Release and Recovery of Saccharides in Their Reduced Form

1. Dissolve the glycosylated protein in 0.1 *M* NaOH containing 0.25 NaBH$_4$ (final concentrations) and incubate at 30°C for 16 h.
2. Add excess acetic acid to neutralize the NaOH and to decompose remaining NaBH$_4$.
3. The sample is desalted by either a) passing through a small column of Dowex 50 (H$^+$) resin in water to remove cations (exchanging sodium ions from the NaOH and NaBH$_4$ for resin hydrogen ions), followed by evaporating the column eluate of neutral components to dryness several times from acidic methanol (1% acetic acid in methanol) to remove volatile methyl borate esters derived from excess NaBH$_4$, or b) passing through a mixed-bed ion-exchange resin column such as Amberlite MB-3 to remove salts and residual protein or peptides, but with potential loss of acidic phosphorylated saccharides on the anion-exchange component of the mixed-bed resin. The volume of either ion-exchange resin needed can be estimated from the known resin capacities being from 1 to 2 millequivalents/mL wet resin. A 5- to 10-fold molar excess of resin capacity is generally utilized for these desalting procedures.

3.1.2. Release and Recovery of Nonreduced Saccharides

1. Omit the $NaBH_4$ from the initial NaOH solution and omit the acidic methanol evaporation step in the above procedure *(19,20)* (*see* **Note 4**).

3.2. Reductive Labeling of Saccharides

3.2.1. Reduction With NaB^3H_4

1. Dissolve the sample of nonreduced saccharides in a minimal volume (0.1 mL or less) of 0.01 *M* NaOH and add high radiospecific activity NaB^3H_4 (0.1–1.0 mCi of 100 mCi/µmol NaB^3H_4). Allow initial reduction to occur for several hours at room temperature.
2. Add unlabeled $NaBH_4$ to 0.1 *M* final concentration and continue incubation for several hours to assure complete reduction of saccharides.
3. Remove excess reagents with recovery of the radiolabeled saccharides as previously described for reductive alkaline β-elimination. Appropriate caution is taken for release of tritium gas upon acidification of the reaction mixture.

3.2.2. Reductive Amination With 2-Aminobenzamide

Fluorescent labeling of released nonreduced saccharides is carried out by reductive amination with 2-aminobenzamide (AB). This is accomplished by use of kits, including detailed instructions, available from ProZyme or Sigma-Aldrich.

3.3. Size Analysis of Saccharides

3.3.1. Size Exclusion Column Chromatography

1. Oligomers of reduced or nonreduced saccharides are separated by size exclusion chromatography (gel filtration) on columns of BioGel P-2 or BioGel P-4 *(23)*. For preparative chromatography, and depending on the size range of oligosaccharides to be resolved, columns from 1- to 2-cm in diameter and 1- to 2-m in length are prepared and run in the descending mode at room temperature with eluting solvent of 0.01 *M* ammonium acetate or 0.1 *M* acetic acid:1% propanol. Flow rates can be controlled by physical pumping of eluting solvent, or simply by maintaining a constant pressure head with the height of the eluting solvent chamber. Size calibration is done with standard oligomers of mannose (monomer through pentamer), either native, reduced, or AB labeled. Eluted saccharides are detected with phenol-sulfuric acid *(23)*, radioactivity, or fluorescence (*see* **Note 5**).
2. The presence of esterified phosphate on any of the oligosaccharides will significantly alter the elution position from BioGel sizing columns as compared with the corresponding neutral oligosaccharide. Preliminary fractionation by HPLC on a weak anion exchange resin, such as DEAE- or amino-, will separate the neutral saccharides from the acidic saccharides *(25,26)*. Conversion of the phosphorylated saccharides to the neutral saccharides is done by mild acid hydrolysis in 0.05 *M* HCl at 100°C for 1 h to remove the α1-linked mannose residue from the phospho-diester group, followed by phosphomonoesterase (alkaline phosphatase) treatment to remove the remaining phosphomonoester.

3.4. HPLC Analysis of Saccharides

Resolution and quantitation of neutral saccharides is accomplished by HPAEC *(27)*. The Dionex system employs a CarboPac PA column and a pulsed amperometric detector (PAD), and allows separation and detection of monomer through hexamer (or higher) of reduced or nonreduced saccharides of mannose. Monosaccharides other then mannose are also detected. Chromatographic conditions as provided by the supplier are followed and generally utilize a sodium acetate gradient in 0.1 *M* NaOH for elution. The following procedure employs two CarboPac PA1 standard columns (4 × 250 mm) connected in tandem. Quantities of saccharides at the picomole level are detected.

1. Equilibrate the column with 0.1 *M* NaOH.
2. Apply the sample (10–50 µL) dissolved in 0.1 *M* NaOH.
3. Elute the column with a linear gradient of sodium acetate from 0 to 0.0 *M*, all in 0.1 *M* NaOH, over the first 10 min of a 30-min analysis. (*See* **Fig. 1** for typical elution profiles of reduced and nonreduced saccharides obtained from *P. pastoris* cell wall "glycopeptide A" *[10]*).

3.5. Glycosidase Analysis

Several exoglycosidases are commercially available and are supplied with detailed conditions for incubations with substrates. Most useful have been Jack Bean α-1,2/3/6-mannosidase, *Aspergillus saitoi* α-1,2-mannosidase, and *Xanthomonas manihotis* α-1,2/3-mannosidase for characterizing the α-glycosidic linkages in the oligosaccharides. Any β-glycosidic linkages will be resistant to hydrolysis by these enzymes.

1. Incubate enzyme with substrate under specified conditions.
2. Stop the reaction by heating in a boiling water bath for 2 to 3 min.
3. Remove any precipitate by centrifugation.
4. Analyze the saccharide products by HPLC analysis as previously described. (*See* **Fig. 2** for elution profiles of saccharides, obtained from *P. pastoris* cell wall mannan, before and after treatment with Jack bean α-mannosidase and with *A. saitoi* α-1,2-mannosidase *[10]*). The component in **Fig. 2C** eluting near 18 mins thus reveals the presence of a disaccharide containing an α-mannosidic linkage (hydrolyzed by Jack bean α-mannosidase) that is not a specific α-1,2-linkage (not hydrolyzed by *A. saitoi* α-1,2-mannosidase).

3.6. Structure Analysis by Physical Methods

Nuclear magnetic resonance (NMR) spectroscopy and mass spectrometry (MS) have become of great utility in determining the stereochemistry (α or β) of the glycosidic linkages within an oligosaccharide and in determining the

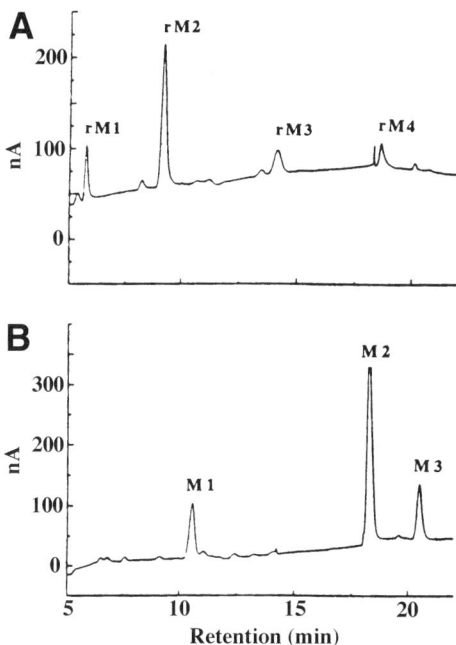

Fig. 1. HPAEC-PAD analyis of saccharides obtained from *P. pastoris* cell wall gly-copeptide A. (**A**) Profile of reduced saccharides obtained by alkaline β-elimination performed in the presence of $NaBH_4$. Peaks eluting at 6, 9, 14, and 19 min correspond respectively to monomer, dimer, trimer, and tetramer saccharide alcohols of mannose. (**B**) Profile of saccharides obtained by alkaline β-elimination performed in the absence of $NaBH_4$. Peaks eluting at 11, 18, and 21 min correspond respectively to monomer, dimer, and trimer saccharides of mannose.

masses of the parent oligosaccharide and resulting fragments. Detailed methodology on the application of these two techniques to structural determination of carbohydrates can be found in numerous other literature sources (*see [28,29]* for reviews and other leading references). With particular reference to application of NMR spectroscopy to saccharides found on expressed glycoproteins from *P. pastoris*, the early study reported by Trimble and coworkers *(30)* on the structures of N-linked oligosaccharides on the *P. pastoris*-expressed *Saccharomyces* invertase can be consulted. Oligosaccharides released from the secreted invertase and separated by size and HPAEC, were characterized by proton NMR to yield structures resulting from the various metabolic reactions in the biosynthetic pathway. The recent report by Trimble and coworkers *(16)* utilizes mass spectrometry, along with proton NMR spectroscopy, to characterize the N- and O-linked saccharides on a human bile salt-stimulated lipase secreted

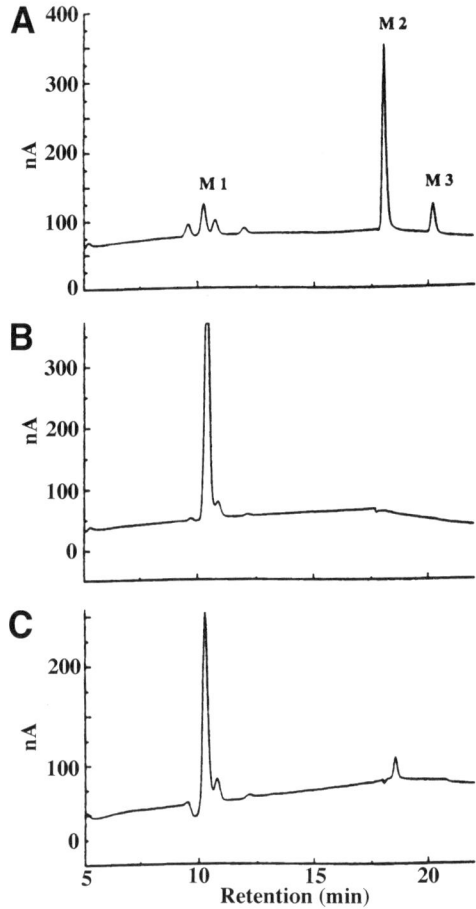

Fig. 2. HPAEC-PAD analysis of saccharides obtained from *P. pastoris* cell wall mannan and digested with α-mannosidases. (**A**) Profile of saccharides obtained by alkaline
β-elimination (in absence of NaBH$_4$), showing monomer, dimer, and trimer saccharides
of mannose eluting respectively at 11, 18, and 21 min. (**B**) Profile after treatment of the
components in (**A**) with Jack bean α-mannosidase, showing only monomer at 11 min.
(**C**) Profile after treatment of the components in (**A**) with *A. saitoi* α-1,2-mannosidase,
showing monomer at 11 min and residual dimer at 18 min.

by *P. pastoris*. In addition to the characterization of a phosphorylated O-linked
saccharide, β-glycosidic linkages were shown to be present in the O-linked
pentasaccharide and hexasaccharide having structures determined to be
Manβ1,2Manβ1,2Manα1,2Manα1,2[Manα1,2]$_{0-1}$Man-ol. The physiological
relevance of these particular protein-bound oligosacchardes in *P. pastoris*,

including potential presence of biosynthetic β-mannosyltransferases, remains to be elucidated.

3.7. Release of O-Linked Saccharides by Other Methods

Other chemical methods, in particular hydrazinolysis, have been utilized to release saccharides from glycoproteins. Both N- and O-linked saccharides will be released, but conditions can be adjusted to preferentially release the O-linked saccharides. Enzymes that will catalyze removal of intact O-linked saccharides from native glycoproteins are not available (*see* **Note 6**).

4. Notes

1. These items are available from various sources as New England Biolab, Inc., Calbiochem-Novabiochem Corp., ProZyme, and Sigma-Aldrich Co., and are supplied with detailed conditions for use.
2. Many methods that are of general use in characterizing glycoproteins and their carbohydrate components are found in the *Methods in Enzymology* series. In particular, *Guide to Techniques in Glycobiology*, Volume 230, can be consulted *(18)*.
3. Contaminating cellular mannan, containing N- and O-linked saccharides, has been reported to be largely removed by trapping on concanavalin A-agarose *(19)*.
4. A procedure for selectively releasing and recovering N-linked and O-linked saccharides from glycoproteins is described in detail by Verostek and coworkers *(22)*. Briefly, the denatured glycoprotein is first treated with endo-glycosidases (endo H and/or PNGase) to release N-linked saccharides, followed by precipitation of the released saccharides and residual protein with 80% acetone at –20°C. Selective extraction from the precipitate of the released saccharides is achieved with 60% ice-cold aqueous methanol. The remaining precipitate, containing any protein with O-linked saccharides, is then subjected to reductive β-elimination and processed for recovery of the reduced saccharides as described above. This procedure is stated to be applicable to starting glycoprotein concentrations from μg to tens of mg/mL, with greater then 90% recovery of the saccharide components.
5. The phenol-sulfuric colorimetric procedure utilized is based on the original published method *(24)*. Mannose (0.18 mg/mL (1.0 μmol/mL in water) is used as a standard. Various volumes (0.05–0.5 mL) of standard or unknown samples are added to Pyrex test tubes, followed by addition of water to make final volumes of 0.5 mL. To each tube is then added 0.5 mL of phenol solution (5% *[w/v]* in water) and the contents mixed. Using a glass syringe, 2.5 mL of concentrated sulfuric acid is then added by rapid injection to obtain thorough mixing. Caution is taken to avoid any splattering of the very hot resulting solution. After standing at room temperature for 30 min, absorbance at 490 nm of each tube is determined. This procedure gives results in terms of micromoles monosaccharide (mannose) present in samples of oligosaccharides, and can be converted to micromoles oligosaccharide when the chain length is known.

6. Other methods for releasing O-linked saccharides from proteins have been reported and may be of particular use. These are summarized as follows:

a. Hydrazine: Saccharides linked to asparagine (N-linked) and serine or threonine (O-linked) are released as intact oligosaccharides with hydrazine and can be further reduced or labeled with AB. Preferential release of O-linked saccharides by hydrazinolysis at 65°C vs release of both O-linked and N-linked saccharides at 85°C, has been reported by Patel et al. *(31)*. This technique has been applied to a study of the saccharides on recombinant mouse gelatinase B expressed in *P. pastoris (11)*, showing that hydrazinolysis at 65°C released saccharides from monomer through pentamer (recovered as AB-labeled saccharides) that were degraded to monomer (with some possible residual trimer) with Jack Bean α-mannosidase.

b. Trifluoromethanesulphonic acid: Degradation of both N- and O-linked saccharides on glycoproteins, with retention of protein structure and, depending on conditons used, retention of the N- and/or O-linked monosaccharide, has been reported *(32)* by use of this anhydrous acid (*see* kit available through Europa Bioproducts or QA-Bio). Recombinant *Toxoplasma gondi* surface antigen I, mutated to delete the N-glycosylation site and expressed in *P. pastoris,* was concluded to be highly O-glycosylated by release of saccharides with this acid treatment *(14)*.

c. Mannosidase: Hydrolysis of the Man-O-Ser/Thr glycosidic bond by Jack Bean α-mannosidase is generally considered not to occur in intact proteins. This lack of activity results from inaccessibility of the glycosidic linkage (in intact proteins) to the enzyme active site, as the O-glycosidic linkages in synthetic Man-α-O-Ser/Thr-Gly (N-tert-butoxycarbonyl-O-α-D-mannosylpyranosyl-seryl/threonyl-glycine methyl amide) glycopeptides have been reported to be hydrolyzed *(33,34)*. Thus the use of Jack Bean α-mannosidase is a potential tool to identify the hexose unit as mannose in α-glycosidic linkage in only very small glycopeptides.

References

1. Lin Cereghino, J. and Cregg, J. M. (2000) Heterologous protein expression in the methylotrophic yeast *Pichia pastoris. FEMS Microbiol. Rev.* **24,** 45–46.
2. Gemmill, T. R. and Trimble, R. B. (1999) Overview of *N-* and *O*-linked oligosaccharide structures found in various yeast species. *Biochim. Biophys. Acta* **1426,** 227–237.
3. Wildt, S. and Gerngross, T. U. (2005) The humanization of *N*-glycosylation pathways in yeast. *Nat. Rev. Microbiol.* **3,** 119–128.
4. Gentzsch, M. and Tanner, W. (1996) The *PMT* gene family: protein *O*-glycosylation in *Saccharomyces cerevisiae* is vital. *EMBO J.* **15,** 5752–5759.
5. Gentzsch, M. and Tanner, W. (1997) Protein-*O*-glycosylation in yeast: protein-specific mannosyltransferases. *Glycobiology* **7,** 481–486.
6. Girrbach, V. and Strahl, S. (2003) Members of the evolutionarily conserved PMT family of protein *O*-mannosyltransferases form distinct protein complexes among themselves. *J. Biol. Chem.* **278,** 12,554–12,562.

7. Ecker, M., Mrsa, V., Hagen, I., Deutzmann, R., Strahl, S., and Tanner, W. (2003) *O*-Mannosylation precedes and potentially controls the *N*-glycosylation of a yeast cell wall glycoprotein. *EMBO Rep.* **4**, 628–632.

8. Strahl-Bolsinger, S., Gentzsch, M., and Tanner, W. (1999) Protein *O*-mannosylation. *Biochim. Biophys. Acta* **1426**, 297–307.

9. Jigami, Y. and Odani, T. (1999) Mannosylphosphate transfer to yeast mannan. *Biochim. Biophys. Acta* **1426**, 335–345.

10. Duman, J. G., Miele, R. G., Liang, H., et al. (1998) *O*-Mannosylation of *Pichia pastoris* cellular and recombinant proteins. *Biotechnol. Appl. Biochem.* **28**, 39–45.

11. Van den Steen, P., Rudd, P. M., Proost, P., et al. (1998) Oligosaccharides of recombinant mouse gelatinase B variants. *Biochim. Biophys. Acta* **1425**, 587–598.

12. Bewley, M. C., Tam, B. M., Grewal, J., et al. (1999) X-ray crystallography and mass spectroscopy reveal that the N-lobe of human transferrin expressed in *Pichia pastoris* is folded correctly but is glycosylated on serine-32. *Biochemistry* **38**, 2535–2541.

13. Mochizuki, S., Hamato, N., Hirose, M., et al. (2001) Expression and characterization of recombinant human antithrombin III in *Pichia pastoris*. *Protein Express. Purif.* **23**, 55–65.

14. Letourneur, O., Gervasi, G., Gaia, S., Pages, J., Watelet, B., and Jolivet, M. (2001) Characterization of *Toxoplasma gondii* surface antigen I (SAGI) secreted from *Pichia pastoris*: evidence of hyper *O*-glycosylation. *Biotechnol. Appl. Biochem.* **33**, 35–45.

15. Boraston, A. B., Sandercock, L. E., Warren, R. A. J., and Kilburn, D. G. (2003) *O*-Glycosylation of a recombinant carbohydrate-binding module mutant secreted by *Pichia pastoris*. *J. Mol. Microbiol. Biotechnol.* **5**, 29–36.

16. Trimble, R. B., Lubowski, C., Hauer III, C. R., et al. (2004) Characterization of *N*- and *O*-linked glycosylation of recombinant bile salt-stimulated lipase secreted by *Pichia pastoris*. *Glycobiology* **14**, 265–274.

17. O'Leary, J. M., Radcliffe, C. M., Willis, A. C., Dwek, R. A., Rudd, P. M., and Downing, A. K. (2004) Identification and removal of *O*-linked and non-covalently linked sugars from recombinant protein produced using *Pichia pastoris*. *Protein Express. Purif.* **38**, 217–227.

18. Lennarz, W. J. and Hart, G. W. (eds.) (1994) *Guide to Techniques in Glycobiology*. Methods in Enzymology, vol. 230. Academic Press, San Diego, CA.

19. Palczewska, M., Batta, G., and Groves, P. (2003) Concanavalin A-Agarose removes mannan impurities from an extracellularly expressed *Pichia pastoris* recombinant protein. *Cell. Mol. Biol. Lett.* **8**, 783–792.

20. Nakajima, T. and Ballou, C. E. (1974) Characterization of the carbohydrate fragments obtained from *Saccharomyces cerevisiae* mannan by alkaline degradation. *J. Biol. Chem.* **249**, 7679–7684.

21. Ballou, C. E. (1990) Isolation, characterization, and properties of *Saccharomyces cerevisiae mnn* mutants with nonconditional protein glycosylation defects. *Methods Enzymol.* **185**, 440–470.

22. Verostek, M. F., Lubowski, C., and Trimble, R. B. (2000) Selective organic precipitation/extraction of released *N*-glycans following large-scale enzymatic deglycoslyation of glycoproteins. *Anal. Biochem.* **278,** 111–122.

23. Yamashita, K., Mizuochi, T., and Kobata, A. (1982) Analysis of oligosaccharides by gel filtration. *Methods Enzymol.* **83,** 105–126.

24. Dubois, M., Gilles, K. A., Hamilton, J. K., Rebers, P. A., and Smith, F. (1956) Colorimetric method for determination of sugars and related substances. *Anal. Chem.* **28,** 350–356.

25. Miele, R. G., Castellino, F. J., and Bretthauer, R. K. (1997) Characterization of the acidic oligosaccharides assembled on the *Pichia pastoris*-expressed recombinant kringle 2 domain of human tissue-type plasminogen activator. *Biotechnol. Appl. Biochem.* **26,** 79–83.

26. Hirose, M., Kameyama, S., and Ohi, H. (2002) Characterization of *N*-linked oligosaccharides attached to recombinant human antithrombin expressed in the yeast *Pichia pastoris.* *Yeast* **19,** 1191–1202.

27. Townsend, R. R. and Hardy, M. R. (1991) Analysis of glycoprotein oligosaccharides using high-pH anion exchange chromatography. *Glycobiology* **1,** 139–147.

28. Duus, J. O., Gotfredsen, C. H., and Bock, K. (2000) Carbohydrate structural determination by NMR spectroscopy: modern methods and limitations. *Chem. Rev.* **100,** 4589–4614.

29. Dell, A. and Morris, H. R. (2001) Review: Glycoprotein structure determination by mass spectrometry. *Science* **291,** 2351–2356.

30. Trimble, R. B., Atkinson, P. H., Tschopp, J. F., Townsend, R. R., and Maley, F. (1991) Structure of oligosaccharides on *Saccharomyces SUC2* invertase secreted by the methylotrophic yeast *Pichia pastoris.* *J. Biol. Chem.* **266,** 22,807–22,817.

31. Patel, T., Bruce, J., Merry, A., et al. (1993) Use of hyrdazine to release in intact and unreduced form both *N*- and *O*-linked oligosaccharides from glycoproteins. *Biochemistry* **32,** 679–693.

32. Edge, A. S. B. (2003) Review article. Deglycosylation of glycoproteins with trifluoromethanesulphonic acid: elucidation of molecular structure and function. *Biochem J.* **376,** 339–350.

33. Ibatullin, F. M., Newstroev, K. N., Golubev, A. M., and Firsov, L. M. (1993) Cleavage of *O*-glycosyl linkages in glycopeptides. *Biokhimiya* **58,** 852–856 (English translation).

34. Ibatullin, F. M., Golubev, A. M., Firsov, L. M., and Neustroev, K. N. (1993) A model for cleavage of *O*-glycosidic bonds in glycoproteins. *Glycoconjugate J.* **10,** 214–218.

9

Modification of the *N*-Glycosylation Pathway to Produce Homogeneous Human-Like Glycans Using GlycoSwitch Plasmids

Wouter Vervecken, Nico Callewaert, Vladimir Kaigorodov, Steven Geysens, and Roland Contreras

Abstract

Glycosylation is an important issue in heterologous protein production for therapeutic applications. Glycoproteins produced in *Pichia pastoris* contain high mannose glycan structures that can hamper downstream processing, might be immunogenic, and cause rapid clearance from the circulation. This chapter describes a method that helps solving these glycosylation-related problems by inactivation of *OCH1*, overexpression of an HDEL-tagged mannosidase, and overexpression of a Kre2/GlcNAc-transferase I chimeric enzyme. Different plasmids are described as well as glycan analysis methods.

Key Words: *Pichia pastoris*; glycosylation; *N*-glycans; glyco-engineering; GlycoSwitch; OCH1; α-1,2-mannosidase; GlcNAc-transferase I; DSA-FACE; protein production; glycan analysis.

1. Introduction

N-linked glycosylation of proteins in eukaryotes is performed in the organelles of the secretory pathway. It starts as a cotranslational process in the endoplasmic reticulum (ER), where a dolichyl-linked oligosaccharide is transferred *en bloc* to an Asn residue of a nascent protein molecule at a specific recognition site (Asn-Xxx-Ser/Thr). This oligosaccharide, which consists of 14 sugar residues ($Glc_3Man_9GlcNAc_2$ where Glc = glucose, Man = mannose, GlcNAc = *N*-acetylglucosamine) (*see* **Fig. 1** for the structure) is modified by removal and addition of sugar residues as the glycoprotein moves through the secretory pathway (*see* *[1]* for a review). After the removal of the three glucose residues and a mannose residue, the glycoprotein is transferred to the Golgi. Oligosaccharides of this type, called high-mannose glycans, contain only

From: *Methods in Molecular Biology, vol. 389:* Pichia *Protocols, Second Edition*
Edited by: J. M. Cregg © Humana Press Inc., Totowa, NJ

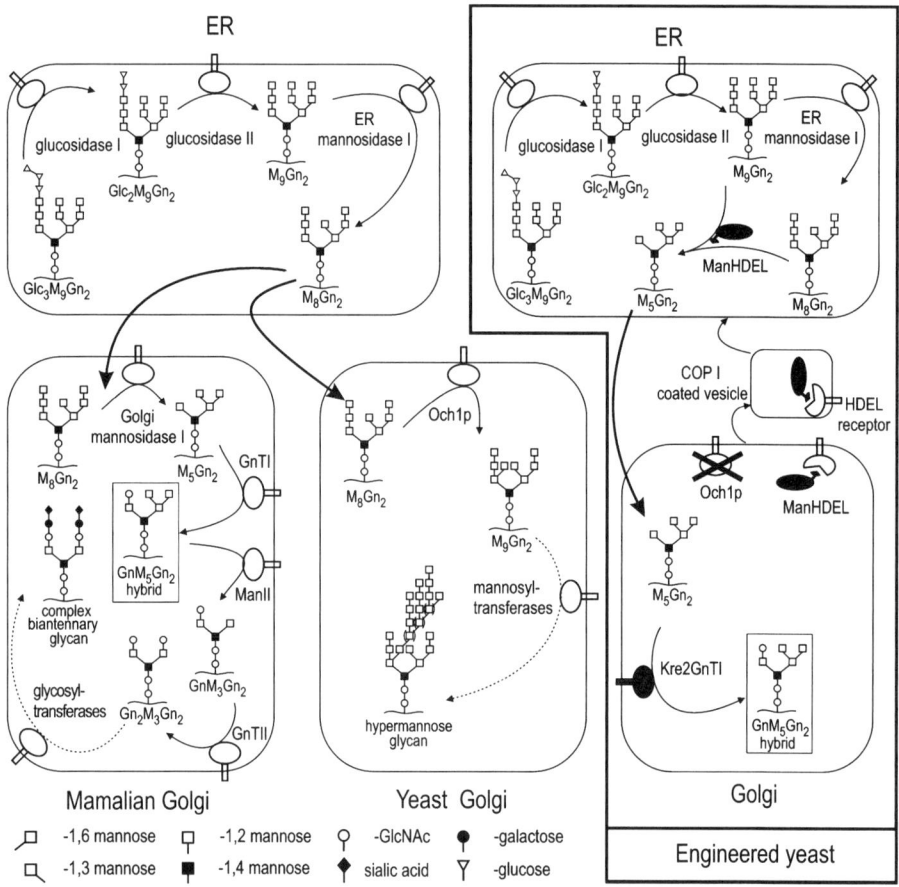

Fig. 1. Schematic representation of the glycosylation pathway in human and yeast and the steps taken for glyco-engineering. The upper part depicts the process that is localized in the ER. A $Glc_3Man_9GlcNAc_2$ structure on the glycoproteins is converted to $Man_9GlcNAc_2$ by two glucosidases. Then the glycans are trimmed to $Man_8GlcNAc_2$ by ER-mannosidase I. Glycoproteins with this structure are transported to the Golgi (lower part). The subsequent steps in human and yeast are different. In human cells the glycans are trimmed to a $Man_5GlcNAc_2$ structure by Golgi mannosidase I and elongated to $GlcNAcMan_5GlcNAc_2$ by GlcNAc-transferase I (GnTI). Further processing involves trimming by Mannosidase II (ManII), which leads to a $GlcNAcMan_3GlcNAc_2$ structure that is elongated by GlcNAc-transferase II (GnTII) and several other glycosyltransferases, to ultimately form a complex type oligosaccharide. In the yeast Golgi, the $Man_8GlcNAc_2$ structure can be further elongated by the addition of an α-1,6-mannose residue by Och1p. This mannose can be further elongated by several mannosyltransferases to form a hypermannose glycan. In the engineered yeast this initiating mannosyltransferase is inactivated, thereby abolishing hypermannosylation. Additionally, an ER retained α-1,2-mannosidase is overexpressed. The enzyme is tagged with an HDEL

mannose as terminal sugars. In the Golgi apparatus, the modification of the protein-linked oligosaccharide differs among different species. In mammals, the glycans are trimmed first to a $Man_5GlcNAc_2$ and later modified to complex type structures that contain GlcNAc, galactose (Gal), fucose (Fuc), and sialic acid in different structural arrangements and combinations. The most common form of complex type *N*-glycans is a biantennary structure (*see* **Fig. 1**). There are also hybrid structures that combine features of both high mannose and complex type glycans (*see* **Fig. 1**).

Mammalian cells mainly produce these complex type sugars, whereas yeast cells produce only high-mannose oligosaccharides that can be extended by the formation of an outer chain. This outer chain is initiated by the α-1,6-mannosyl-transferase Och1p. The addition of this α-1,6-mannose serves as a starting point for further elongation. Different mannosyltransferases can elongate this initiating mannose. In this way, large heterogeneous outer chains are formed, finally containing not only mannoses but also phosphate and GlcNAc residues. This leads to formation of hyperglycans or hypermannose-type glycans, which in *Pichia pastoris* can contain from 30 to 50 glycan residues (i.e., 5 to 8 kDa/glycan). These heterogenous outer chains translate into heterogeneity at the protein level, which can hamper downstream processing of heterologously produced glycoproteins. Furthermore, therapeutic glycoproteins produced in yeast are rapidly cleared from the bloodstream because lectins on cells of the reticuloendothelial system recognize the terminal mannose (*2*). In addition, glycans from yeast serve as epitopes that induce antibody production. Consequently, humanization of the glycans is necessary in cases where long-term blood residence is required, in order to reduce doses and the risk of developing a humoral immune response. Furthermore, regulatory authorities increasingly require that glycosylation patterns of therapeutics be consistent from batch to batch.

Described here are tools to homogenize and humanize the *N*-glycans of glycoproteins produced in *P. pastoris*. The steps involve the inactivation of *OCH1* to abolish hyperglycosylation, the overexpression of an ER-retained mannosidase, and the overexpression of a Golgi-retained human GlcNAc-transferase, as we have recently described (*3*). In addition to describing the different steps taken to reach this goal, two carbohydrate electrophoresis methods—FACE (fluorophore-assisted carbohydrate electrophoresis and DSA-FACE (DNA sequencer-assisted FACE)—for analyzing glyco-engineered strains are also discussed.

Fig. 1. (*Opposite page*) sequence that leads to retrieval of the protein from the Golgi to the ER. This enzyme removes all α-1,2-mannose residues, producing a $Man_5GlcNAc_2$ structure, which can then be elongated by GnTI. The concomitant expression of a chimeric enzyme consisting of the GnTI catalytic domain and the targeting signals of the yeast Golgi localized protein Kre2p results in the production of $GlcNAcMan_5GlcNAc_2$.

2. Materials

2.1. Strains and Plasmids for OCH1 Inactivation

1. Plasmid pGlycoSwitchM8 (*see* **Fig. 2**) (*see* **Note 1**).
2. *P. pastoris* strain GS115 (*his4*).
3. Vector pPICZB and antibiotic Zeocin (both from Invitrogen, Carlsbad, California, USA).
4. Primer 1: 5′-AAGGAGTTAGACAACCTGAAGTCTA-3′.
5. Primer 2: 5′-AGATCTTTAGTCCTTCCAACTTCCTT-3′.

2.1.1. Expression of α-1,2-Mannosidase

1. Plasmids pGlycoSwitchM5 (*see* **Fig. 3A**) and pPIC9ManHDEL (*see* **Fig. 3B**).
2. AOX1 primers 5′: 5′-GACTGGTTCCAATTGACAAGC-3′ and 3′: 5′- GCAAAT GGCATTCTGACATCC-3′.

2.1.2. Expression of Kre2GnTI

1. Plasmids pPIC6Kre2GnTI (*see* **Fig. 3C**) and pGAPKre2GnTI-Blast (*see* **Fig. 3D**).
2. Primer 3: 5′-GTCCCTATTTCAATCAATTGAA-3′.
3. Antibiotic blasticidin (Fluka, Buchs, Switzerland).

2.2. Glycan Analysis Products

1. PNGase F (NE BioLabs, Beverly, MA).
2. Bovine pancreas RNAse B derived and malto-oligosaccharide reference sugars (Prozyme, San Leandro, CA).

2.2.1. Fluorophore-Assisted Carbohydrate Electrophoresis (FACE)

1. Mini polyacrylamide gel electrophoresis system (*see* **Note 2**).
2. Buffers supplied with PNGase F from NEB:
   a. 10X denaturing buffer: 5% sodium dodecyl sulfate (SDS) and 10% 2-mercapto-ethanol.
   b. 10X G7 buffer: 0.5 *M* sodium phosphate (pH 7.5) and 10% Nonidet P40.
3. Carbograph columns and resin (Alltech Associates, Deerfield, IL).
4. 80% acetonitrile + 0.1% trifluoroacetic acid (TFA).
5. 25% acetonitrile + 0.05% TFA.
6. ANTS (8-aminonaphtalene-1,3,6-trisulfonic acid) (Molecular Probes, Eugene, CA; also available from several other suppliers). It can be stored at −80°C at 64 mg/mL in acetic acid/H_2O 3:17 *(v/v)*.
7. 1 *M* $NaCNBH_3$ (sodium cyanoborohydride) in DMSO (62 mg/mL, freshly prepared). $NaCNBH_3$ is toxic and should be handled in a hood. It is also hygroscopic and should therefore be kept in a container secluded from the air.
8. 50 mL acrylamide stock solution: 60% acrylamide/1.6% bis-acrylamide.
9. TEMED.

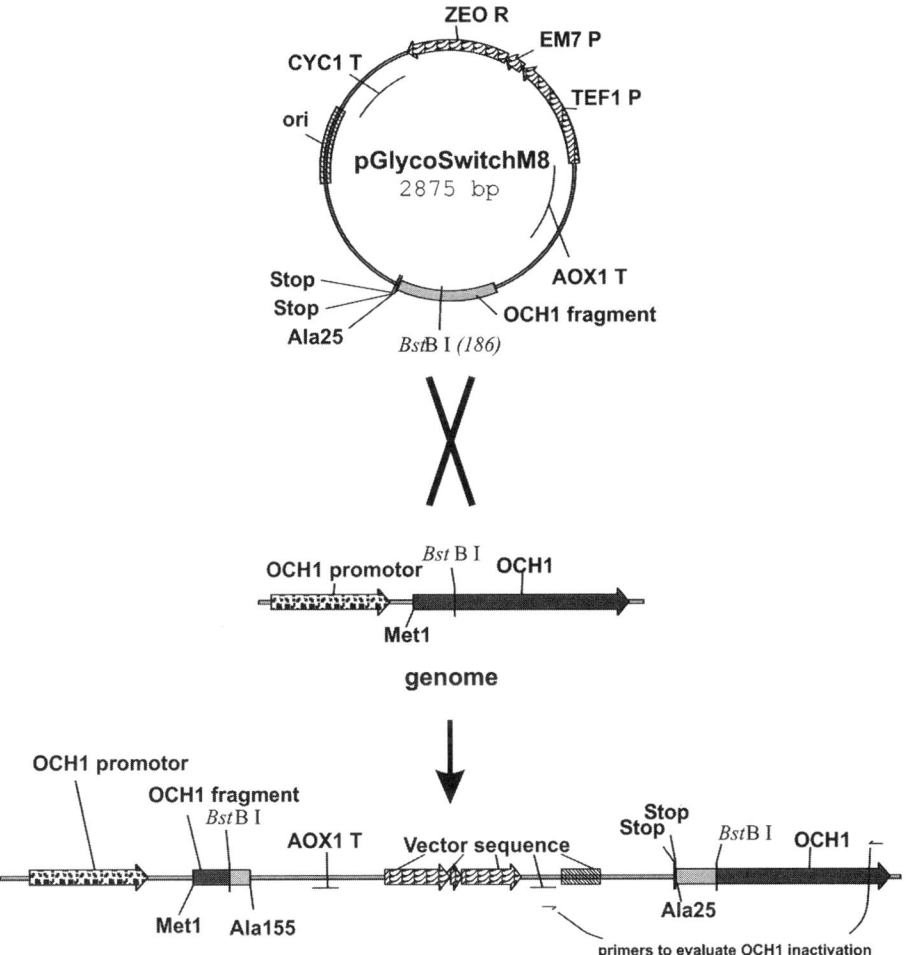

genome after OCH1 inactivation

Fig. 2. *OCH1* inactivation plasmid pGlycoSwitch M8. After linearization of the plasmid with *Bst*B I and transformation to *Pichia*, a single cross-over event at the *OCH1* locus inactivates the gene. An *OCH1* fragment that is situated downstream of the *OCH1* promoter can only result in the synthesis of a truncated polypeptide with 155 amino acids. This fragment does not contain the catalytic part of *OCH1*. In addition, after introduction of the plasmid in the genome, there remains an *OCH1* fragment that is promoterless and lacks the first 24 amino acids. There are also two in frame stop codons just before the *OCH1* sequence that prevent the expression of a functional protein from RNA that might be transcribed from any cryptic promoter that may lie upstream. Reproduced with permission from **ref. 3**.

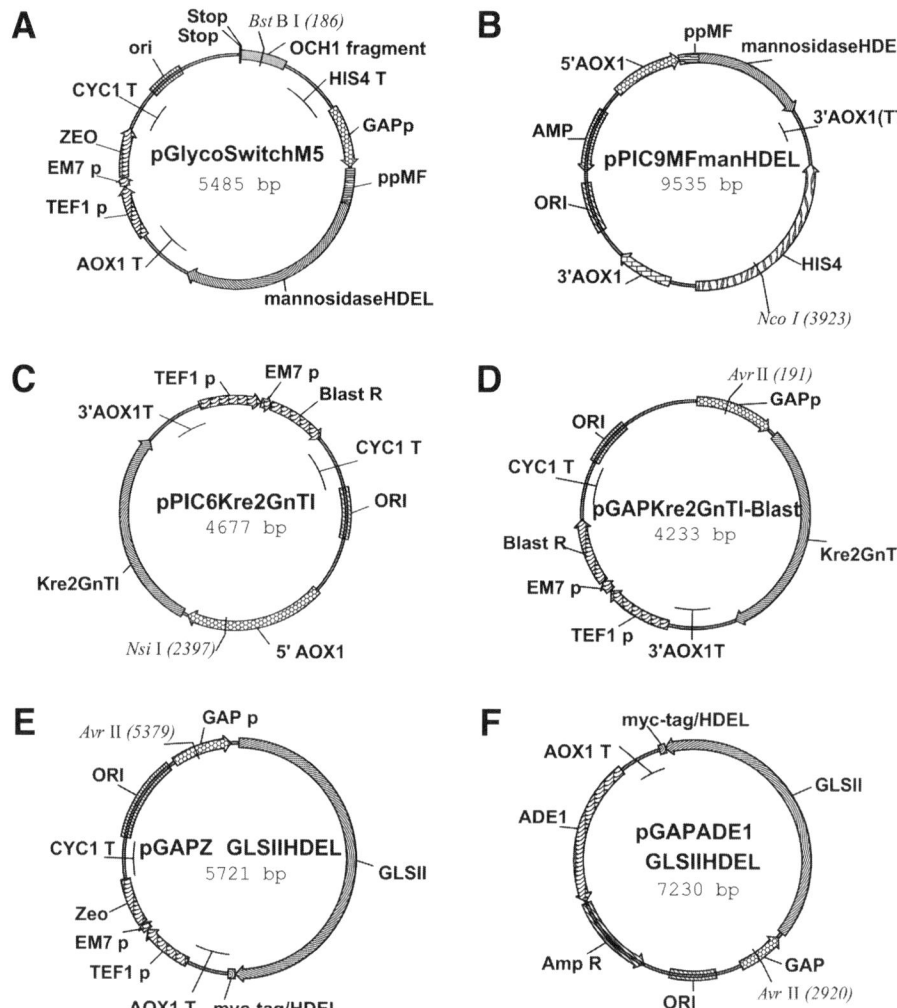

Fig. 3. Circular map of the plasmids used in the glyco-engineering of *Pichia*. The indicated unique restriction sites are used for linearization of the plasmids before transformation.

10. 10% ammoniumperoxidosulfate (APS).
11. 1.5 *M* Tris-HCl (pH 8.8).
12. 10X running buffer: 1.92 *M* glycine, and 0.25 *M* Tris-HCl (pH 8.5).

2.2.2. DNA-Sequencer Assisted-FACE (DSA-FACE) Glycan Analysis Procedure

1. Applied Biosystems 377 DNA sequencer (*see* **Note 3**).
2. Vacuum manifold (Millipore, Bedford, CA).

3. RCM buffer: 8 M urea, 360 mM Tris-HCl (pH 8.6), and 3.2 mM ethylene diamine tetraacetic acid (EDTA).
4. Multiscreen-Immobilon-P plates (Millipore, Bedford, CA).
5. 0.1 M dithiothreitol (DTT) (prepared monthly, the stock is stored at –20°C).
6. 0.1 M iodoacetic acid.
7. 1% polyvinylpyrrolidone 360 (Sigma-Aldrich, Bornem, Belgium).
8. 10 mM Tris-acetate (pH 8.3).
9. 1.2 M citric acid.
10. 1 M NaCNBH$_3$ in dimethyl sulfoxide (DMSO) (freshly made).
11. 20 mM APTS (Molecular Probes, Eugene, CA) in 1.2 M citric acid (store at –20°C).
12. Multiscreen-Durapore HV 0.45-µm membrane-lined 96-well plates (Millipore, Bedford, CA).
13. Tapered-well microtiter plates (V-shaped bottom).
14. 100 µL Multiscreen column loader (Millipore, Bedford, CA).
15. Sephadex G10 (Amersham Pharmacia Biotech AB, Uppsala, Sweden).
16. ROX-labeled Genescan™ 500 standard mixture (Perkin Elmer, Foster City, CA, USA).
17. 19:1 mixture of acrylamide:bisacrylamide (BioRad, Hercules, CA).
18. Running buffer: 89 mM Tris-borate and 2.2 mM EDTA.

2.3. Preparation of Mannoproteins

1. 0.9% NaCl.
2. 20 mM Na citrate (pH 7.0).
3. Methanol.

3. Methods

3.1. Inactivation of OCH1 Gene With pGlycoSwitchM8

To abolish hyperglycosylation, the *OCH1* gene must be inactivated. A strategy was developed to accomplish this in a simple, efficient homologous recombination step (*see* **Fig. 2**). A plasmid was designed containing a 391-bp fragment of the *OCH1* gene starting at bp +75. Two in-frame stop codons were inserted upstream of this fragment to prevent potential start of translation of a truncated *OCH1* after correct integration of the plasmid in the *P. pastoris* genome. Before transformation, the plasmid is linearized by digestion of the unique *Bst*B I site in the *OCH1* fragment. Upon correct integration in the *OCH1* locus, a promotor-less truncated fragment that has 2 in-frame stop codons is created. A short ORF (Met1-Ala155) can be translated, but it lacks the catalytic domain (*see* **Fig. 2** for an overview). Thus, inactivation of *OCH1* is achieved through a single cross over event following a very efficient transformation step (more than half of the antibiotic-resistant clones have a correct integration).

3.1.1. Plasmid Construction

1. Before transformation, digest plasmid pGlycoSwitchM8 with *Bst*B I to direct the construct to the *OCH1* locus.

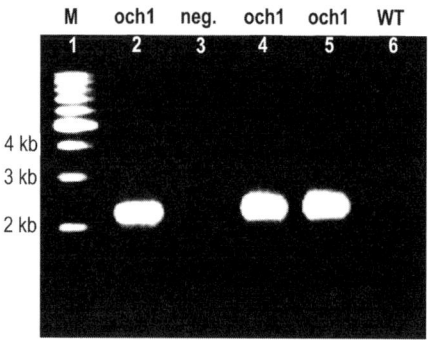

Fig. 4. PCR on genomic DNA from Zeocin resistant clones after transformation with pGlycoSwitchM8 plasmid. *Lanes 2, 4,* and *5* display the correct band of 2.1 kb. *Lane 3* represents a Zeocin resistant clone in which *OCH1* is not inactivated (incorrect integration). *Lane 6* is the WT strain. *Lane 1* contains a DNA marker ladder.

2. Use either the electroporation or whole cell-polyethylene glycol (PEG)$_{1000}$ transformation method to transform *P. pastoris* strain GS115.
3. Plate the transformation mixture on YPDS plates containing 50 µg/mL Zeocin.
4. Zeocin resistant clones are analyzed on the genomic level by PCR using primer 1 and primer 2. This should amplify a DNA fragment of 2101 bp. **Fig. 4** shows the amplified fragments separated by agarose gel electrophoresis of different Zeocin resistant clones after introduction of pGlycoSwitchM8. Lanes indicated with *och1* are from clones with the correctly integrated plasmid.

3.1.2. Glycan Analysis

Several methods can be used to evaluate the change in glycosylation patterns. However, it should be emphasized that most glycan analysis procedures (e.g., mass spectrometry, DSA-FACE, HPAEC-PAD) are not well suited for the detection of hypermannose structures. As *P. pastoris* hypermannose structures are very heterogeneous in terms of mass, charge, and sugar composition, the high-resolution of the high-molecular-weight components of the above-mentioned techniques becomes a disadvantage. The exponentially increasing number of isomers is partially resolved, resulting in a high background signal which is very hard to quantify. As a result, the degree of hyperglycosylation is often underestimated.

Therefore, for the analysis of hyperglycosylation we chose the older classical FACE method described by Jackson et al. *(4)*. In this technique, the high molecular weight hyperglycans are not resolved, but appear as one smearing band on the gel (*see* **Fig. 5**) *T. reesei* α-1,2-mannosidase was heterologously expressed as a secreted protein in the WT GS115 and in the *och1* inactivated strain. The smearing band indicated as hypermannose structure (*see* **Fig. 5B**,

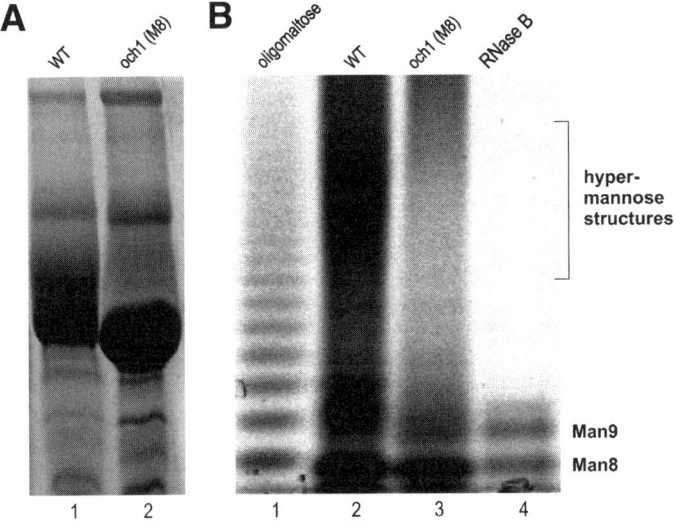

Fig. 5. Inactivation of *OCH1* on the glycoprotein level. (**A**) SDS-PAGE of secreted mannosidase heterologously expressed in wild type GS115 *(lane 1)* and in the GS115 GlycoSwitchM8 strain *(lane 2)*. The smearing above the mannosidase is absent from the *och1* inactivated strain. (**B**) FACE analysis of *N*-glycans from the same mannosidase. *Lane 1:* oligomaltose reference; *lane 2:* glycans from mannosidase produced in the WT strain; *lane 3:* glycans from mannosidase produced in an och1 inactivated strain, *lane 4:* reference glycans from RNase B. The hypermannose structures present in the WT strain are absent. Reproduced with permission from **ref. 3**.

lane 2) is absent from the *och1* strain. In addition to the absence of hyperglycosylation, changes can also be observed in the low-molecular-weight glycan structures upon inactivation of *OCH1*. The latter changes can also be evaluated using the DSA-FACE procedure described below, in which case only the glycans that are not extensively elongated after Och1p activity are observed (*see* **Fig. 6**, *panels 2* and *3*).

3.1.2.1. Deglycosylation Using Peptide: N-Glycosidase F (PNGase F)

1. Precipitate the glycoprotein sample with 3 volumes ice cold ethanol.
2. Denature between 20 and 100 µg of glycoprotein in 100 µL denaturing buffer (10X: 5% SDS, 10% 2-mercaptoethanol) at 100°C for 20 min.
3. Add one-tenth of a volume of each 10X G7 buffer (0.5 *M* sodium phosphate, [pH 7.5]) and 10% Nonidet P-40, and mix. NP-40 is required for PNGase activity under these strongly denaturing conditions.
4. Add 1 to 5 µL (500-2500 Units, unit definition NEB) PNGase F and incubate overnight at 37°C.

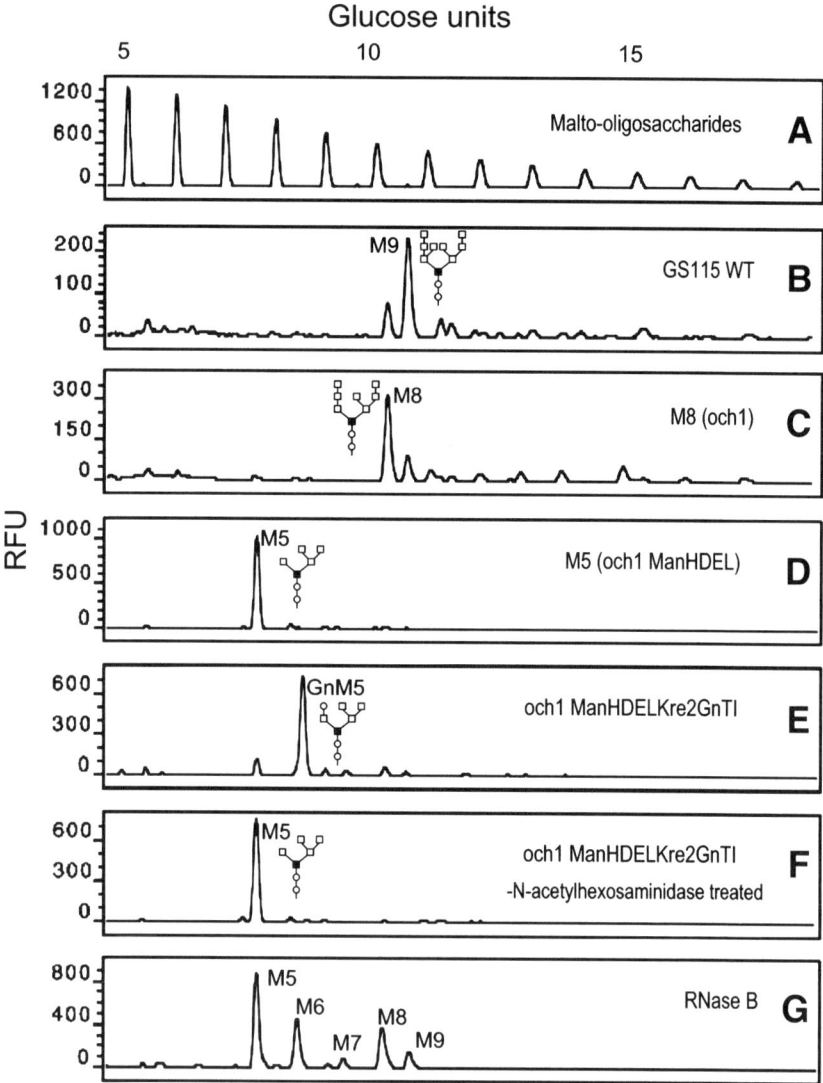

Fig. 6. DSA-FACE analysis of *N*-linked carbohydrates from mannoproteins from several glyco-engineered *Pichia* strains. *Lane 1:* malto-oligosaccharides reference. Next panels contain *N*-glycans from: *lane 2:* wild-type strain GS115. The main peak is $Man_9GlcNAc_2$ presumably the Och1p modified $Man_8GlcNAc_2$ ER-exit structure. *Lane 3: och1* inactivated strain. Main peak: $Man_8GlcNAc_2$. *Lane 4: och1* strain expressing ManHDEL. Main peak: $Man_5GlcNAc_2$. *Lane 5: och1* inactivated strain expressing ManHDEL and KreGnTI. Main peak: $GlcNAcMan_5GlcNAc_2$. *Lane 6:* Same as *lane 5* but glycans treated with β-*N*-acetylhexosaminidase. The $GlcNAcMan_5GlcNAc_2$ peak shifts to the $Man_5GlcNAc_2$ position, indicating that terminal GlcNAc was present. *Lane 7:* Reference glycans from bovine RNase B ($Man_{5-9}GlcNAc_2$). Adapted with permission from **ref. (3)**.

3.1.2.2. GLYCAN PURIFICATION

The glycans are purified by solid phase extraction on Porous Graphitized Carbon. Detergents and proteins bind with very high affinity to this material, whereas glycans can be selectively eluted with acetonitrile-containing solutions (*see* **Note 4**).

1. Wash a Carbograph column once with 2 mL of 80% acetonitrile + 0.1% TFA and once with 10 mL water (*see* **Note 5**).
2. Load the sample diluted with water to 200 µL and wash with 5 to 10 mL of water.
3. Elute with 2 mL of 25% acetonitrile + 0.05% TFA (neutral and charged sugars are eluted together).
4. Place in an evaporator to reduce the volume to 100 µL, then transfer to a 200 µL polymerase chain reaction (PCR) tube and evaporate to dryness.

3.1.2.3. ANTS LABELING OF THE SUGARS

ANTS (8-aminonaphtalene-1,3,6-trisulfonic acid) is a fluorescent reagent that reacts with the reducing end of the isolated sugars. The product of the reaction is fluorescent and has three extra negative charges. As only one ANTS molecule reacts one oligosaccharide, quantitative analysis is possible.

1. Add 5 µL of 0.15 M ANTS (64 mg/mL in acetic acid/H_2O 3:17 *[v/v]*) and 5 µL of 1 M NaCNBH$_3$ in DMSO (62 mg/mL, freshly prepared) to the dried sugars. Incubate overnight at 37°C.
2. Precipitate the sugars by adding 10 µL of H_2O and 6 volumes of cold acetone. Keep at –20°C for 30 min and centrifuge at 18,000g for 20 min at –10°C.
3. Discard the supernatant and leave the pellet to dry completely. Dissolve the pellet in 5 to 8 µL of 25% glycerol.

3.1.2.4. ELECTROPHORETIC FRACTIONATION OF THE SUGARS

1. The labeled (and charged) sugars are now separated by PAGE using 25–30% gels. Hyperglycosylation is better detected with 25% gels, which are also easier to cast.
2. The separating gel contains 8.3 mL of 60%/1.6% acrylamide/bisacrylamide, 5 mL of 1.5 M Tris (pH 8.8), 6.6 mL of H_2O, 120 µL of 10% APS (freshly prepared), and 20 µL of TEMED (*see* **Note 6**).
3. The stacking gel is composed of 650 µL of 60%/1.6% acrylamide/bisacrylamide, 2.5 mL of 1.5 M Tris (pH 8.8), 6.5 mL of H_2O, 60 µL 10% of APS and 20 µL of TEMED.
4. While running the gel, put the apparatus in an ice bath and use precooled running buffer (4°C) to prevent in-gel diffusion and possible breakdown of the ANTS molecules.
5. Apply 100 V during stacking. After the glycans enter the separating gel, raise the voltage to 180 V.

3.1.2.5. VISUALIZATION OF THE OLIGOSACCHARIDES

ANTS fluorescence is excited at 370 nm, with an emission centered at 500 nm. Place the gel, still sandwiched between the glass plates, on an ultraviolet (UV) transilluminator.

3.2. Expression of ER-Retained α-1,2-Mannosidase: From M8 to M5

Trimming $Man_8GlcNAc_2$ to $Man_5GlcNAc_2$ requires expression of an α-1,2-mannosidase that is retained in the secretory pathway.

A prerequisite for α-1,2-mannosidase functionality is the absence of glucose residues on the α-1,3-branch of its substrate, which means the presence a functional glucosidase II. These extra glucose residues have been observed on glycans of *Trichoderma reesei (5)* and *P. pastoris (6)*. If such glucose residues are present, which is rarely the case, it is advised first to overexpress a glucosidase II. In **Fig. 3E,F**, two plasmids are represented containing the GlsII catalytic α-subunit from *S. cerevisiae,* tagged at the C-terminus with an HDEL sequence for retrieval of the protein from the Golgi to the ER. The HDEL sequence interacts with an HDEL receptor in the Golgi apparatus, which leads to sequestration of the protein in the ER in a COPI dependent way.

As an α-1,2-mannosidase we selected a mannosidase that had been PCR amplified from a *T. reesei* cDNA library *(7)* and added here to a C-terminal HDEL-tag to retain the protein in the ER *(8)*.

We evaluated two promoters to control the open reading frame of the mannosidase (i.e., *AOX1* and *GAP* gene promoters). When the constitutive *GAP* promoter was used to drive expression of the mannosidase, we obtained complete N-glycan trimming to $Man_5GlcNAc_2$. The same was true with the *AOX1* promotor upon methanol induction. The manHDEL sequence was cloned in frame with the pre-pro sequence of the yeast α mating factor for targeting to the secretory pathway.

Plasmid pPIC9ManHDEL (*see* **Fig. 3B**), which contains the MFmannosidase HDEL under control of the *AOX1* promoter, can be transformed into an *och1* strain, to yield the $Man_5GlcNAc_2$ structure. However, the most efficient way of converting wild type *N*-glycosylation to $Man_5GlcNAc_2$ is to use pGlycoSwitchM5. A GAP-driven mannosidaseHDEL expression cassette was cloned into the above-described plasmid pGlycoSwitchM8 to yield this pGlycoSwitchM5. Plasmid pGlycoSwitchM5 (*see* **Fig. 5A**) allows achievement of $Man_5GlcNAc_2$ in only one transformation step.

3.2.1. Transformation

1. Digest pGlycoSwitchM5 with *Bst*B I and transform GS115 using one of the methods described in Chapter 3.
2. Select transformants by plating the transformation mixture on YPDS plates containing 50 µg/mL Zeocin.

3. Screen for proper integration of the construct in the *P. pastoris* genome by PCR, using primers 1 and 2 as described for pGlycoSwitchM8 (**Subheading 3.1.2.**). An amplicon of 2101 bp should be generated.
4. Digest pPIC9MFManHDEL with *Nco* I and transform the pGlycoSwitchM8-engineered strain that was created as described in **Subheading 3.1**.
5. Plate the transformation mixture on RDB–His medium.
6. Screen for the presence of the construct in the genome by PCR using 3′- and 5′-*AOX1* primers, which should amplify a fragment of 2004 bp.

3.2.2. Mannoprotein Preparation

If no heterologous glycoprotein is expressed, we found it convenient to prepare mannoproteins from the glyco-engineered strains to analyze the change in *N*-glycosylation *(9)*. It must be emphasized that if the *AOX1* controlled mannosidase is expressed (plasmid pPIC9MFManHDEL) the strain must be grown for 24 h in medium containing methanol.

1. Grow an overnight culture of the transformed strain in 10 mL.
2. Pellet the cells by centrifugation and wash with a 0.9% NaCl solution.
3. Wash the cells twice with water and resuspend them in 1.5 mL of 0.02 *M* sodium citrate (pH 7.0) in a microcentrifuge tube.
4. Autoclave the cells for 90 min at 121°C.
5. Vortex and spin down the cellular debris.
6. Collect the supernatant. Precipitate the mannoproteins overnight with 4 volumes methanol at 4°C with rotary motion.
7. Collect the precipitate by centrifugation. Allow the pellet to dry and dissolve it in 50 µL of water.

3.2.3. DSA-FACE: Carbohydrate Analysis by Standard DNA Sequencing Equipment

DSA-FACE *(10)* is extremely sensitive, has high throughput, and separates most isomers. Less than 1 µg of glycosylated protein is sufficient to obtain size profiles of the nondigested and glycosidase-digested *N*-glycans. It is well suited for analysis of oligosaccharides of up to at least 25 glucose units. The procedure is schematically represented in **Fig. 7**.

3.2.3.1. MICROTITER 96-WELL PLATE DEGLYCOSYLATION PROCEDURE

The deglycosylation protocol has been elaborated in detail by Papac et al. *(11)* (*see* **Note 7**).

1. Dilute the glycoprotein in 2 volumes RCM buffer to a final volume of at least 50 µL, and incubate for 1 h at 55°C.
2. Wet the polyvinylidene fluoride (PVDF) membrane at the bottom of the wells of a Multiscreen-IP plate with 300 µL methanol, then wash it 3× with 300 µL water and once with 50 µL of RCM buffer. Finally add 15 µL RCM buffer.

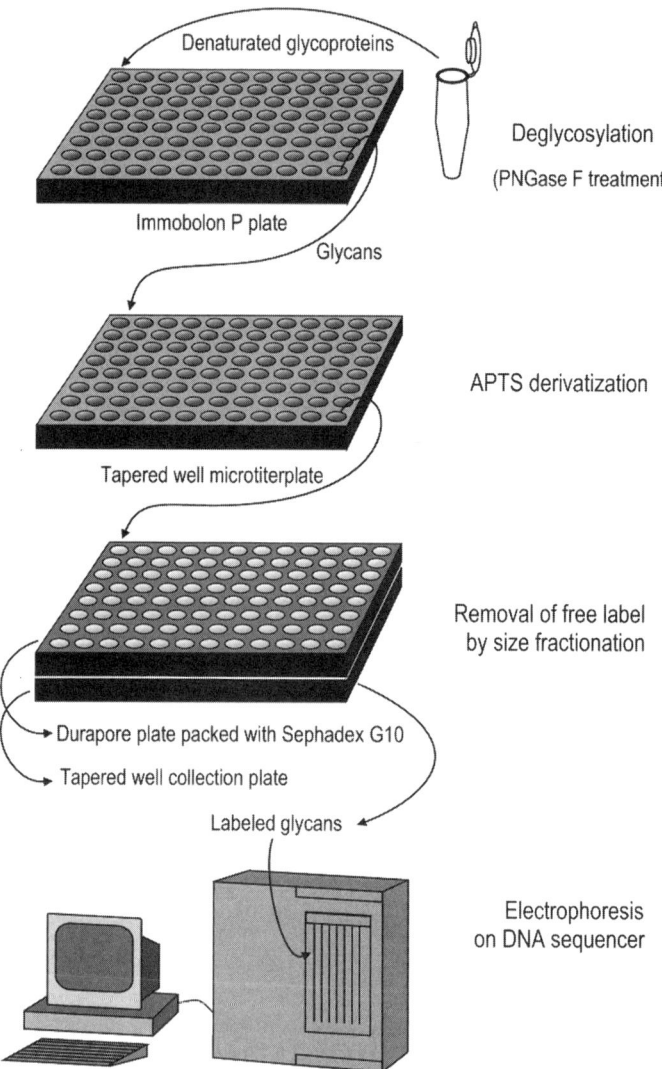

Fig. 7. Schematic overview of the DSA-FACE procedure.

3. Load the glycoprotein in the wells.
4. Place the plate on a vacuum manifold and apply gentle vacuum to bind the protein to the membrane.
5. Wash twice with 50 µL RCM buffer.
6. Reduce the bound protein by adding 50 µL of 0.1*M* DTT in RCM buffer and incubating at 37°C for 1 h. Remove the reducing solution and wash the wells three times with 300 µL of water, always removing the washing fluid through the membrane on the vacuum manifold.

7. Perform carboxymethylation by adding 50 µL of 0.1 *M* iodoacetic acid in RCM buffer and incubating for 30 min at room temperature in the dark.
8. Remove the solution by washing three times with 300 µL of water.
9. Block the remaining protein binding capacity of the wells by incubation with 100 µL of 1% polyvinylpyrrolidone (PVP) 360 in water at room temperature for 1 h.
10. Wash three times as described above.
11. Prepare a 1:5 dilution of PNGase F in 10 m*M* Tris-acetate (pH 8.3) (store at –20°C). Dilute this mixture 1:20 in 10 m*M* Tris-acetate (pH 8.3), and add 20 µL to the wells. Allow to digest for 3 h at 37°C. (Finally, 100 units are added in 20 µL; unit definition NEB).
12. Transfer the solution to a tapered-well microtiter plate (*see* **Note 8**).

3.2.3.2. APTS DERIVATIZATION REACTION

1. Evaporate the deglycosylation mixture to dryness.
2. Add 1 µL of a 1:1 mixture of 20 m*M* APTS in 1.2 *M* citric acid and 1 *M* NaCNBH$_3$ in DMSO to each well.
3. Vortex the plate carefully and centrifuge it briefly, and then incubate it upside down at 37°C overnight, tightly wrapped in parafilm.
4. The following morning, quench the reaction by the addition of 4 µL of water.

3.2.3.3. 96-WELL SEPHADEX G10 POSTDERIVATIZATION CLEANUP

1. Pack the wells of a Multiscreen-Durapore membrane-lined 96-well plate with 200 µL dry Sephadex G10 using a resin loader with a capacity of 100 µL. First, add 100 µL of the resin and wet it with 100 µL water, then load the rest of the resin and wet it (*see* **Note 9**).
2. Prior to sample loading, add 2 × 150 µL water to the resin, then centrifuge for 10 s at 750*g* in an Eppendorf table-top centrifuge equipped for handling 96-well plates.
3. After loading the sample on the resin that had been centrifuged to dryness, elute by adding 10 µL of water and centrifuging the plate for 10 s at 750*g*. (Optimal centrifugation conditions depend on the properties of the centrifuge). Collect the eluate in another tapered-well microtiter plate. Elute two more times, so that 35 µL is collected in the tapered well microtiter plate.
4. Evaporate the eluate to dryness and reconstitute in 5 µL water, then repeat **step 3** (it is once again treated as a sample; *see* **Note 8**). Successful cleanup is hallmarked by the detection of only faint fluorescence of the eluate on a standard UV-light box.
5. After evaporation, reconstitute the derivatized glycans in 5 µL of water.

3.2.3.4. PREPARATION OF THE SAMPLES FOR GEL LOADING

1. To 1 µL of each sample, add 0.5 µL of the ROX-labelled Genescan™ 500 standard mixture (*see* **Note 10**). Save the rest for eventual exoglycosidase digestions.
2. After addition of the internal standard, add 0.5 µL of deionized formamide to facilitate sample loading.

3.2.3.5. Gel Electrophoresis and Data Analysis

We performed our experiments on an Applied Biosystems 377A DNA-sequencer equipped with an external cooling bath kept at 23°C (easily connectable to the sequencer according to the ABI PRISM 377 DNA sequencer user bulletin "Modifications for subambient temperature operations") (*see* **Note 3**). Because carbohydrates diffuse rapidly, we load alternate lanes to avoid cross-contamination and to ease the lane tracking process. The gel contains 10% of a 19:1 mixture of acrylamide:bisacrylamide in the standard DNA-sequencing running buffer (89 mM Tris-borate, and 2.2 mM EDTA). Polymerization is catalyzed by the addition of 200 µL of a 10% ammonium peroxidosulfate solution in water and 20 µL of TEMED. The gels are of the standard 36-cm well-to-read length. The gels are prerun at 3000 V for 1 h, the wells are thoroughly rinsed with the sequencing buffer, and 1.6 µL of the samples is loaded. Electrophoresis is performed at 3500 V and data are collected for 5 h, which is sufficient to separate glycans of up to 25 glucose units in size. Data analysis is performed using the Genescan 3.1 software. The fluorescence-overlap correction matrix employed for DNA sequencing using BigDye^{TM} dye terminators can normally be used here. The fluorescence of APTS-derivatized carbohydrates and rhodamine-labeled oligonucleotides is readily resolved.

Figure 6, *panel 1* shows DSA-FACE analysis of the effects of the inactivation of *OCH1* and the overexpression of mannosidaseHDEL on the mannoprotein *N*-glycans. *Panel 2* depicts GS115 WT glycans where the major peak is Man$_9$GlcNAc$_2$. After inactivation of *OCH1*, the main peak becomes Man$_8$GlcNAc$_2$ (*panel 3*), whereas the overexpression of mannosidaseHDEL in the *och1* background results only in Man$_5$GlcNAc$_2$ (*panel 4*).

3.3. Expression of Golgi Targeted GlcNAc-Transferase I

The next step in the formation of human-like *N*-glycans is the overexpression of GlcNAc-transferase I (GnTI), which adds a GlcNAc residue to the Man$_5$GlcNAc$_2$ structure (*see* **Fig. 1**). To target GnTI to the Golgi we fused the catalytic domain to the yeast Kre2p *N*-terminal part responsible for the localization of the yeast protein to the Golgi; the plasmid contains a chimera composed of the first 100 amino acids from Kre2p and the last 327 amino acids from human GlcNAc-transferase I. This chimera was cloned under control of the *AOX1* promoter of plasmid pPIC6A to form pPIC6Kre2GnTI. This plasmid can be transformed in a strain that produces a Man$_5$GlcNAc$_2$ structure on its glycoproteins and is selectable on blasticidin. An available alternative construct is pGAPKre2GnTI-Blast, in which the fusion construct is under control of the constitutive *GAP* promoter. It is advisable to use the *GAP* controlled constructs when constitutive heterologous glycoprotein expression is desired, and to use

AOX1 driven glyco-engineering enzymes when the heterologously expressed protein is under control of the *AOX1* promoter.

3.3.1. Transformation

1. Digest pGAPKre2GnTI-Blast with *Avr*II and transform into a pGlycoSwitchM5-engineered strain.
2. Select on YPDS plates containing 500 µg/mL blasticidin (*see* **Note 11**).
3. Screen for the presence of the construct by PCR using the 3'*AOX1* primer and primer 3. An amplified DNA fragment of 1620 bp should be generated.
4. Digest plasmid pPIC6Kre2GnTI (*see* **Fig. 3C**) with *Nsi* I before transformation to a pGlycoSwitchM8 + pPIC9MFManHDEL-engineered strain.
5. Plate the transformation mix on YPDS plates containing 500 µg/mL blasticidin.
6. Screen for the presence of the construct by PCR using the *AOX1* primers. A fragment of 1620 bp should be amplified.

3.3.2. Glycan Analysis

Glycan analysis can be performed on expressed glycoproteins as well as on mannoproteins. Oligosaccharides from mannoproteins are prepared as described above (*see* **Subheading 3.2.3.1.**), and for purified glycoproteins there is a choice between the procedures described in **Subheadings 3.1.2.1.** and **3.2.3.1.** (*see* **Note 12**). Analysis is preferably performed by DSA-FACE as described in **Subheading 3.2.3**.

A peak about one glucose unit larger than $Man_5GlcNAc_2$ appears in the electropherogram. Additional proof of the presence of terminal GlcNAc can be obtained by digestion of the oligosaccharides with β-*N*-acetylhexosaminidase. After digestion, the peak containing GlcNAc peak shifts to the position of $Man_5GlcNAc_2$.

1. Transfer 1 µL of the cleaned-up derivatized *N*-glycans (*see* **Subheading 3.2.3.3.**, **step 5**) to a 250 µL PCR tube or to a tapered-well microtiter plate.
2. Add 10 µL of 20 m*M* sodium acetate (pH 5.5) containing 30 U/mL Jack bean β-*N*-acetylhexosaminidase, and digest overnight at 37°C.
3. Continue with **Subheading 3.2.3.4**.
4. The result of Kre2GnTI expression on the glycan level is represented in **Fig. 6**, *panel 5*. Treatment of the glycans with *N*-acetylhexosaminidase is represented in **Fig. 6**, *panel 6*.

4. Notes

1. The plasmids described in this chapter (*see* **Figs. 2** and **3**), and in Vervecken et al. *(3)* can be obtained from us. Please address your request to Roland Contreras, Fundamental and Applied Molecular Biology, Ghent University—VIB, Technologiepark 927, B-9052 Gent-Zwijnaarde, Belgium. E-mail: Roland. Contreras@dmbr.UGent.be. This glyco-engineering technique is Intellectual Property

of Flanders Interuniversity Institute for Biotechnology and Ghent University. Patent pending in exclusive license to Research Corporation Technologies.

2. Using a casting system composed of clear glass plates on both sides and thick spacers (1.5 mm) simplifies detection afterwards. The protein electrophoresis system Mini-PROTEAN 3 from BioRad worked well in our hands.

3. Capillary sequencers have replaced the gel-based systems. Our recent experiments have shown that capillary sequencers are at least equally well suited for glycan analysis. The sensitivity is up to 20× higher. The procedures suggested by the manufacturer for DNA fragment analysis can be used.

4. Previously, proteins were separated from released sugars by differential precipitation. In our hands, these methods often resulted in an extensive loss of oligosaccharide material, and thus we included this small chromatographic step. In addition, the purified carbohydrates obtained can be used directly in matrix-assisted laser desorption/ionization time-of-flight mass spectrometry (MALDI-TOF MS) analysis.

5. The Carbograph graphite resin can also be purchased as a powder. The described purification of oligosaccharides also works in batch using this powder. Put about 150 mg powder in a microcentrifuge tube for samples of 20 to 100 μg oligosaccharide material. Perform the same washing steps, and use the binding and elution solutions as described for the column procedure.

6. Polymerization should not proceed too rapidly; if gas bubbles appear between the gel and the glass plates, use less APS and/or TEMED.

7. The deglycosylation procedure described here looks somewhat difficult, but it is the most robust method and has the highest throughput. Alternatively, the deglycosylation procedure described for the FACE protocol can also be used, but a larger glycoprotein sample will be needed. Perform the procedure of **Subheading 3.1.2.2** and proceed with **Subheading 3.2.3.2**. A combination of the above described mannoprotein preparation and *N*-glycan isolation as described for the FACE analysis is not encouraged. The preparation of mannoproteins involves the precipitation of all carbohydrates, including a huge amount of cell wall glucan. This glucan does not bind to the PVDF membranes at the bottom of the Immobilon-P plates and is washed away, whereas it remains in the FACE procedure and disturbs the glycan pattern. The use of the liquid phase deglycosylation of mannoproteins instead of the PVDF-plate method requires purification of protein bound sugars from the free carbohydrates.

8. Alternatively, if no evaporator is present that can handle 96-well plates, the solutions can be transferred to small 200 μL PCR tubes.

9. It is necessary to load the whole plate with resin at once, even if only a few samples are to be analyzed. The unused wells can be used for later analyses. It is difficult to reproducibly load the resin by pipetting the water-suspended slurry in the wells. We therefore stick to the resin loader procedure.

10. Alternatively, we used a mixture containing 250 fmol each of a rhodamine-labeled 6-,18-, 30-, and 42-meric oligonucleotide (consisting of repeats of the basic sequence 5′-TAC-3′, synthesized and PAGE-purified by Invitrogen).

11. The use of blasticidin is not as convenient as other antibiotics. We have the experience that sometimes three selection rounds are needed to obtain single clones. If after the first plating no individual clones can be picked, it is advised to take some of the confluent growing cells and put them back on selective medium. Eventually resistant clones will appear.

12. If the liquid phase procedure for deglycosylation of the glycoproteins is used and DSA-FACE to detect them, leave the procedure after **Subheading 3.1.2.2.**, step 4 and proceed with **Subheading 3.2.3.2.**

Acknowledgments

WV holds a fellowship of IWT, NC is a Post-doctoral Researcher of FWO-Flanders. Research was funded by Research Corporation Technology (Tucson, Arizona, USA), FWO-Flanders, and a GOA grant of Ghent University (grant 12051099). Annelies Van Hecke is acknowledged for technical assistance.

References

1. Kornfeld, R. and Kornfeld, S. (1985) Assembly of asparagine-linked oligosaccharides *Annu. Rev. Biochem.* **54,** 631–664.

2. Lee, S. J., Evers, S., Roeder, D., et al. (2002) Mannose receptor-mediated regulation of serum glycoprotein homeostasis. *Science* **295,** 1898–1901.

3. Vervecken, W., Kaigorodov, V., Callewaert, N., Geysens, S., De Vusser, K., and Contreras, R. (2004) *In vivo* synthesis of mammalian like hybrid type N-glycans in *Pichia pastoris. Appl. Environ. Microbiol.* **70,** 2639–2646.

4. Jackson, P. (1990) The use of polyacrylamide-gel electrophoresis for the high-resolution separation of reducing saccharides labelled with the fluorophore 8-aminonaphthalene-1,3,6-trisulphonic acid. Detection of picomolar quantities by an imaging system based on a cooled charge-coupled device. *Biochem. J.* **270,** 705–713.

5. Maras, M., De Bruyn, A., Schraml, J., et al. (1997) Structural characterization of N-linked oligosaccharides from cellobiohydrolase I secreted by the filamentous fungus *Trichoderma reesei* RUTC 30. *Eur. J. Biochem.* **245,** 617–625.

6. Martinet, W., Saelens, X., Deroo, T., et al. (1997) Protection of mice against a lethal influenza challenge by immunization with yeast-derived recombinant influenza neuraminidase. *Eur. J. Biochem.* **247,** 332–338.

7. Maras, M., Callewaert, N., Piens, K., et al. (2000) Molecular cloning and enzymatic characterization of a *Trichoderma reesei* 1,2-α-D-mannosidase. *J. Biotechnol.* **77,** 255–263.

8. Callewaert, N., Laroy, W., Cadirgi, H., et al. (2001) Use of HDEL-tagged *Trichoderma reesei* mannosyl oligosaccharide 1,2- alpha-D-mannosidase for N-glycan engineering in *Pichia pastoris. FEBS Lett.* **503,** 173–178.

9. Ballou, C. E. (1990) Isolation, characterization, and properties of *Saccharomyces cerevisiae* mnn mutants with nonconditional protein glycosylation defects. *Methods Enzymol.* **185,** 440–470.

10. Callewaert, N., Geysens, S., Molemans, F., and Contreras, R. (2001) Ultrasensitive profiling and sequencing of N-linked oligosaccharides using standard DNA-sequencing equipment. *Glycobiology* **11,** 275–281.
11. Papac, D. I., Briggs, J. B., Chin, E. T., and Jones, A. J. (1998) A high-throughput microscale method to release N-linked oligosaccharides from glycoproteins for matrix-assisted laser desorption/ionization time-of-flight mass spectrometric analysis. *Glycobiology* **8,** 445–454.

10

N-Linked Glycan Characterization of Heterologous Proteins

Huijuan Li, Robert G. Miele, Teresa I. Mitchell, and Tillman U. Gerngross

Abstract

Our laboratory has focused on the re-engineered of the secretory pathway of *Pichia pastoris* to perform glycosylation reactions that mimic processing of N-glycans in humans and other higher mammals *(1,2)*. A reporter protein with a single N-linked glycosylation site, a His-tagged Kringle 3 domain of human plasminogen (K3), was used to identify combinations of optimal leader/catalytic domain(s) to recreate human N-glycan processing in the *Pichia* system. In this chapter we describe detailed protocols for high-throughput purification of K3, enzymatic release of N-glycans, matrix-assisted laser desorption ionization time-of-flight and high-performance liquid chromatography analysis of the released N-glycans. The developed protocols can be adapted to the characterization of N-glycans from any purified protein expressed in *P. pastoris*.

Key Words: *Pichia pastoris*; N-linked glycosylation; N-glycans; protein production; Kringle 3 domain; human plasminogen.

1. Introduction

Glycosylation is a common posttranslational modification found on proteins expressed in *Pichia pastoris*. This modification may affect the structure and function of proteins expressed in *P. pastoris*, and has a significant impact on serum-stability of such proteins in higher mammals. N-linked glycosylation is generated through a common pathway, which starts in the endoplasmic reticulum (ER) with the transfer of a core oligosaccharide to an asparagine residue in the sequence Asn-X-Ser/Thr within the polypeptide chain. This highly conserved core oligosaccharide structure is first assembled in the cytosol, and then flipped to the inside of the ER where it is transferred to the protein. Following transfer, three glucose moieties and one terminal α-1,2-mannose are removed in the ER, before the partially processed glycoprotein is transported to the Golgi

From: *Methods in Molecular Biology, vol. 389:* Pichia *Protocols, Second Edition*
Edited by: J. M. Cregg © Humana Press Inc., Totowa, NJ

apparatus. In the Golgi a number of α-1,6- mannosyl, and α-1,2-mannosyl-transferases, act on the N-glycans, resulting in high-mannose structures ranging from Man 8 to Man 14 with some degree of hypermannosylation (Man > 15) *(3)*. Like in *Saccharomyces cerevisiae,* these glycans may contain varying amounts of charges resulting from the transfer of phosphomannose, but lack the α-1,3-mannose found in *S. cerevisiae.* Efforts to engineer glycosylation in *P. pastoris* were first described by Koji et al., *(4)*, Maras et al., *(5)*, and Callewaert et al. *(6)*. More recently our laboratory reported the successful production of human glycoproteins in engineered strains of *P. pastoris (1,2)*. This process involved extensive reengineering of the glycosylation pathway including gene knockouts as well as functional expression and localization of several mannosi-dases, sugar nucleotide transporters and glycosyltransferases. One of the earli-est challenges was the proper localization of active α-1,2-mannosidase in the yeast's secretory pathway. To overcome this we created an extensive library of protein fusions between fungal targeting peptides (mostly leaders of type II membrane proteins) and catalytic domains of α-1,2-mannosidases, and then set up a screen to select for those leader/catalytic domain fusions that effectively mediated mannose trimming in vivo. This screen is based on a high-throughput protein purification scheme, followed by enzymatic N-glycan release and matrix-assisted laser desorption ionization time-of-flight (MALDI-TOF) analysis. Protocols described here can be adapted to the characterization of N-glycans from any purified protein expressed in *P. pastoris.*

2. Materials

2.1. Chemicals, Oligosaccharide Standards, and Enzymes

All reagents should be of analytical grade. Oligosaccharide standards were from EMD Biosciences (San Diego, CA, formerly Calbiochem). Sources of enzymes: N-glycosidase F (PNGaseF), β-*N*-Acetylhexosaminidase and Jack Bean mannosidase from Glyko (Novato, CA) or New England BioLabs (Berverly, MA).

2.2. Buffers, Stock Solutions, and Equipment

1. 10X Binding buffer: 50 m*M* imidazole, 5M NaCl, and 200 m*M* Tris-HCl (pH 7.9) (*see* **Note 1**).
2. 10X Wash buffer: 300 m*M* imidazole, 5 *M* NaCl, 200 m*M* and Tris-HCl (pH 7.9).
3. 4X Elution buffer: 2 *M* imidazole, 2 *M* NaCl, and 80 m*M* Tris-HCl (pH 7.9).
4. 4X Strip buffer: 400 m*M* ethylene diamine tetraacetic acid (EDTA), 2 *M* NaCl, and 80 m*M* Tris-HCl (pH 7.9).
5. 4X Charge buffer: 200 m*M* $NiSO_4$.
6. HisBind resin (EMD Biosciences, San Diego, CA; formerly Novagen).
7. 96-well lysate clearing plate (Wizard SV96, Promega Corp, Madison, WI).

8. Beckman 96-well plate manifold (Beckman Coulter, Palatine, IL).
9. (optional) Beckman BioMek 2000 laboratory robot (Beckman Instruments, Fullerton, CA).
10. RCM buffer: 8 M Urea, 360 mM Tris, and 3.2 mM EDTA (pH 8.6).
11. 0.1 M dithiothreitol (DTT) in RCM buffer (prepare freshly).
12. 0.1 M iodoacetic acid in RCM buffer (prepare freshly).
13. 1% Polyvinylpyrrolidone (PVP-360) in water (prepare freshly).
14. 10 mM NH_4HCO_3 (pH 8.3).
15. MAIPN4510 96-well plate (pore size 0.45 μm, Millipore, Bedford, MA).
16. Voyager DE PRO MALDI-TOF Mass Spectrometer (Applied Biosystems, Foster City, CA).
17. 2-Aminobenzamide (2-AB).
18. Dimethyl sulfoxide (DMSO).
19. Glacial acetic acid.
20. Sodium cyanoborohydride.
21. Water, high-performance liquid chromatography (HPLC) grade.
22. Acetonitrile, HPLC grade.
23. Heating block set to 65°C (avoid water baths).
24. 96-Well collecting plates.
25. Centrifugal evaporator (Thermo Savant, Holbrook, NY).
26. 50 mM ammonium formate (pH 4.4) (6.62 mL of formic acid in 3.5 L water, adjust pH to 4.4 with NH_4OH).
27. 0.6 mL and 1.5 mL siliconized tube (Fisher Scientific, Pittsburgh, PA).
28. N-linked oligosaccharide standard ($Man_8GlcNAc_2$) (EMD Biosciences, San Diego, CA; formerly Calbiochem).
29. 50 mM ammonium acetate (pH 5.0).

3. Methods

3.1. 96-Well Ni-affinity Purification of K3

His-tagged K3 was purified from the medium by Ni-affinity chromatography utilizing a 96-well format on a Beckman BioMek 2000 laboratory robot. The robotic purification is an adaptation of the protocol provided by Novagen for the HisBind resin.

Robotic protocol was as follows:

1. Manually prepare 96-well Ni-affinity plate by transfering 200 μL HisBind slurry to each well to yield 100 μL of packed resin.
2. Wash 3× with 150 μL of water.
3. Wash 5× with 150 μL of 1X Charge buffer.
4. Wash 3× with 150 μL of 1X Binding buffer.
5. Load samples 10×200 μL (Prior to loading, culture supernatant was buffered with phosphate bufered saline (PBS) (60 mM phosphate, 16.2 mM KCl, and 822 mM NaCl) to pH 6.5–7.0 by adding 0.6 mL of PBS to 0.9 mL of culture supernatant) (*see* **Note 2**).

6. Wash resin 10× with 150 μL of 1X Binding buffer.
7. Wash resin 6× with 150 μL of 1X Wash buffer.
8. Elute by washing with 4 × 150 μL of 1X Elution buffer and evaporate to dryness in a centrifugal evaporator (Thermo Savant, Holbrook, NY). At this point, the plate is ready for N-glycan release (*see* **Subheading 3.2**).
9. Strip resin 4× with 150 μL of 1X Strip buffer and keep in water containing 0.05% sodium azide at 4°C. This resin can be reused approx 10 min.

3.2. Release of N-Linked Glycans

The glycans are released and separated from the glycoproteins by a modification of a previously reported method by Papac et al. *(7)*. The protocol was adapted as follows.

1. Wet 96-well MultiScreen IP (Immobilon-P membrane) plate (Millipore, Bedford, MA) with 100 μL of methanol per well, and drain by applying a gentle vacuum using the Beckman 96-well plate manifold.
2. Wash once with 100 μL of water followed by 100 μl of RCM buffer and drain by applying a gentle vacuum after each addition.
3. Dissolve the dried protein from the Ni-affinity purification above in 200 μL of RCM buffer and transfer to the Immobilon-P membrane plate. Note: Each well should contain about 20 μg of protein, and can be from other sources than Ni-affinity purification.
4. Drain the wells and wash twice with RCM buffer (2 × 300 μL). Reduce the protein by addition of 50 μL of 0.1M DTT in RCM buffer for 1 h at 37°C. Place the plate in a humidified container.
5. After incubation, apply gentle vacuum to drain RCM/DTT buffer. Wash wells 3× with 300 μL of water. Carboxymethylate the protein by addition of 60 μL of 0.1M iodoacetic acid in RCM buffer for 30 min in the dark at room temperature in a humidified container.
6. After incubation, apply gentle vacuum to drain RCM/iodoacetic acid buffer. Wash the wells three times with 300 μL water and block the membranes by the addition of 100 μL of 1% PVP-360 in water for 1 h at room temperature in a humidified container.
7. After incubation, apply gentle vacuum to drain PVP. Drain the wells and wash 4× with 300 μL of water and deglycosylate the protein by the addition of 25 to 30 μL of 10 mM NH$_4$HCO$_3$ (pH 8.3) containing one milliunit of PNGase F (Glyko) or 10 units of PNGase F (New England Biolab). After 16 h at 37°C, collect solution containing the glycans by centrifugation and evaporate to dryness. Samples are now ready for MALDI-TOF analysis (*see* **Note 3**).

3.3. Matrix Assisted Laser Desorption Ionization Time of Flight Mass Spectrometry

Molecular mass of the glycans can be determined using a Voyager DE PRO linear MALDI-TOF (Applied Biosystems, Foster City, CA) mass spectrometer using delayed extraction. The dried glycans from above are dissolved in 15 μL

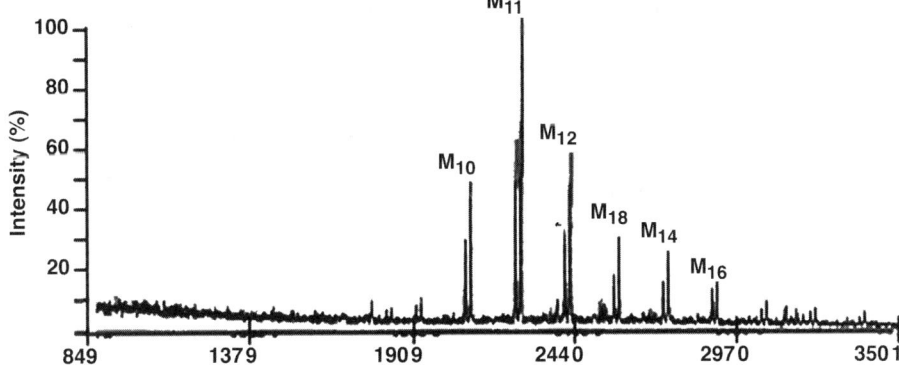

Fig. 1. Positive-ion MALDI-TOF mass spectrum of N-linked glycans released from K3. K3 was produced in wild type *P. pastoris* strain BK64 *(5)*.

of water per well, and 0.6 µL were spotted on a stainless steel sample plate, to which 0.6 µL of S-DHB matrix (10 mg/mL of dihydroxybenzoic acid dissolved in 20% acetonitrile containing 0.1% TFA mixed with 1 mg/mL of 5-methoxysalicilic acid dissolved in 50% acetonitrile containing 0.1% TFA in 9:1 ratio) were added, and the mixture was allowed to dry. Automatic spotting was performed by a Symbiot I system (Applied Biosystems, Foster City, CA). Ions were generated by irradiation with a pulsed nitrogen laser (337 nm) with a 4 ns pulse time. The instrument was operated in the delayed extraction mode with a 125 ns delay and an accelerating voltage of 20 kV. The grid voltage was 93.00%, guide wire voltage was 0.1%, internal pressure was less than 5×10^{-7} torr. Spectra were generated from the sum of 100 to 200 laser pulses and acquired with a 500 MHz digitizer. $Man_5GlcNAc_2$ or $Man_8GlcNAc_2$ oligosaccharides were used as an external molecular weight standard. All spectra were generated with the instrument in the positive ion mode. **Figure 1** shows a MALDI-TOF spectrum of N-glycans released from K3 produced in a wild type *P. pastoris* strain.

3.4. 2-AB Labeling Reaction

Fluorescent labeling of glycans was performed following a previously reported method *(8,9)* using the Glyko signal 2-AB labeling kit, from QA Bio (San Mateo, CA). (*see* **Note 4**). The protocol was adapted as follows.

1. Add 50 µL glacial acetic acid to 150 µL DMSO and mix by pipetting.
2. Add 100 µL of this mixture to 2-AB dye (5 mg) and mix until the dye is dissolved.
3. Add the entire contents of the above mixture to the sodium cyanoborohydride (6.3 mg) and mix by pipetting until the reductant is completely dissolved. This is the labeling reagent. If the reductant is difficult to dissolve, then gently warm the vial for up to 3 min at 65°C and mix by pipetting.

4. Add 5 μL of labeling reagent to each dried glycan sample (200 ng to 20 μg), close the microcentrifge tube, mix thoroughly, and then gently tap to ensure the labeling solution is at the bottom of the tube.

5. Place the tubes in a heating block at 65°C and incubate for 2 to 3 h. The incubation must be performed in a dry environment. The samples must be completely dissolved in the labeling solution to ensure efficient labeling. The samples can be briefly vortexted or tapped 30 mins after the start of the 65°C incubation to ensure complete dissolution of the samples.

6. After the incubation period, remove the samples, centrifuge briefly, then allow to cool completely to room temperature or freeze for later analysis. These samples are now ready for final cleanup to remove excess labeling reagent.

7. Prepare a 96-well cleanup plate (Wizard SV96, Promega Corp, Madison, WI) by washing each well with 3 mL acetonitrile, 3 mL water followed by 3 mL 30% acetic acid. After draining, wash with 1 mL of acetonitrile. If flow is slow, slight pressure can be applied with a rubber bulb to the top of the well in order to assure normal flow (*see* **Note 5**).

8. Spot each labeled sample onto a freshly washed disc in the 96-well cleanup plate ensuring that the disc is still wet with acetonitrile when sample is applied. If a disc has dried out it can be rewetted by washing with additional acetonitrile (0.5 mL) prior to loading the sample. When applying the sample to the cleanup plate, try to spread the sample over the entire area of the disc.

9. Leave for 15 min to allow glycans to adsorb to the disc.

10. Rinse each tube with 100 μL acetonitrile and transfer to the corresponding disc.

11. Wash each disc with 1 mL acetonitrile, followed by 3 × 1 mL 96% acetonitrile/4% water. Discard these washes into a suitable waste container.

12. Recover the glycans by eluting 3 × with 0.4 mL HPLC grade water into a 96-well plate, allowing each aliquot to drain before the next is applied.

13. Evaporate the samples to dryness in a centrifugal evaporator and reconstitute the glycans in HPLC-grade water. The samples are now ready for HPLC or MALDI-TOF analysis.

3.5. HPLC Analysis of 2-AB Oligosaccaride Derivatives

1. HPLC detection requires a fluoresecence detector set at an excitation wavelength of 330 nm and an emission wavelength of 420 nm. We use both Hitachi F1050 and Hitachi F1080 detectors (Tokyo, Japan).

2. Prepare the following buffers: Solvent A: 100% acetonitrile. Solvent B: 50 mM ammonium formate (pH 4.5).

3. Fluorescence-labeled samples are analyzed by HPLC using an Econosil NH_2 4.6 × 250 mm, 5 micron bead, amino-bound silica column (Altech, Avondale, PA). The flow rate is 1.0 mL/min and the column is maintained at 30°C. After eluting isocratically (68% A:32% B) for 3 min, a linear solvent gradient (68% A:32% B to 40% A:60% B) is employed over 27 min to elute the glycans. The column is equilibrated with solvent (68% A:32% B) for 20 min between runs. (*see* **Note 6**).

4. The HPLC program is as follows:

Time/min	% A	% B	Comments
0	68	32	Initial isocratic region
3	68	32	Start of salt gradient
30	40	60	End of salt gradient
33	68	32	Start gradient back to initial conditions
40	68	32	Back to initial conditions

The column is equilibrated with solvent (68% A:32% B) for 20 min between runs. **Figure 2** shows standard N-linked oligosaccharide $Man_8GlcNAc_2$ (A) and $Man_5GlcNAc_2$ (B) after 2-AB labelling and analysis by the above described HPLC program.

3.6. Selected Glycosidase Digestions of N-glycans

3.6.1. Jack Bean Mannosidase and α-1,2 Mannosidase Digest of $Man_8GlcNAc_2$

1. Reconstitute the standard N-linked oligosaccharide $Man_8GlcNAc_2$ (20 μg) in 100 μL HPLC grade water. Aliquot 10 μL to a 0.6 mL siliconized tube.
2. Evaporate the sample to dryness.
3. Add 10 μL of 50 mM ammonium acetate (*see* **Note 7**).
4. Add Jack Bean mannosidase (0.03 units) or α-1,2 mannosidase from *Trichoderma reseei* (0.03 mU, a gift from Dr R. Contreras, Unit of Fundamental and Applied Molecular Biology, Department of Molecular Biology, Ghent University, Ghent, Belgium).
5. Incubate the sample in enzyme for 16 to 24 h at 37°C.
6. Evaporate the sample to dryness. Reconstitute the sample in 10 μL of water. The sample is now ready for MALDI-TOF analysis (*see* **Fig. 3**).

3.6.2. β-N-Acetylhexosaminidase (Jack Bean) Digest of $GlcNAc_2Man_3GlcNac_2$

1. Aliquot 200 ng of $GlcNAc_2Man_3GlcNac_2$ produced from *Pichia* strain YSH44 (*2*) to a 0.6-mL siliconized tube.
2. Evaporate the sample to dryness.
3. Add 10 μL of 50 mM ammonium acetate.
4. Add Jack Bean β-N-acetylhexosaminidase (0.06 units).
5. Incubate the sample in enzyme for 16 to 24 h at 37°C.
6. Evaporate the sample to dryness. Reconstitute the sample in 10 μL of water. The sample is now ready for MALDI-TOF analysis (*see* **Fig. 4**).

4. Notes

1. It is useful to have stocks of 0.5 M Tris-HCl, and 4M imidazole to make the reagents.
2. It is important to adjust the pH of the culture supernatant to between 6.5 and 7.0 in order to ensure optimum binding of the His-tagged protein.

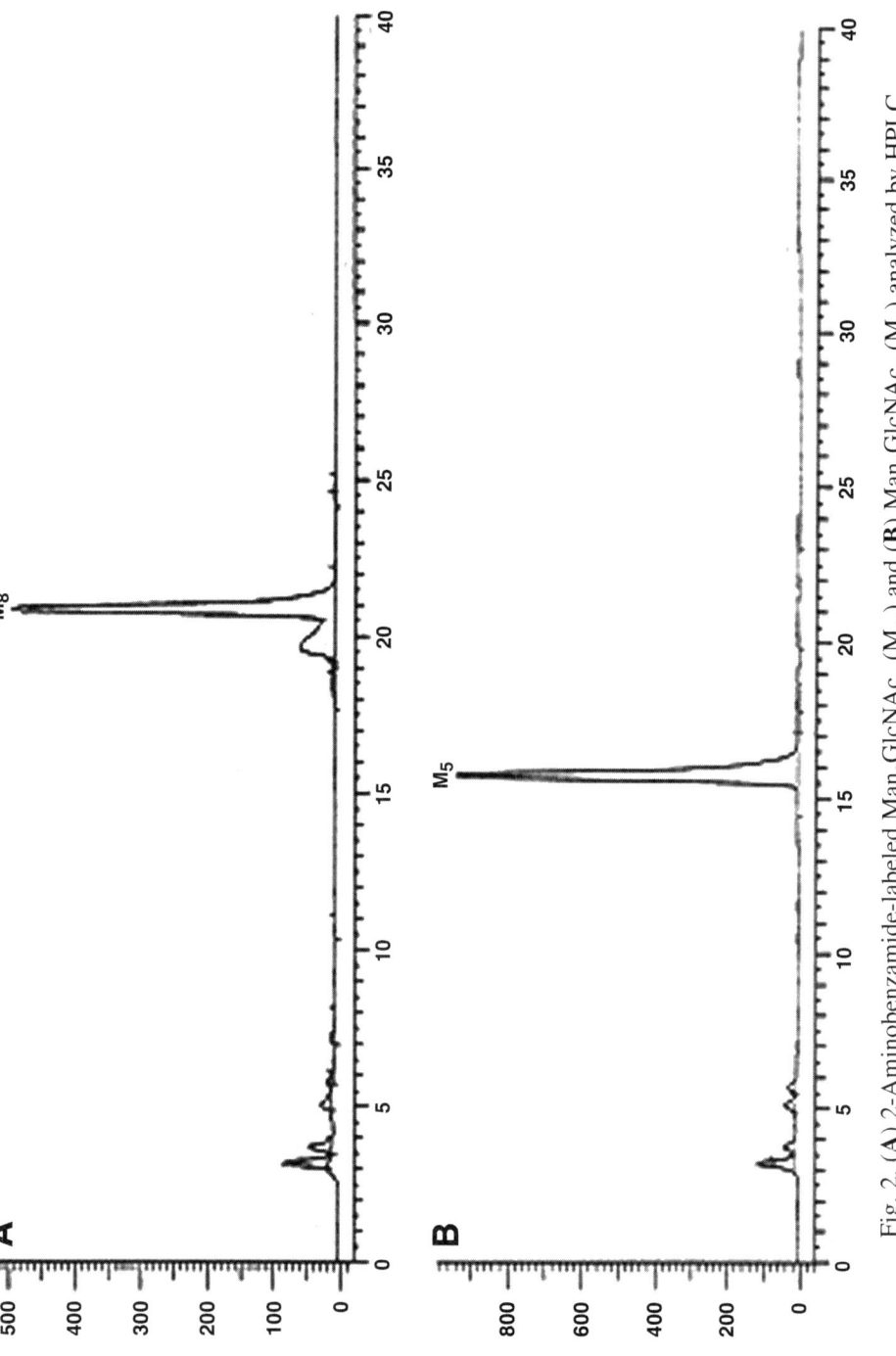

Fig. 2. (**A**) 2-Aminobenzamide-labeled $Man_8GlcNAc_2$ (M_8) and (**B**) $Man_5GlcNAc_2$ (M_5) analyzed by HPLC.

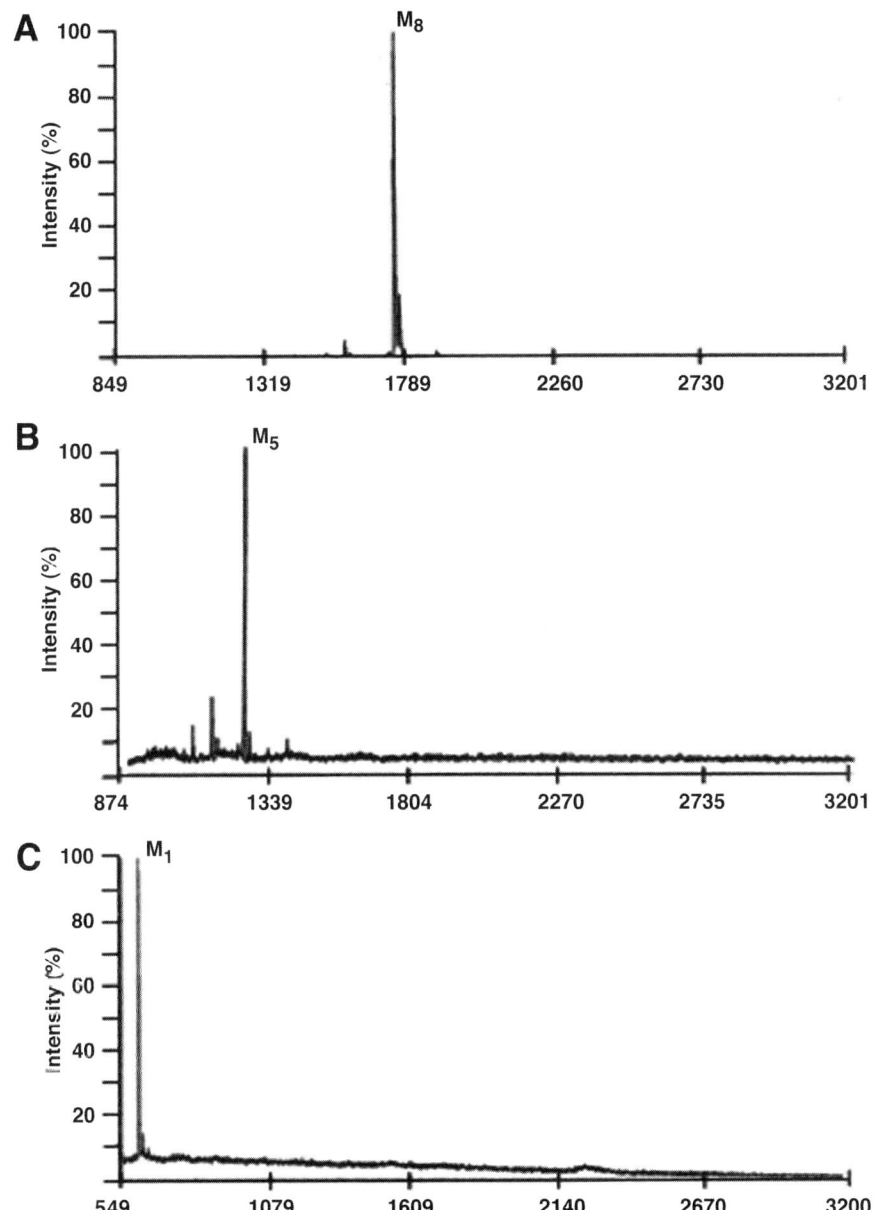

Fig. 3. (**A**) Positive-ion MALDI-TOF mass spectrum of standard M_8. (**B**) M_8 after α-1,2 mannosidase treatment. (**C**) M_8 after Jack Bean mannosidase treatment.

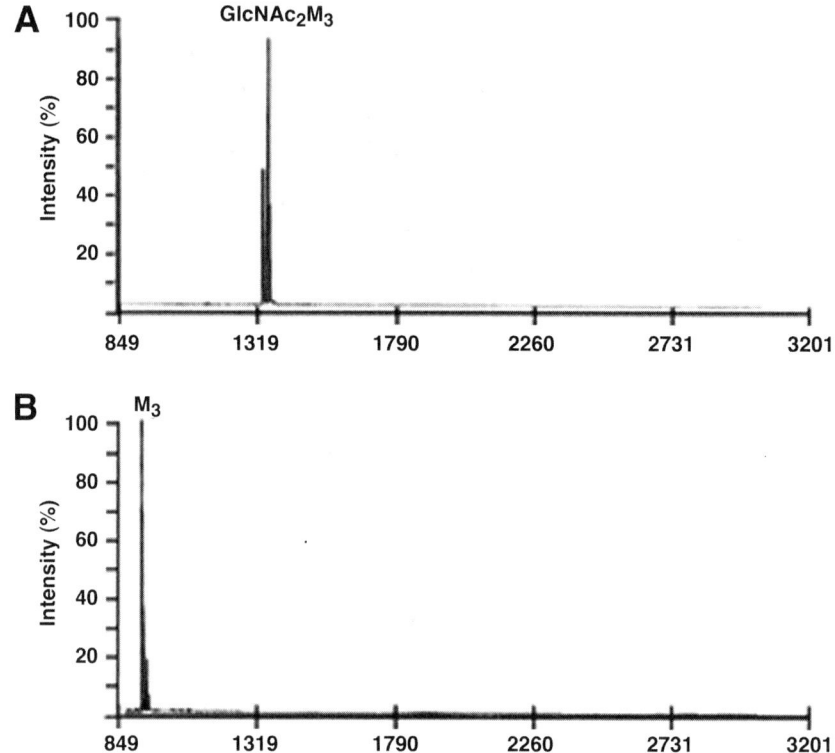

Fig. 4. Positive-ion MALDI-TOF mass spectra of N-linked glycans. (**A**) N-glycans released from K3 in *P. pastoris* strain YSH44 (**6**). (**B**) Glycans from YSH44 after β-*N*-acetylhexosaminidase treatment.

3. High concentrations of salt in the purified protein sample did not seem to interfere with PNGaseF digest on the 96-well plate. However extensive washing of the membranes with HPLC grade water is necessary before incubating the protein with the enzyme in volatile buffer of 10 mM NH$_4$HCO$_3$. In this way, the collected glycans will be salt-free after evaporation.

4. 2-AB labeling reaction mixture has to be prepared freshly prior to use. For high efficiency, nonselective labeling of glycans:

   a. The temperature of the labeling reaction must be maintained at 65°C for the entire labeling reaction.
   b. The glycans must be salt free and contain no other contaminants.
   c. The glycans must be dissolved completely in the labeling mixture.

5. Removal of excess labeling reagent can be performed using the GlycoClean S cleanup cartridges (Glyko) or a 96-well lysate clearing plate as described in **Subheading 3.4.**

6. Buffers for HPLC analysis need to be degassed; ideally a helium gas line is used to continuously purge each buffer. Econosil NH2 columns have a tendency to deteriorate over time; it is therefore critical to run known standards prior to the analysis of unknown samples.
7. Volatile buffer (50 m*M* ammonium acetate [pH 4.0–6.5]) was used for glycosidase digest of glycans. This buffer is ideal for enzymes with a pH optimum in the range of 4.0 to 6.5 and obviates the need for dialysis before MALDI-TOF analysis.

References

1. Choi, B-K., Bobrowicz, P., Davidson, R. C., et al. (2003) Use of combinatorial genetic libraries to humanize N-Linked glycosylation in the yeast *Pichia pastoris. Proc. Natl. Acad. Sci. USA* **100,** 5022–5027.
2. Hamilton, S. R., Bobrowicz, P., Bobrowicz, B., et al. (2003) Production of complex human glycoproteins in yeast. *Science* **301,** 1244–1246.
3. Bretthauer, R. K. and Castellino F. J. (1999) Glycosylation of *Pichia*-derived proteins. *Biotechnol. Appl. Biochem.* **30,** 193–200.
4. Koji, M., Narutoshi, S., and Tomoyasu, R. (1996) Sugar chain-elongating protein and DNA derived from yeast *Pichia* genus. *Jpn Patent JP* 8336387.
5. Maras, M., De Bruyn A., Vervecken, W., et al. (1999) *In vivo* synthesis of complex N-glycans by expression of human N-acetylglucosaminyltransferase I in the filamentous fungus *Trichoderma reesei. FEBS letters* **452,** 365–370.
6. Callewaert, N., Laroy, W., Cadirgi, H., et al. (2001) use of HDEL-tagged *Trichoderma reesei* mannosyl oligosaccharide 1,2-α-mannosidase for N-glycan engineering in *Pichia pastoris. FEBS letters* **503,** 173–178.
7. Papac, D. I., Briggs, J. B., Chin E. T., and Jones A. J. S. (1998) A high-throughput microscale method to release N-linked oligosaccharides from glycoprpteins for matrix-assisted laser desorption/ionization time-of flight mass spectrometric analysis. *Glycobiology.* **8,** 445–454.
8. Bigge, J. C., Patel, T. P., Bruce, J. A., Goulding, P. N., Charles, S. M., and Parekh, R. B. (1995) Nonselective and efficient fluorescent labeling of glycans using 2-amino benzamide and anthranilic acid. *Anal. Biochem.* **230,** 229 238.
9. Townsend, R. R., Lipniunas, P. H., Bigge, C., Ventom, A., and Parekh R. (1996) Multimode high-performance liquid chromatography of fluorescently labeled oligosaccharides from glycoproteins. *Anal. Biochem.* **239,** 200–207.

11

Heavy Labeling of Recombinant Proteins

Eric Rodriguez

Abstract

Because of the cost of isotopic chemicals and heterologous proteins to produce, an economical $^{15}N/^{13}C$ isotopic labeling method is critically needed. Four protocols have been tested for the expression of Ovine interferon-τ in *Pichia pastoris*. ^{13}C-glucose in place of ^{13}C-glycerol as well as the need for $^{15}N/^{13}C$-sources were evaluated during the growth phase. Sequential addition of $^{15}NH_4Cl$ and ^{13}C-methanol were also evaluated at different ratio. Our results demonstrate that $^{15}N/^{13}C$ isotopes are not required throughout the initial growth period but are necessary at low concentration a few hours prior to the methanol induction period. We have evaluated the cost of the use of isotopes $^{15}NH_4Cl$, ^{13}C-glucose and ^{13}C-methanol in our optimised P4 protocol conditions. The cost was one-third that of the standard method using $^{15}NH_4Cl$ and ^{13}C-glucose throughout the entire growth period and was even lower using ^{13}C-glycerol.

Key Words: $^{15}N/^{13}C$ Isotopic labeling; *Pichia pastoris*; protein expression; methanol induction; ovine interferon-τ; NMR.

1. Introduction

Isotopic labeling of heterologous proteins for NMR experiments needs a cost-effective expression system that uses limited amounts of expensive isotopic chemical products such as $(^{15}NH_4)_2SO_4$, ^{13}C-glucose, ^{13}C-glycerol or ^{13}C-methanol. The methylotrophic yeast *Pichia pastoris* has been developed into a highly successful system for the production of large quantities of a variety of secreted heterologous proteins in their correctly folded native states *(1)*. The yield of expressed proteins in *P. pastoris* depends critically on growth conditions and attainment of high-cell densities by fermentation. These factors have been shown to improve protein yields by 5- to 100-fold. *P. pastoris* is therefore the expression system of choice for NMR. However, because of the cost of isotopic compounds and the difficulty in producing some proteins, optimized culture and fermentation conditions are often critically needed. ^{13}C-glycerol is one of the

From: *Methods in Molecular Biology, vol. 389:* Pichia *Protocols, Second Edition*
Edited by: J. M. Cregg © Humana Press Inc., Totowa, NJ

more expensive sources of ^{13}C-label. An alternative carbon source that might be more economical than glycerol is glucose. However, the strong inducible alcohol oxidase I (*AOX1*) promoter is severely repressed by glucose. In fact, suddenly switching the carbon source from glucose to methanol results in cell death because the cells cannot convert the methanol without alcohol oxidase, which is produced from the *AOX1* gene, and the methanol can become toxic. Because the protein of interest is not produced during the growth phase supplementation with medium containing $(^{15}NH_4)_2SO_4$ and ^{13}C-glucose may not be required. Our goal was to identify an effective and economical method to label proteins expressed in *P. pastoris*. Here, we report a comparison of four $^{15}N/^{13}C$ isotopic labeling methods, two of which were utilized previously *(2,3)*. The expression of a 20-kDa recombinant protein, namely the Ovine interferon-tau (rOvIFN-τ) with anti-viral *(4)* and antiproliferative *(5)* properties is used to illustrate the relative efficiencies of the four heavy labelling methods in *P. pastoris (6)*.

2. Materials

2.1. Host Strain, Vector, and Antibiotics

1. Zeocin resistance-based plasmids (Invitrogen, Carlsbad, CA).
2. OvIFN-τ cDNA *(7)*.
3. *P. pastoris* strain X-33 (Mut+ His+) (Invitrogen).
4. Zeocin (Invitrogen): Used as a stock solution of 100 mg/mL in sterile water. The stock solution is added to media solutions that are <60°C. Zeocin stock solutions are stored at –20°C until use.

2.2. Culture Media

All carbon sources are from Fisher, Pittsburg, PA.

1. YPD medium: 10 g/L yeast extract, 20 g/L peptone, and 20 g/L dextrose.
2. YNB medium: 6.7 g/L yeast nitrogen base without without amino acids and ammonium sulfate (YNB w/o amino acids and AS), 11.6 g/L monobasic potassium phosphate, and 2.7 g/L dibasic potassium phosphate.
3. Buffered glycerol-complex medium (BMGY): 1% yeast extract, 2% peptone, 1.34% YNB w/o amino acids and AS, 4×10^{-5}% Biotin, 1% methanol, and 100 m*M* sodium phosphate buffer (pH 6.0) (Invitrogen). All the carbon sources are from Fisher, Pittsburgh, PA.
4. Buffered methanol-complex medium (BMMY): 1% yeast extract, 2% peptone, 1.34% YNB w/o amino acids and AS, 4×10^{-5}% Biotin, 0.5% methanol, and 100 m*M* sodium phosphate buffer (pH 6.0).
5. Shake-flask cultures are grown in 0.5 to 1 L volumes in 4-L baffled flasks (Fisher). Small cultures are grown in borosilicate glass culture tubes (18 × 125 mm) (Fisher).
6. Base solution: 4.4 *M* KOH and 7.4 *M* NaOH.

7. FM21 medium *(3)*: YNB w/o amino acids and AS (0.34%), Casamino acids (1%), biotin (2 mg/L).
8. FM22 medium *(2)*: 1g/L $CaSO_4 \cdot 2 H_2O$, 14,28 g/L K_2SO_4, 11.7 g/L $MgSO_4 \cdot 7 H_2O$, 400 µg/L biotin, and 1 mg/L PTM1 salts) (Invitrogen).
9. FM23 medium: 0.8% YNB w/o amino acids and AS, 2 mg/L biotin, 1.2% $(NH_4)_2SO_4$, 0.3% K_2HPO_4, 0.28% KH_2PO_4, and 2 mg/L PTM1 salts.
10. D-glucose (Sigma).
11. Biotin (Invitrogen).

    a. Glycerol (Fisher).
    b. Methanol (Fisher).

2.3. Analytical Materials

2.3.1. Tricine SDS-PAGE

1. 12% Acrylamide separating gels in Tris-HCl buffer (pH8.5), 0.1% sodium dodecyl sulfate (SDS), and 2% crosslinker.
2. 4% Acrylamide stacking gels in 1 *M* Tris-HCl (pH 8.5), 0.1% SDS, and 2% crosslinker.
3. Anode running buffer: 0.2 *M* Tris-HCl (pH 8.8).
4. Cathode running buffer: 0.1 *M* Tris-HCl, 0.1 *M* Tricine, and 0.1% SDS (pH 8.0).
5. 2X Sample buffer: 4% SDS, 12% glycerol, 0.1 *M* Tris-HCl (pH 6.8,) 0.004% Coomassie brillant blue G, and 0.02% β-mercaptoethanol.
6. Coomassie Brillant Blue R-250 (Sigma).
7. Acetic acid (Fisher).
8. Methanol (Fisher).

2.3.2. Purification Columns, Buffers, and Reagents

1. Q-Sepharose gel (Pharmacia, Peapack, NJ).
2. HS100 Sephacryl gel (Pharmacia).
3. 1 *M* Tris-HCl (pH 7.6) (Sigma).
4. Sodium chloride (Sigma).
5. 5-kDa cut-off dialysis bags (Sigma).
6. Sodium azide (Sigma).

2.3.3. Heavy Isotopes, Chemicals for NMR Sample Preparation, NMR, and MS Instruments

1. ^{15}N-ammonium chloride (99% ^{15}N, Cambridge Isotope Laboratories, MA).
2. $^{13}C_6$-D-glucose (99% ^{13}C, Isotech, Miamisburg, OH).
3. ^{13}C-methanol (99.2% ^{13}C, Isotech).
4. Tris-d_{11} (Sigma).
5. D_2O (Fisher).
6. Bruker AM-600 MHz NMR instrument.
7. Matrix-assisted laser desorption ionization time-of-flight mass spectrometry (MALDI-TOF MS) (PerSeptive Biosystems, Framingham, MA).

2.4. Antiviral Assay

1. MDBK cells (BioWhittaker, Verviers, Belgium).
2. Vesicular stomatitis virus (VSV).
3. 96-well plates (Fisher).
4. Incubator at 37°C.

3. Methods

The methods described below outline: (i) the expression of the recombinant OvIFN-τ in *P. pastoris*, (ii) rOvIFN-τ ^{15}N/^{13}C isotopic labeling protocols, (iii) protein electrophoresis and purification, (iv) analysis of ^{15}N/^{13}C incorporation into rOvIFN-τ using MS and NMR analysis, (v) antiviral assay of rOvlFN-τ activity.

3.1. Expression of Recombinant Ovine Interferon-τ

The recombinant Ovine Interferon-tau (rOvIFN-τ) was expressed in *P. pastoris* as previously described with slight modifications *(7)*. The cDNA for OvIFN-τ was fused to DNA sequence encoding the prepro region of the *Saccharomyces cerevisiae* mating hormone, α-mating factor (MF) into a Zeocin resistance-based plasmid, pPICZ α (Invitrogen). The MF prepro sequence is an 89 amino acid polypeptide, which can direct peptides fused to it through the yeast secretory pathway. *P. pastoris* strain (Mut+ His+) was transformed with the vector linearized with the restriction enzyme *Pme*I. This directs integration of the entire *Pme*I-linearized expression vector into the *AOX1* locus of the *P. pastoris* genome by an additive homologous recombination event. Transformants are selected on zeocin (100 μ/mL) YPD agar plates. In order to screen for the best rOvIFN-τ producer, each transformant was grown in GMGY medium for 24 h. Aliquots (50 μL) of culture medium from each transformant culture taken during methanol induction period were evaluated for rOvIFN-τ presence by Tricine-PAGE, MS, NMR analysis, and an antiviral assay.

3.2. rOvIFN-τ ^{15}N/^{13}C Isotopic Labeling Protocols

In support of NMR studies aimed at determining the structure of the rOvIFN-τ, the protein was uniformly labeled with heavy isotopes of nitrogen (^{15}N) and carbon (^{13}C). The enriched chemicals used are: ^{15}N-ammonium chloride (99%), ^{13}C$_6$-D-glucose (99% ^{13}C) and ^{13}C-methanol (99.2% ^{13}C). Because the recombinant proteins in *P. pastoris* are only expressed during the methanol-induction phase, it is not necessary to add ^{15}N-ammonium chloride and ^{13}C$_6$-D-glucose during the entire growth-phase. Four protocols were tried and are described in **Table 1**. Two protocols, P3 and P4 were designed and compared with slight modification of protocols P1 and P2 previously described *(2,3)*.

Table 1
**^{15}N/^{13}C Isotopic Labeling Protocols in *Pichia pastoris*

	P1	P2	P3	P4
Growth phase (0–30 h)	G1 medium: FM21 medium^a + ^{15}NH$_4$Cl (1%) + ^{13}C-glucose (2%)	G2 medium: FM22 medium^b + ^{15}NH$_4$Cl (0.5%) + ^{13}C-glucose (5%)	G3 medium: FM23 medium^c + ^{15}NH$_4$Cl (1.2%) + ^{13}C-glucose (3%)	G4 medium: FM23 medium^c + NH$_4$Cl (1.2%) + D-glucose (3%)
Growth phase (at 24 h)				Addition of D-glucose (2%)
Growth phase (30–36 h)	G1 medium: FM21 medium^a + ^{15}NH$_4$Cl (1%) + ^{13}C-glucose (2%)	G2 medium: FM22 medium^b + ^{15}NH$_4$Cl (0.5%) + ^{13}C-glucose (5%)	G3 medium: FM23 medium^c + ^{15}NH$_4$Cl (1.2%) + ^{13}C-glucose (3%)	G4 medium: FM23 medium^c + ^{15}NH$_4$Cl (0.02%) + ^{13}C-glucose (0.1%)
Induction phase (4 days)	I1 medium: FM21 mediuma + ^{15}NH$_4$Cl (1%) + ^{13}C-methanol (2%)	I2 medium: FM22 medium^b + ^{15}NH$_4$Cl (0.5%) Addition of ^{13}C-methanol (0.5%) every 12 h	I3 medium: FM23 medium^c + ^{15}NH$_4$Cl (1.2%) Addition of ^{13}C-methanol as described in **Fig. 1**	I4 medium: FM23 mediumc + ^{15}NH$_4$Cl (1.2%) Addition of ^{13}C-methanol as described in **Fig. 1**
^{15}N/^{13}CrOvIFN-τ	98 mg/L	226 mg/L	85 mg/L	260 mg/L

P1 to P4 represent the four protocols compared in this study. Media recipes were prepared in a 100 mL final volume and filtered through a 0.22 μm membrane.

The pH was adjusted to 5.5 with a base solution (KOH 4.4*M*, NaOH 7.4*M*). After 30 h of grow at 30°C, each culture was centrifuged (6000*g* for 5 min) and suspended with the corresponding second step growth-phase prewarmed medium. The growth phase was continued for 6 h and stopped by centrifugation (6000*g* for 5 min). Yeasts were washed with a 0.2% glycerol solution, centrifuged, then suspended and cultured according to the respective induction protocol (*see* **Note 1**). For protocols P3 and P4, ^{13}C-methanol (0.2%) was added at a low concentration during the adaptation phase and incrementally supplemented (0.4%) every 12 h up to 1.4% from 72 to 96 h of the time-course (*see* **Fig. 1**).

^{a}FM21 medium (**3**): 0.34% YNB w/o amino acids and 0.34% AS, 1% Casamino acids, and 2 mg/L biotin.

^{b}FM22 medium (**2**): 1 g/L CaSO$_4$.2 H$_2$O, 14.28 g/L K$_2$SO$_4$, 11.7 g/L MgSO$_4$.7 H$_2$O, biotin (400 μg/L), and PTM1 salts (1 mL/L) (Invitrogen).

^{c}FM23 medium: 0.8% YNB w/o amino acids and AS, 2 mg/L biotin, 1.2% (NH$_4$)$_2$SO$_4$, 0.3% K$_2$HPO$_4$, 0.28% KH$_2$PO$_4$, and 2 mL/L PTM1 salts.

Abbr: YNB w/o amino acids and AS, yeast nitrogen base without amino acids and ammonium sulfate.

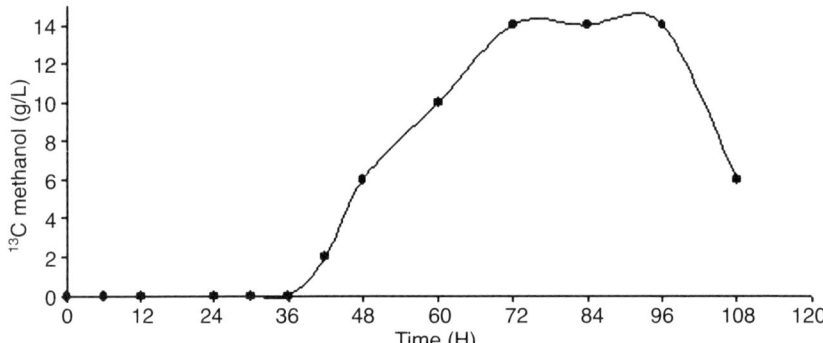

Fig. 1. Profile of $^{13}C_6$-methanol induction for protocols P3 and P4. *P. pastoris* were induced with $^{13}C_6$-methanol as described in the Materials and Methods section.

1. Throughout the study, *P. pastoris* cultures were agitated at 250 rpm at 30°C.
2. Four different growth media were prepared (G1 to G4, see details in **Table 1**). *P. pastoris* transformant was cultured for 30 h in each respective four growth media (G1 to G4).
3. For protocol P4, after 24 h of growth, 2% of D-glucose was added to the growth media.
4. After 30 h of growth at 30°C, each culture was centrifuged (6000*g* for 5 min) and suspended with the corresponding second step growth-phase pre-warmed medium (G1 to G4). Medium G4 is a modification of medium G4 by adding $^{15}NH_4Cl$ (0.02%) and ^{13}C-glucose (0.1%) to FM23 medium. The growth phase was continued for 6 h and stopped by centrifugation (6000*g* for 5 min).
5. At the end of the growth period, each culture was centrifuged briefly, washed with a 0.2% glycerol solution, centrifuged and suspended in the protein induction medium (I1 to I4, *see* details in **Table 1**). The transformant was then cultured for 3 d in presence of $^{15}NH_4Cl$ (1% for I1, 0.5% for I2, 1.2% for I3, and 1.2% for I4) and ^{13}C-methanol (2% for I1; 0.5% every 12 h for I2; and 0.2% from 36 to 42 h, then addition of 0.4% every 12 h to a total of 1.4% (assuming no evaporation of methanol) from 72 to 96 h of the time course of protocols P3 and P4.
6. Because a small portion of *P. pastoris* cells lyse at high cell concentrations, the amount of intracellular proteases in the medium can be substantial. To reduce this problem, cells were shifted to 0.8% of ^{13}C-methanol during the last 12 h of culturing.

   a. The pH was controlled and adjusted under agitation, when necessary, to 5.4 with a sterilized base solution of 4.4 *M* KOH and 7.4 *M* NaOH.

3.2.1. Expression and Yield of $^{15}N/^{13}C$ rOvIFN-τ

The $^{15}N/^{13}C$ double-labeled rOvIFN-τ was expressed and secreted into the culture medium (*see* **Fig. 2**). After 4 d of methanol induction, protein quantification results demonstrate 98, 226, 85, and 260 mg of $^{15}N/^{13}C$ double-labeled rOvIFN-τ/L of final culture when P1, P2, P3 and P4 protocols were used, respectively.

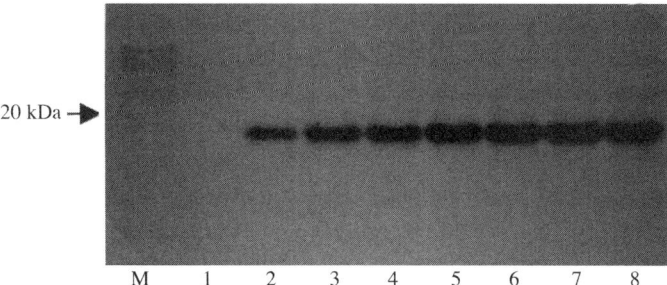

20 kDa ➤

M   1   2   3   4   5   6   7   8

Fig. 2. Tricine-PAGE of purified $^{15}N/^{13}C$ rOvIFN-τ expressed during ^{13}C-methanol induction (42, 60, 72, 84, 96, 108, 120 and 132h) in shake flask (protocol P4). *Lane M* indicates the prestained protein molecular marker (BioRad, Hercules, CA). *Lanes 1–8* show purified $^{15}N/^{13}C$ rOvIFN-τ expressed at 42, 60, 72, 84, 96, 108, 120 and 132h respectively.

3.2.2. Growth Curves and $^{15}N/^{13}C$ rOvIFN-τ Labeling

Prior to ^{13}C- methanol induction, yeasts were washed with a 0.2% glycerol solution, centrifuged, then suspended and cultured according to the appropriate induction protocol. Cells were induced with ^{13}C-methanol when early starvation was observed (**Fig. 1**). Protocol P2 containing 5% ^{13}C-glucose at the early exponential phase ensured a much higher rate of cell growth in culture as compared with the remaining 3 protocols, P1, P3 and P4 (*see* **Note 2**). However, supplementation with 2% glucose in P4 during the exponential growth phase significantly increased cell growth in this phase (*see* **Note 3**). The results demonstrate a growth rate to cell density similar to P2 just prior to ^{13}C-methanol induction. The decreased cell density observed for P2 after 72 h may be explained by the toxicity of the 2% ^{13}C-methanol treatment during the induction period. To prevent cell-death of *P. pastoris* after a long-term ^{13}C-methanol exposure period, the rate of ^{13}C-methanol addition was reduced. During the methanol adaptative phase (for 6-h post-growth phase), ^{13}C-methanol was given at a low concentration (0.2%) and the concentration was subsequently increased by 0.4% every 12 h until it reached 1.4%. After 24 h at a high-methanol concentration (1.4%), the ^{13}C-methanol was decreased to 0.6% for the last 12 h. Starting the ^{13}C-methanol induction period at about the same cell density, P4 resulted in a slight increase (16%) of $^{15}N/^{13}C$ labeling over protocol P2. Protocols P1 and P3 resulted in about same growth curves and gave 98 and 85 mg respectively of purified $^{13}C/^{15}N$ double-labeled rOvIFN-τ/L of culture. Protocol P4 is the only growth and induction procedure with 0.1% ^{13}C-glucose and 0.02% 15N-ammonium chloride isotopes added for 6 h at the end of the growth phase. Importantly, when compared with the isotope incorporation in rOvIFN-τ using P3 and the other protocols, the P4 growth and induction protocol resulted in a 99% incorporation of ^{15}N and ^{13}C isotopes into rOvIFN-τ (*see* **Figs. 3** and **4**). We concluded that P4 was the most economical protocol when compared with the others (*see* **Note 4**).

3.3. Protein Electrophoresis and Purification

1. Protein concentration was measured by the Bradford's method *(8)*.
2. The purified rOvIFN-τ was subjected to polyacrylamide gel electrophoresis (PAGE) on a 15% gel using Tricine buffer.
3. Proteins were stained with Coomassie brillant blue R-250 (Sigma).
4. Prior to purification, the protein suspension was extensively washed by filtration through a 3-kDa Amicon membrane using 20 mM Tris-HCl (pH 7.6).

3.3.1. Anion-Exchange Chromatography Purification Process

1. The concentrated fraction resulting from a 3-kDa filtration was dialyzed against 20 mM Tris-HCl (pH 7.6), at 4°C for 48 h.
2. Proteins were loaded on an 800 mL Q-Sepharose Fast Flow column (Pharmacia, Uppsala, Sweden) equilibrated with 20 mM Tris-HCl (pH 7.6).
3. Elution was carried out using 20 mM Tris-HCl and 0.35 M NaCl (pH 7.6), at a flow rate of 1 mL/min.

3.3.2. Gel Filtration Purification Process

1. The Q-Sepharose material was concentrated to 30 to 40 mg/mL and then purified over a 1.5 × 120 cm Sephacryl S-100 HR column equilibrated in 20 mM Tris-HCl (pH 7.6).
2. Proteins (2 mL) were eluted at a flow rate of 1 mL/min using 20 mM Tris-HCl, 0.3 M NaCl (pH 7.6). Fractions containing the purified material were identified on 15% PAGE-Tricine and were pooled, dialyzed for 36 h against 10 mM Tris-HCl (pH 7.6), then 1 mM Tris-d$_{11}$, lyophilized, and kept at –80°C in 0.02% sodium azide.
3. After the double purification step, unlabeled and ^{13}C/^{15}N double-labeled rOvIFN-τ were obtained with over 95% purity as observed in Tricine-PAGE electrophoresis, 1D-NMR, and MS analysis (*see* **Figs. 2–4**).

3.4. Analysis of ^{15}N/^{13}C rOvIFN-τ Incorporation Using MS and NMR Analysis

Uniform isotopic labeling of purified rOvIFN-τ was confirmed by MALDI-TOF MS (PerSeptive Biosystems, Framingham, MA) analysis. The acceleration voltage was set at 25 kV. Apomyoglobin was used as standard in the mass spectrometer. Purified ^{15}N/^{13}C-labeled rOvIFN-τ was used for 1D proton and 2D-NMR experiments (*see* **Figs. 3** and **4**).

1. Lyophilized protein (1 mM) was suspended in 1 mM Tris-d$_{11}$ (pH 7.2) with 10% D$_2$O.
2. All NMR experiments were performed at 25°C on a Bruker AM-600 MHz NMR. The purpose of the study was to determine the best way to label rOvIFN-τ with

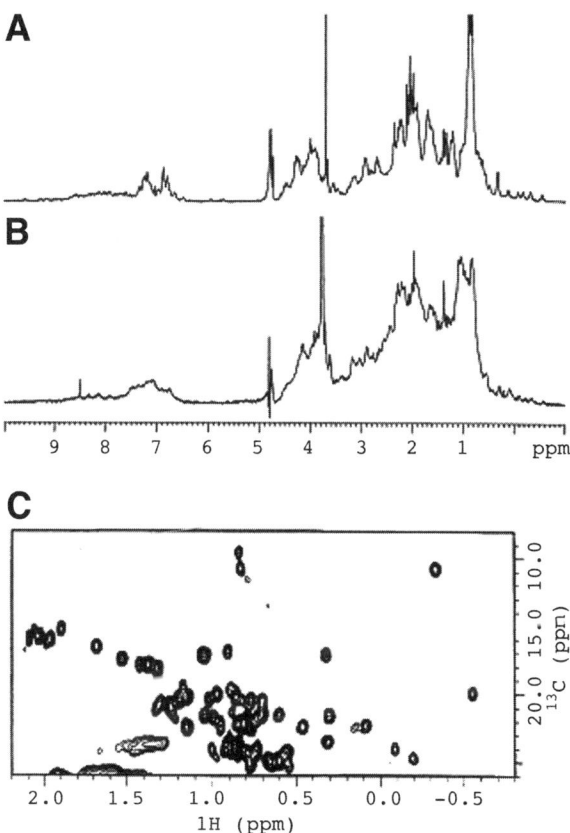

Fig. 3. (**A**) ^{1}H 1D-NMR spectra of the unlabeled and (**B**) ^{15}N/^{13}C rOvIFN-τ and (**C**) 2D-NMR of the ^{15}N/^{13}C rOvIFN-τ at pH 7.2, 1 mM, 25°C.

^{15}N and ^{13}C isotopes at the lowest cost in shake flask experiments. Two protocols P3 and P4 were compared with previous protocols (P1 *[3]* and P2 *[2]*). The results demonstrate that ^{15}N and ^{13}C isotopes are not required at all times during the initial growth period but are necessary at a low concentration just a few hours prior to induction of the protein expression period (*see* **Figs. 3** and **4**). Full incorporation (99%) was observed when 0.1% ^{13}C-glucose and 0.02% ^{15}N-ammonium chloride were added 6 h prior to the ^{13}C-methanol induction phase.

3.5. Antiviral Assay

An antiviral activity assay measuring the protection of Madi-Darby Bovine Kidney cells (MDBK) (BioWhittaker, Verviers, Belgium) against vesicular stomatitis virus (VSV)-induced CPEs (cytopathic effects) was performed as previously described *(4)*. In brief, 0.5 × 10^5 cells were seeded into each well of

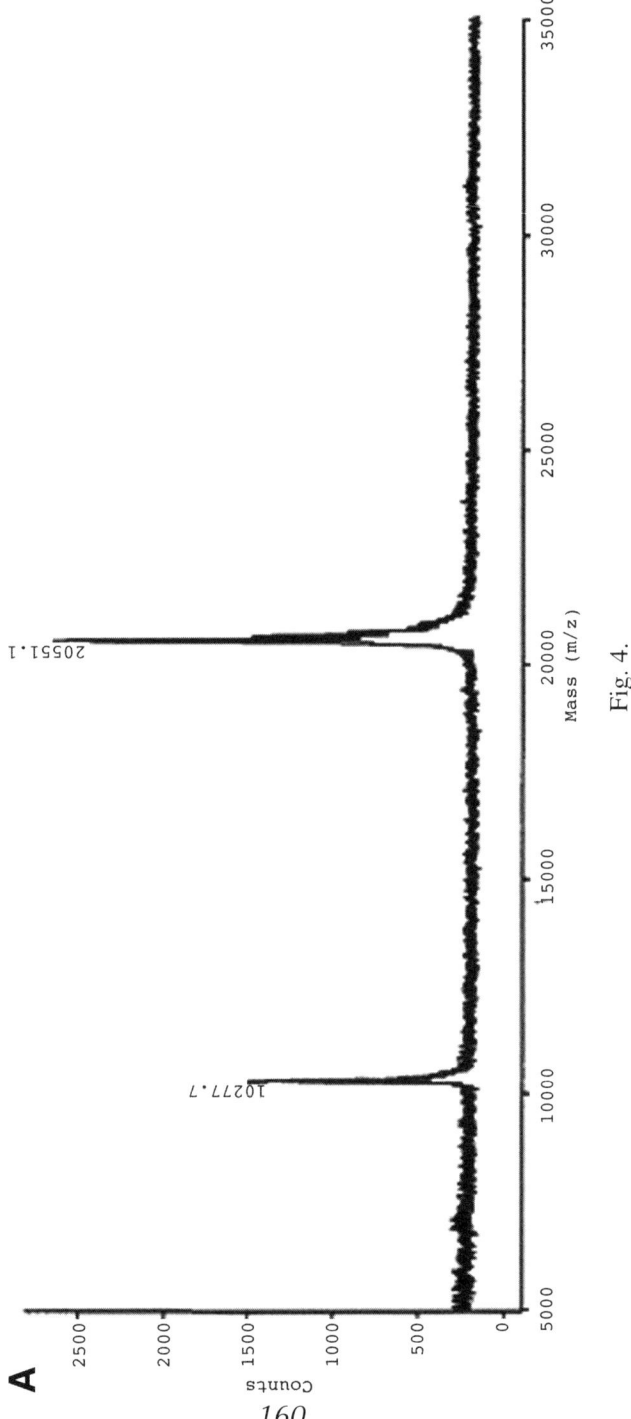

Fig. 4.

160

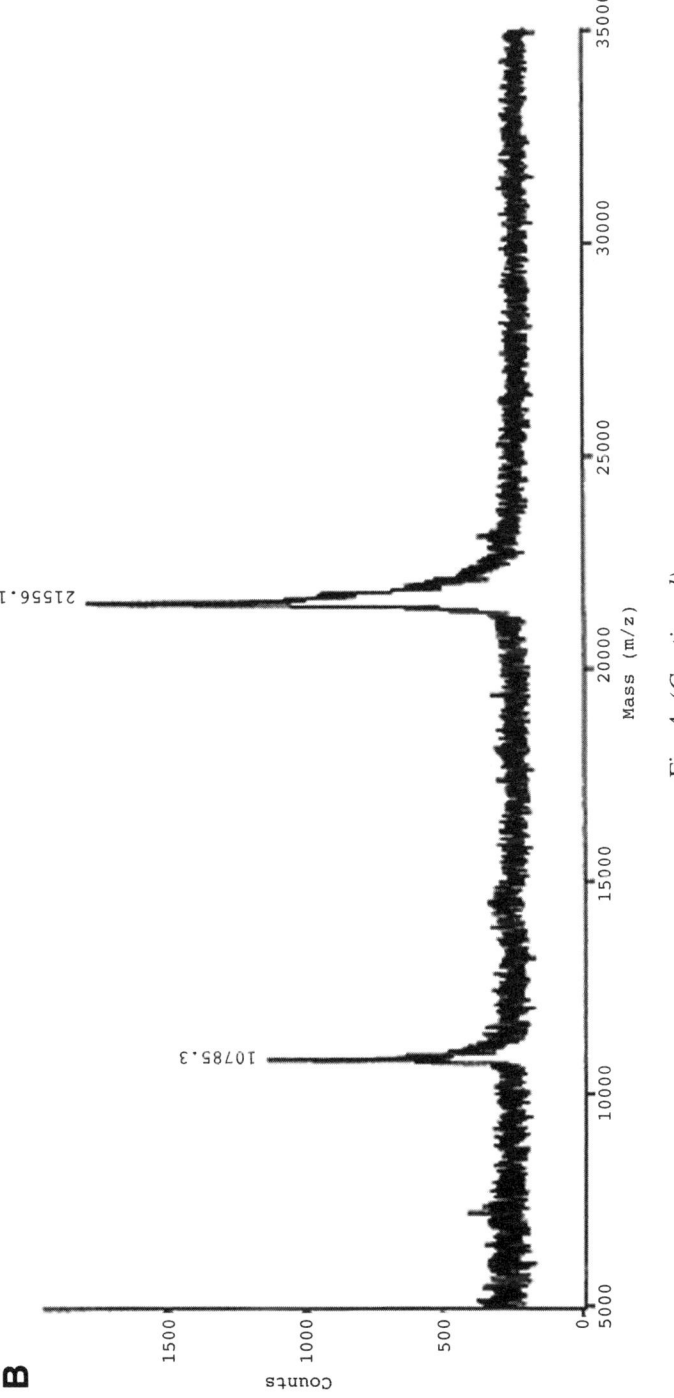

Fig. 4. (*Continued*)

161

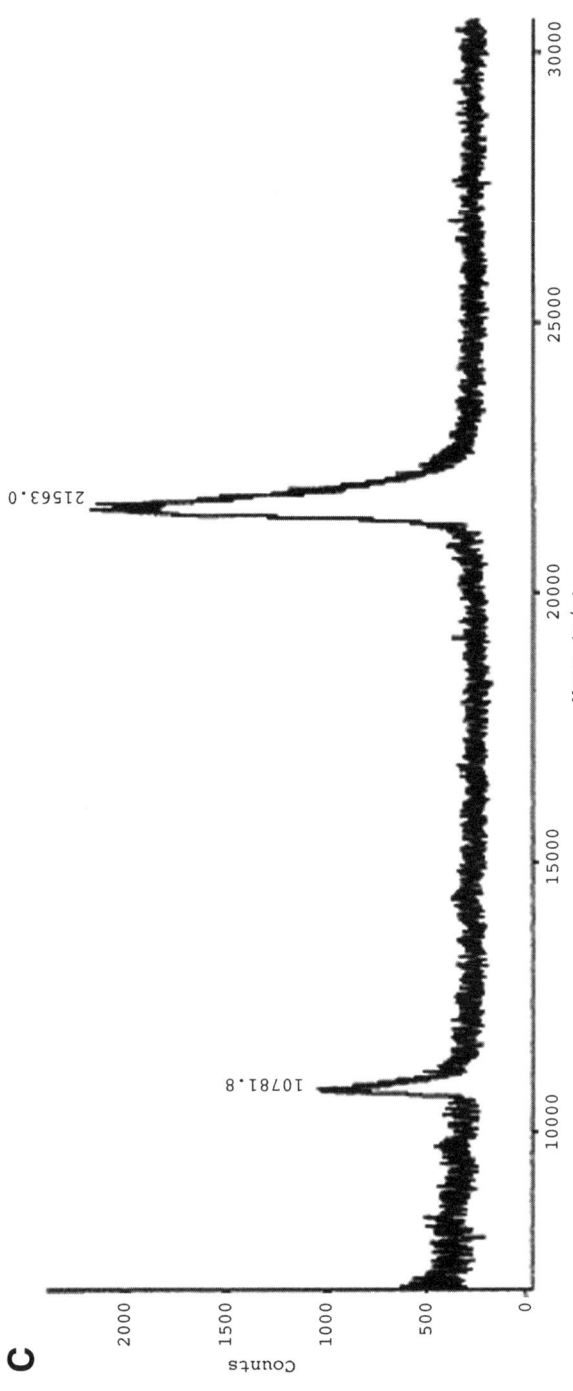

Fig. 4. (**A**) MALDI-TOF MS of unlabeled rOvIFN-τ, (**B**) $^{15}N/^{13}C$-labeled rOvIFN-τ (control using P2 protocol; and C, using P4 protocol).

96-well plates and incubated with two-fold serial dilutions of supernatant samples for 18 h at 37°C. After incubation, the cells were challenged with VSV, and the plates were incubated at 37°C for 18 h. Virus-induced CPEs were assessed by microscopic examination, and results were expressed as the inverse dilution that provided 50% protection of the cells from virus-induced CPEs. The purified ^{15}N/^{13}C double-labeled rOvIFN-τ from flask experiments demonstrated full anti-viral activity against VSV on MDBK cells using a cytopathic effect microplate assay (data not shown). Activities of 1×10^8 and $10^{7.5}$ antiviral units/mL were detected in the purified samples from shake flasks and fermentation experiments, respectively.

4. Notes

1. Prior to ^{13}C-methanol induction period, washing cells with 0.2% glycerol was found beneficial. When the 0.2% glycerol wash step was not performed, the expression level of rOvIFN-τ was significantly lower (41 to 77% less, depending on the protocol used).

2. Sequential addition of carbon source or addition of 5% of ^{13}C-glucose during the exponential growth phase was beneficial in improving cell density in shake flasks (*1,2*). However, aeration is a main limited factor in shake flasks and cannot be obviously controlled. One way is to increase aeration by agitation when 2% of glucose is added during the exponential growth phase. Another way is to culture strains in a fermenter, which allows more control of oxygenation parameters. In order to obtain more labeled protein at a lower cost, fermentation experiments using the P4 protocol may help. First, oxygen transfer is controlled and at much higher levels than during fermentation than in shake flask cultures. Second, a fed-batch phase using D-glucose (24-36 h) following the growth phase gives a much higher cell density, thus giving much more labeled recombinant protein. Third, methanol accumulation, which can be responsible for cell toxicity, can be monitored and better controlled during the induction phase.

3. Isotopic labeling may be slightly different from one protein to another. Time of culturing and concentration of ^{13}C-glucose may be optimized. Addition of ^{13}C-glucose (0.1%) 6 h prior to methanol induction was found to be sufficient to have efficient ^{15}N/^{13}C-labeling of rOvIFN-τ. However, addition of ^{13}C-glucose at different time and/or at a lower concentration was not tested.

4. The cost of this method and previous methods were evaluated. In our experiments, a 100 mL *P. pastoris* culture resulted in the expression and further purification of 26 mg of protein which was sufficiently concentrated to run NMR experiments. Taking into account isotope prices, the cost of our best method (P4) was found to be about one-third of the cost of the classical method that used ^{15}N-ammonium chloride (0.25%) and ^{13}C-D-glucose (3%) during the entire growth phase. The cost of our method is even lower when compared with methods that use ^{13}C-glycerol during the growth phase, because this compound is even more expensive than ^{13}C-D-glucose. It seems that cell lysis and the resulting possible protein degradation

observed when using a high concentration of methanol (>1.5%) can be prevented by reducing the amount of ^{13}C-methanol during the initial adaptation phase (0.2 to 0.6%) and the last 12 h of the methanol induction period (0.6%). This was in contrast to protocol P2 where 2% ^{13}C-methanol was utilized at all times during the induction period. This result is in agreement with a prior study showing that initial and terminal moderate methanol induction can regulate *Escherichia coli* acid phosphatase appA2 mRNA expression in *P. pastoris* (**9**).

Acknowledgments

This work was performed in the NMR facility of the University of Alabama at Birmingham, Birmingham, AL, USA. This works was supervised by N. R. Krishna and supported in part by NCI grants 1R01 CA-84177 and CA-13148 (Cancer Center Shared Facilities). We are grateful to Drs. Robert Krull and Huadong Zeng for NMR measurements and Dr. Ted Sakai for constructive discussions. We also thank Dr. Barbara Torres at the University of Florida, Gainesville, for the antiviral assay.

References

1. Cereghino, J. L. and Cregg, J. M. (2000) Heterologous protein expression in the methylotrophic yeast *Pichia pastoris. FEMS Microbiol. Reviews* **24,** 45–66.
2. Laroche, Y., Storme, V., De Mutter, J., Messens, J., and Lauwerys, M. (1994) High-level secretion and very efficient isotopic labeling of tick anticoagulant peptide (TAP) expressed in the methylotrophic yeast, Pichia pastoris. *Bio/Technology* **12,** 1119–1124.
3. Wood, M. J. and Komives, E. A. (1999) Production of large quantities of isotopically labeled protein in *Pichia pastoris* by fermentation. *J. Biomol. NMR* **13,** 149–159.
4. Pontzer, C. H., Torres, B. A., Vallet, J. L., Bazer, F. W., and Johnson, H. M. (1988) Antiviral activity of the pregnancy recognition hormone ovine trophoblast protein-1. *Biochem. Biophys. Res. Commun.* **152,** 801–817.
5. Pontzer, C. H., Ott, T. L., Bazer, F. W., and Johnson, H. M. (1991) Antiproliferative activity of a pregnancy recognition hormone, ovine trophoblast protein-1. *Cancer Res.* **51,** 5304–5307.
6. Rodriguez, E. and Krishna, N. R. (2001) An economical method for (15)N/(13)C isotopic labeling of proteins expressed in *Pichia pastoris. J. Biochem. (Tokyo).* **130,** 19–22.
7. Van Heeke, G., Ott, T. L., Strauss, A., Ammaturo, D., and Bazer, F. W. (1996) High yield expression and secretion of the ovine pregnancy recognition hormone interferon-tau by Pichia pastoris. *J. Interferon Cytokine Res.* **16,** 119–126.
8. Bradford, M. M. (1976) A rapid and sensitive method for the quantitation of microgram quantities of protein utilizing the principle of protein-dye binding. *Anal. Biochem.* **72,** 248–254.
9. Rodriguez, E., Han, Y., and Lei, X. G. (1999) Cloning, sequencing, and expression of an Escherichia coli acid phosphatase/phytase gene (appA2) isolated from pig colon. *Biochem. Biophys. Res. Commun.* **257,** 117–123.

12

Selenomethionine Labeling of Recombinant Proteins

Anna M. Larsson and T. Alwyn Jones

Abstract

Selenomethionine incorporation is a standard method for determining the phases in protein crystallography by single- or multiwavelength anomalous dispersion. Recombinant expression of selenomethionine-containing protein in non-auxotrophic *Pichia pastoris* strains yield an incorporation of about 50%. The expression of a mutated variant of *Penicillium minioluteum* dextranase in *P. pastoris* is used to illustrate the method utilized to obtain selenomethionyl-substituted protein and to show the phasing power of the acquired anomalous signal. The dextranase structure was solved using the anomalous signal achieved from 50% selenomethionine incorporation.

Key Words: Dextranase; MAD; multiwavelength anomalous dispersion; *Pichia pastoris*; SeMet; selenomethionine.

1. Introduction

Selenomethionine (SeMet) incorporation has become a standard method for determining the phases in protein crystallography by single- or multiwavelength anomalous dispersion (MAD) *(1)*. In *Escherichia coli*, methionine auxotrophic or non-auxotrophic strains are routinely used to achieve a close to complete substitution of methionine to selenomethionine *(2–4)*. Expression of selenomethionine-containing protein in *Saccharomyces cerevisiae* has been shown to yield partial incorporation, where a maximum incorporation of 65% is obtained with a non-auxotrophic strain *(5)*. A similar protocol is used to reach incorporation of about 50% for protein recombinantly expressed in *Pichia pastoris* *(6)*. The expression in *P. pastoris* of a mutated variant of *Penicillium minioluteum* dextranase will be used to illustrate the method utilized to obtain selenomethionyl-substituted protein and to show the phasing power of the acquired anomalous signal.

From: *Methods in Molecular Biology, vol. 389*: Pichia *Protocols, Second Edition*
Edited by: J. M. Cregg © Humana Press Inc., Totowa, NJ

2. Materials

1. *P. pastoris* strain MP36 (his3) *(7)*.
2. pPS-7 vector *(8)*.
3. YNB minimal medium plates: 0.67% yeast nitrogen base, 2% agar, and 2% glucose.
4. YPG growth medium: 1% yeast extract, 2% peptone, 0.1 *M* potassium phosphate buffer (pH 6.0), and 1% glycerol.
5. Saline solution: 0.9% *(w/v)* NaCl.
6. Modified synthetic complete medium: 0.09 mg/mL each of adenine sulfate, uracil, L-tryptophan, L-histidine-HCl, L-arginine-HCl, L-tyrosine, L-leucine, L-isoleucine, and L-lysine-HCl; 0.3 mg/mL each of L-glutamic acid, L-aspartic acid, L-glutamine, and succinic acid; 0.15 mg/mL L-phenylalanine; 0.45 mg/mL L-valine; 0.6 mg/mL L-threonine; 1.2 mg/mL L-serine; 0.34 mg/mL thiamine; 0.12 mg/mL L-cysteine; 0.2 mg/mL L-proline; 0.2 mg/mL L-alanine; 0.01 mg/mL inositol; 1.34% *(w/v)* yeast nitrogen base without amino acids (Difco); and 0.1 mg/mL L-selenomethionine (Calbiochem).
7. Methanol.
8. Reservoir solution: 20% mme polyethylene glycol (PEG) 5000 (Fluka), 0.1 *M* sodium acetate (pH 5.5), 0.1 *M* NaCl and 10 m*M* CaCl$_2$.
9. Cryoprotectant solution: 25% mme PEG 5000, 12.5% glycerol, 0.2 *M* sodium acetate (pH 5.5), 0.2 *M* NaCl and 20 m*M* CaCl$_2$.

3. Methods

The methods described below outline: (i) the construction of the expression system, (ii) the induction of SeMet-labeled protein expression, (iii) the purification and crystallization of the protein, and (iv) the characterization and structure solution of the SeMet labeled protein.

3.1. Expression Construct

The choice of *P. pastoris* strain and vector for expression of the protein might differ and hence the construct will vary. The steps below are a description of the *P. pastoris* strain and vector that we used for expression (*see* **Note 1** for another example.)

3.1.1. P. pastoris *Strain*

The *P. pastoris* strain MP36 *(7)* was used for expression of the dextranase mutant. MP36 is a *his3* deficient strain transformable by *HIS3* gene containing vectors.

3.1.2. pPS-7 Expression Vector

The pPS-7 (*see* **Fig. 1**) expression vector *(8)* is designed to insert the gene by transplacement into the *AOX1* locus giving phenotype methanol-utilization

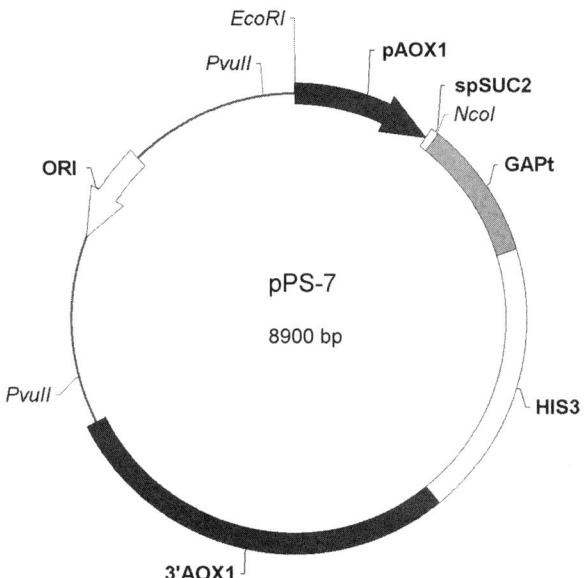

Fig. 1. The pPS-7 vector used for expression of the *dex* gene in *P. pastoris*. The expression cassette contains the promotor from the *AOX1* gene (p*AOX1*) and a fragment of the 3′-end of the *AOX1* gene (3′*AOX1*) for the recombination event. The cassette also consists of the signal peptide from invertase (sp*SUC2*), the terminator for transcription of the gene coding for glyceraldehyde-3-phosphate dehydrogenase (*GAPt*) and the *HIS3* gene as selection marker, all three from *S. cerevisiae*. The vector has a bacterial replication origin (*ORI*).

slow (Mut^S). The expression is induced by methanol and during induction the cell growth is highly reduced for the Mut^S phenotype because the growth relies on the less expressed alcohol oxidase gene, *AOX2*. To obtain secreted protein, the gene is fused to the signal peptide of the sucrose invertase (*SUC2*) gene from *S. cerevisiae*. The termination for the transcription of the gene coding for glyceraldehyde-3-phosphatate dehydrogenase (*GAPt*) and the selection marker *HIS3*, both from *S. cerevisiae*, are also present in the plasmid.

3.1.3. Transformation and Plasmid Selection

To obtain the cassette for transformation into *P. pastoris*, the pPS-7 construction was digested with *Pvu*II and the DNA was extracted with phenol-chloroform. The transformation was done by electroporation according to the *Pichia* Expression Kit Version E Instruction Manual from Invitrogen (www.invitrogen. com). The *HIS3* transformants were selected on yeast nitrogen base (YNB) minimal medium plates incubated at 28°C.

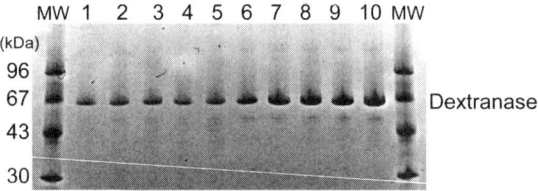

Fig. 2. SDS-PAGE gradient gel (8–25%) of cell culture supernatant concentrated 8×
and run on PhastSystem (Pharmacia). *Lanes 1–5* show the expression after 24, 39, 48,
63, and 72 h postinduction, respectively, of the SeMet protein. *Lanes 6–10* show the
expression after corresponding times of the native protein. The molecular weight of dextranase is 61 kDa. The OD_{600} increased from about 8.5 to 15.5 during the first 24 h after
which it kept constant for both selenomethionine and methionine containing cultures.

3.2. Protein Induction

The next steps in the process are the growth of *P. pastoris* to increase the cell
density followed by induction with methanol to initiate SeMet labeled protein
expression.

3.2.1. Cell Growth

1. Grow a 10 mL preculture in YPG growth medium for about 20 h at room temperature.
2. Add the preculture to 1 L YPG growth medium. Let the cells grow 24 h at 28°C.
3. Harvest the cells by centrifugation (6000*g* for 5 min), discard the supernatant, and
   resuspend the cells in 100 mL saline solution.
4. Repeat **step 3** 2× and resuspend in 300 mL modified synthetic complete medium
   with SeMet the last time.

3.2.2. Induction

1. Add the 300 mL cell suspension to 3 L baffled shaking flask (*see* **Note 2**).
2. Induce expression at 22°C with 1% methanol (*see* **Note 3**).
3. Add 0.5% methanol after 24 h and every 12 h thereafter (*see* **Fig. 2**).

3.2.3. Harvest

1. Terminate expression after 72 h post-induction by centrifuging the cell cultures
   (6000*g* for 5 min).
2. Filter the supernatant containing the secreted protein through a Whatman GF/B filter.

3.3. Protein Purification and Crystallization

The purification and crystallization procedures depend on the protein of
interest. In our case we used the same purification protocol *(6)* as for the
methionine-containing protein. Crystals were obtained under conditions similar

to those for the methionine-containing protein (*6*) and optimized with some minor modifications to the precipitant concentration. The following steps describe the crystallization of dextranase by the hanging-drop vapor-diffusion method (*9*).

3.3.1. Crystallization

1. Prepare the dextranase sample with a protein concentration of 7.5 mg/mL in 0.1 M sodium acetate, (pH 5.5), 0.1 M NaCl, and 10 mM CaCl$_2$. The NaCl is necessary to prevent precipitation of the protein.
2. Add 1 mL reservoir solution into the well of a 24-well crystallization plate.
3. Mix 4 µL protein with 2 µL reservoir solution on a siliconized glass cover slip. To avoid evaporation, seal the cover slip to the crystallization plate with vacuum grease. Store at 25°C.
4. Prepare microseeds by vortexing previously obtained crystals in mother liquor, followed by centrifugation. Microseed the drops after 24 h equilibration by dipping an acupuncture needle into the supernatant of the microseed solution and transfer the microcrystals to the equilibrated drops. Crystals appear within 2 d and are fully grown in a couple of weeks.
5. Prior to data collection, transfer the crystal with a fiber loop to cryoprotectant solution and immediately flash-freeze in liquid nitrogen.

3.3.2. Oxidation

SeMet residues in proteins are more reactive than methionine residues and have a tendency to oxidize (*10*). Problems arise in the phasing when a mixture of oxidized and reduced SeMet is present in the same crystal (*11*). To avoid oxidation, reducing agents like dithiothreitol (DTT), glutathione or β-mercaptoethanol (*11*) are commonly used during the purification and crystallization, or in the cryoprotectant (*12*). Alternatively, deliberate oxidation of all SeMet groups can be stimulated by treatment of the protein with hydrogen peroxide prior to crystallization (*13*). No reduction or oxidation was performed in this example.

3.4. Evaluating the SeMet Labeling

A combination of factors affect the phasing power of the anomalous signal from the selenium atoms in the protein, including the relative amount of methionine that can be substituted in the protein, the degree of SeMet incorporation, the selenium redox state, the crystal quality, and the accuracy of the measured anomalous signal.

3.4.1. Amino Acid Analysis

The mutated version of dextrananse from *P. minioluteum* is a 61-kDa protein and contains 12 methionines. The incorporation of selenomethionine was estimated to be about 50%. The quantitative amino-acid analysis showed this as a decrease in the amount of methionine content relative to the native protein (**Table 1**).

Table 1
Amino Acid Composition of Protein Expressed in Absence (Native)
and Presence of Selenomethionine (SeMet)

Amino acid	No. of residues (theoretical)	No. of residues (native)	No. of residues (SeMet)
A	33	33.8	32.8
C	6	6.4	5.6
D + N	76	77.0	77.7
E + Q	37	38.0	39.0
F	24	24.4	24.6
G	56	57.5	58.0
H	14	13.4	12.0
I	44	41.2	41.9
K	17	16.6	16.1
L	24	24.9	25.5
P	33	31.5	31.9
R	13	13.1	13.1
S	63	63.0	65.0
T	39	39.2	39.4
V	42	40.3	39.6
Y	26	24.9	24.8
M	12	11.9	6.1

3.4.2. Absorption Edge Scan

The presence of selenium in the protein should be established by an absorption edge scan before proceeding with the MAD diffraction experiment. An X-ray absorption spectrum of the crystal with a clear white line was collected near the selenium K absorption edge by measuring the fluorescent signal perpendicular to the beam during an energy scan performed at beamline ID29, ESRF, Grenoble, France (*see* **Fig. 3**). Depending on the redox state of the selenium atoms, the white line is shifted in energy. For the oxidized state it has a slightly higher energy, and the magnitude of the edge increases for a completely oxidized protein *(13)*. In our case the fluorescence spectrum was typical of reduced SeMet.

3.4.3. Anomalous Signal and Phasing

Even at 100% SeMet incorporation, the anomalous signal (i.e., the source of phase information in the MAD method), is a weak signal close to the noise level. Because the SeMet incorporation in *P. pastoris* samples is about 50%, the

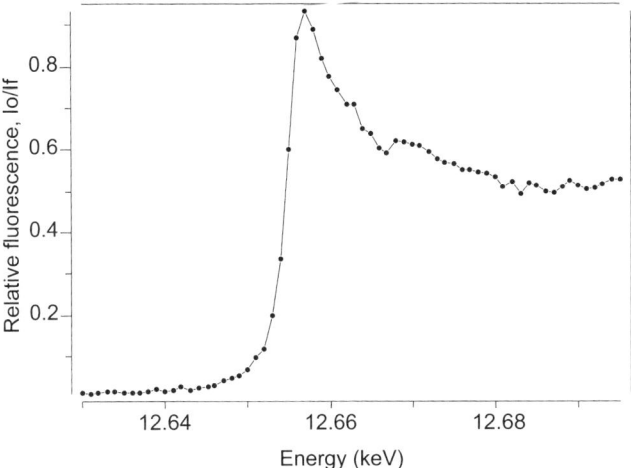

Fig. 3. Selenium *K*-edge fluorescence scan of selenomethionyl dextranase crystals measured at ESRF beamline ID29.

signal is even weaker. The success in solving the structure relies on accurately measuring small differences in reflection intensities, both within the same data set and between the data sets collected at different wavelengths. The crystallographer has to balance collecting sufficient data to produce an accurate anomalous signal vs introducing nonisomorphous differences due to X-ray radiation damage *(14)*.

The dextranase structure was solved using MAD at the selenium *K*-edge *(15)*. The SeMet data was collected on beamline X9A at Brookhaven National Laboratory, a second-generation synchrotron source. The following crystallographic computer programs were used to obtain the electron density map from the intensities of the collected data. Analysis of the anomalous signal with SHELXC *(16)*, (Table 2), indicated that we should restrict the data to 3 Å resolution. In this shell, the correlation coefficients of the anomalous differences have a value of about 30%, the lowest value recommended by Schneider and Sheldrick *(17)*. Phases were refined with autoSHARP *(18)* using 9 selenium sites, producing an overall figure-of-merit of 0.49. Figure-of-merit describes a "confidence level" for the calculated phases. After solvent flattening with DM *(19)*, the electron density map, (*see* **Fig. 4**), could be clearly interpreted.

4. Notes

1. An alternative strain that has been used for SeMet labeling of recombinantly expressed mannanase in *P. pastoris (20)* is the wild-type X-33 strain (Invitrogen).

Table 2
Anomalous Signal

		$\Delta F/\sigma(\Delta F)$			Correlation coefficients (%)		
Dmin (Å)	Dmax (Å)	Hrem	Peak	Infl	Hrem/Peak	Hrem/Infl	Peak/Infl
19.9	8.0	1.62	2.23	2.06	88.4	87.4	91.3
8.0	6.0	1.50	2.17	2.00	78.4	78.7	85.9
6.0	5.0	1.30	1.77	1.71	72.4	70.3	82.7
5.0	4.0	1.07	1.41	1.39	65.0	60.9	77.7
4.0	3.5	1.03	1.27	1.28	46.2	45.3	66.1
3.5	3.0	1.03	1.17	1.16	30.6	28.2	46.0
3.0	2.8	1.03	1.11	1.13	22.6	21.4	33.4
2.8	2.6	1.02	1.10	1.11	15.6	10.8	24.1
2.6	2.4	1.06	1.11	1.08	6.9	5.1	17.5
2.4	2.2	1.04	1.09	1.09	5.1	5.8	10.8
2.2	2.0	1.02	1.04	1.02	4.1	1.4	8.6

$\Delta F/\sigma(\Delta F)$: Signal-to-noise ratio for the anomalous differences.
Correlation coefficient between anomalous differences measured at the two wavelengths.

The medium used for the expression was: 100 mM potassium phosphate (pH 6.0); 1.34% YNB without amino acids (Difco); 0.3 μg/mL biotin, 3 μg/mL each of folic acid, riboflavin, niacinamide, thiamine, and calcium pantothenate; 6 μg/mL pyridoxal; and 0.1 mg/mL each of L-selenomethionine, L-histidine, and L-tryptophan. The induction was done at 17°C with a final concentration of 0.5% methanol and 0.25% NH$_3$, both added every 12 h.

2. The large flask with baffles provides the best surface-to-volume ratio for optimal oxygenation.

3. The glycosylation-free mutant of dextranase may be unstable at 28°C and requires 22°C for synthesis because no protein was obtained when the protein was synthesized at 28°C.

Acknowledgments

We thank Marie Sundqvist and Dr. David Eaker at the Amino Acid Analysis Laboratory, Department of Biochemistry, Uppsala University, Sweden, for analysis and quantification of the amino-acid composition. We are grateful to Drs. José Cremata and Bianca García at the Center for Genetic Engineering and Biotechnology, Cuba, for supervising the construction of the glycosylation-free dextranase clone. We are also thankful to Dr. Jerry Ståhlberg at the Department of Molecular Biology, Swedish University of Agricultural Science, Sweden, for

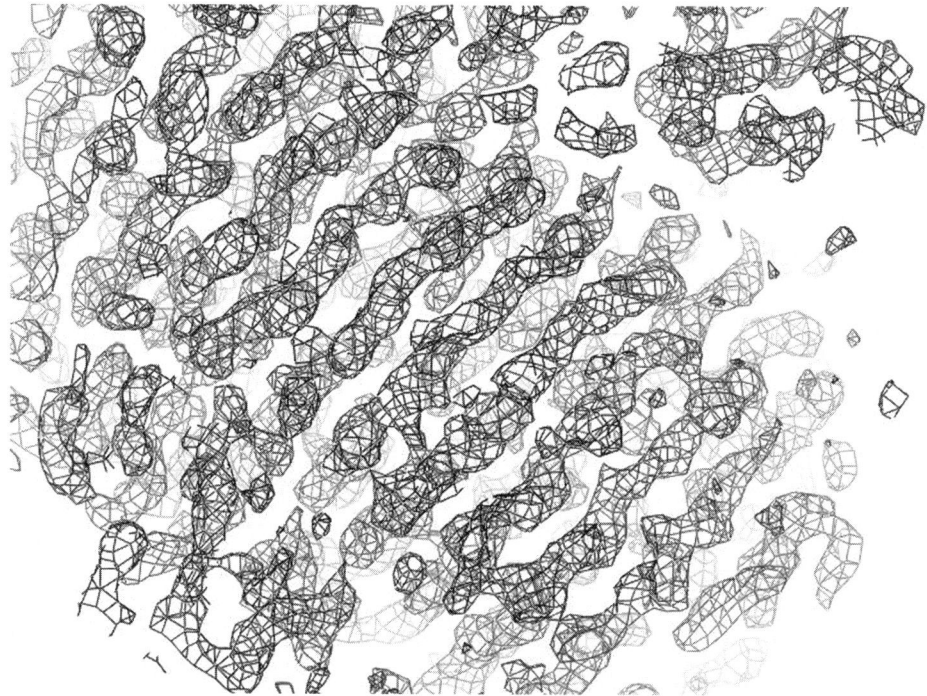

Fig. 4. Electron density map at 3 Å resolution after refinement of 9 selenium sites and phasing with autoSHARP followed by phase improvement with DM. The density is a $2F_O–F_C$ map at 0.3 electrons/Å^3. The tube-like volumes correspond to β-strands. The figure was prepared using the O program *(21)*.

supporting ideas and instructions during the purification steps. The work was supported by Swedish Science Research Council (VR).

References

1. Hendrickson, W. A. (1991) Determination of macromolecular structures from anomalous diffraction of synchrotron radiation. *Science* **254,** 51–58.
2. Hendrickson, W. A., Horton, J. R., and LeMaster, D. M. (1990) Selenomethionyl proteins produced for analysis by multiwavelength anomalous diffraction (MAD): a vehicle for direct determination of three-dimensional structure. *EMBO J.* **9,** 1665–1672.
3. Van Duyne, G. D., Standaert, R. F., Karplus, P. A., Schreiber, S. L., and Clardy, J. (1993) Atomic structures of the human immunophilin FKBP-12 complexes with FK506 and rapamycin. *J. Mol. Biol.* **229,** 105–124.
4. Doublié, S. (1997) Preparation of selenomethionyl proteins for phase determination. *Methods Enzymol.* **276,** 523–530.

5. Bushnell, D. A., Cramer, P., and Kornberg, R. D. (2001) Selenomethionine incorporation in *Saccharomyces cerevisiae* RNA polymerase II. *Structure* **9,** R11–R14.

6. Larsson, A. M., Ståhlberg, J., and Jones, T. A. (2002) Preparation and crystallization of selenomethionyl dextranase from *Penicillium minioluteum* expressed in *Pichia pastoris. Acta Cryst.* D**58,** 346–348.

7. Yong, V., González, M. E., Herrera, L., and Delgado, J. (1992) El gen *HIS3* complementa una mutación *his* de *Pichia pastoris. Biotecnología Aplicada* **9,** 55–61.

8. Yong, V., Herrera, L., Margolles, E., et al. (1991) Method for the expression of heterologuos genes in the yeast *Pichia pastoris*, expression vectors and transformed microorganisms. European Patent Application. Pub. No. 0438 200 A1.

9. McPherson, A. (1982) *Preparation and Analysis of Protein Crystals*. John Wiley & Sons, New York, NY.

10. Shepherd, L. and Huber, R. E. (1969) Some chemical and biochemical properties of selenomethionine. *Can. J. Biochem.* **47,** 877–881.

11. Smith, J. L. and Thompson, A. (1998) Reactivity of selenomethionine - dents in the magic bullet? *Structure* **6,** 815–819.

12. Thomazeau, K., Curien, G., Thompson, A., Dumas, R., and Biou, V. (2001) MAD on threonine synthase: the phasing power of oxidized selenomethionine. *Acta Cryst.* D**57,** 1337–1340.

13. Sharff, A. J., Koronakis, E., Luisi, B., and Koronakis, V. (2000) Oxidation of selenomethionine: some MADness in the method! *Acta Cryst.* D**56,** 785–788.

14. Ravelli, R. B. and McSweeney, S. M. (2000) The "fingerprint" that X-rays can leave on structures. *Structure Fold Des.* **8,** 315–328.

15. Larsson, A. M., Andersson, R., Ståhlberg, J., Kenne, L., and Jones, T. A. (2003) Dextranase from *Penicillium minioluteum*: reaction course, crystal structure, and product complex. *Structure* **11,** 1111–1121.

16. Sheldrick, G. M. (2003) SHELX (http://shelx.uni-ac.gwdg.de/SHELX/).

17. Schneider, T. R. and Sheldrick, G. M. (2002) Substructure solution with SHELXD. *Acta Cryst.* D**58,** 1772–1779.

18. Bricogne, G., Vonrhein, C., Paciorek, W., et al. (2002) Enhancements in autoSHARP and SHARP, with applications to difficult phasing problems. *Acta Cryst.* A**58** (Supplement), C239.

19. Cowtan, K. (1994) An automated procedure for phase improvement by density modification. *Joint CCP4 and ESF-EACBM Newsletter on Protein Crystallography* **31,** 34–38.

20. Xu, B., Munoz, I. G., Janson, J. C., and Ståhlberg, J. (2002) Crystallization and X-ray analysis of native and selenomethionyl β-mannanase Man5A from blue mussel, *Mytilus edulis*, expressed in *Pichia pastoris. Acta Cryst.* D**58,** 542–545.

21. Jones, T. A., Zou, J.-Y., Cowan, S. W., and Kjeldgaard, M. (1991) Improved methods for building protein models in electron density maps and the location of errors in these models. *Acta Crystallogr.* A**47,** 110–119.

13

Selective Isotopic Labeling of Recombinant Proteins Using Amino Acid Auxotroph Strains

James W. Whittaker

Abstract

Labeling proteins with stable isotopes is important for many analytical and structural techniques, including NMR spectroscopy and mass spectrometry. Nonselective labeling, which uniformly labels all amino acids in the protein, may be accomplished with readily available wild-type expression hosts. However, there are often advantages to labeling a specific amino acid, and residue-selective labeling generally requires the use of an expression strain that is auxotrophic for the amino acid in order to efficiently incorporate the isotopic label. The behavior of an auxotrophic strain may be complicated by the regulatory properties of the biosynthetic pathway, by secondary nutritional requirements resulting from disruption of a biosynthetic pathway, and from acquired sensitivity to environmental factors resulting from build-up of metabolic intermediates. As a result, it is important to characterize the phenotype of the each auxotrophic strain in order to optimize its performance as an expression host for selective labeling of proteins. The application of aromatic auxotroph strains of *Pichia pastoris* to labeling tyrosines in a recombinant protein (galactose oxidase) will be used to illustrate selective-labeling methods.

Key Words: Amino acid labeling; tyrosine; phenylalanine; tryptophan; auxotroph; isotope; isotopic labeling; NMR; EPR; mass spectrometry; spectroscopy.

1. Introduction

Interest in methods for stable isotope labeling of proteins has grown rapidly in recent years, driven by the dramatic expansion of biological NMR spectroscopy (with applications in structural biology) *(1,2)* and mass spectrometry (applied to proteomics research) *(3)*. These techniques utilize relatively rare stable isotopes, including ^2H, ^{13}C, and ^{15}N as spectroscopic probes and mass tags, substituting for the majority natural abundance isotopes (^1H, ^{12}C, and ^{14}N) in biomolecules *(4)*. Because the rare isotopes exhibit the same chemistry as the more common isotopic forms, living cells generally do not distinguish between

From: *Methods in Molecular Biology, vol. 389:* Pichia *Protocols, Second Edition*
Edited by: J. M. Cregg © Humana Press Inc., Totowa, NJ

labeled and unlabeled molecules, although D_2O is known to produce both kinetic and structural isotope effects in biological systems *(5)*.

Because most of the atoms in a protein do not exchange with their environment, stable isotope labeling of proteins requires that the labeled amino acid precursors be available for incorporation into the growing polypeptide during protein biosynthesis. When uniform labeling of the protein is desired, the organism is typically adapted to grow on isotopically enriched culture medium, using D_2O in place of normal water, $^{15}NH_3$ as nitrogen source, or ^{13}C-glucose as carbon source. Biosynthetic pathways in the host cell then convert these elementary compounds into complex biomolecules, assembling isotopically labeled amino acids from scratch and inserting them into the protein. In most research applications, the protein of interest will be expressed in a recombinant system where high-level production is possible. *Escherichia coli* is the most commonly used heterologous expression host, but the methylotrophic yeast *Pichia pastoris* has a number of special advantages, particularly for expression of eukaryotic proteins and secretory expression *(6,7)*, and has become one of the standard expression systems for structural genomics. Methods for uniform labeling of recombinant protein by *Pichia pastoris* have been developed, allowing preparation of protein globally labeled with ^2H, ^{13}C or ^{15}N separately or in combination *(8–12)*, and these approaches have been recently reviewed *(13)*.

As the name implies, uniform labeling is unselective, and all amino acids in a protein are similarly labeled. In some cases, it is useful to selectively label amino acids in a protein for enhanced resolution and sensitivity in the analysis of complex NMR or mass spectra. In contrast to uniform labeling methods, residue-specific labeling generally requires the use of auxotroph strains having a strict requirement for the amino acid. Because the auxotroph lacks the biosynthetic machinery to synthesize a given amino acid, dilution of exogenously supplied label with endogenous pools of unlabeled amino acid is avoided.

Genetically well-studied organisms, like *E. coli*, already have a full panoply of amino acid biosynthetic mutants available for expressing labeled protein, although there are often restrictions on which expression vectors may be used in these strains. In *Saccharomyces cerevisiae*, a genome-wide gene-knockout project has made all possible biosynthetic pathway mutants available through targeted disruption and deletion *(14)*. The existence of this complete collection of designer mutants shows the power of targeted gene disruption in creating novel expression hosts when sufficient genomic sequence data is available. Strains created by gene disruption have a well-defined genotype and are genetically stable, providing important advantages over strains isolated as the result of nontargeted mutagenesis, which may revert with significant frequency *(15,16)*. The emerging *Pichia pastoris* genome can be expected to make engineering of new auxotrophic strains routine, enhancing the value of this important

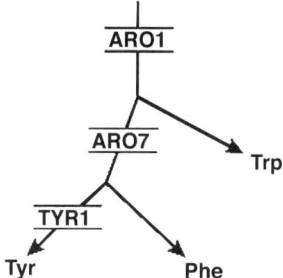

Fig. 1. Pathway for biosynthesis of aromatic amino acids in yeast *(20)*.

expression host. A number of auxotrophic strains of *Pichia pastoris* are already available, including histidine (*HIS4*) *(17)*, arginine (*ARG4*) *(18)*, cysteine (*CYS4*) *(19)*, and aromatic amino acid (*TYR1, ARO7, ARO1*) *(20)* auxotrophs. Techniques that will be generally useful in characterizing auxotrophic strains and protocols for isotope incorporation will be illustrated below in terms of labeling tyrosines in recombinant galactose oxidase *(21,22)*.

A brief review of aromatic amino acid biosynthesis will provide an introduction to these approaches. The organization of the aromatic amino acid biosynthetic pathway of the budding yeast, *S. cerevisiae*, is summarized in **Fig. 1** *(23,24)*. Genomic analysis of other yeasts suggests that the basic features of this pathway are highly conserved *(25,26)*. The first committed step of the pathway is catalyzed by the product of the *ARO1* gene (coding for the arom pentafunctional enzyme). Disrupting *ARO1* is predicted to lead to an absolute nutritional requirement for all three aromatic amino acids (tyrosine, tryptophan, and phenylalanine). Disruption of *ARO7* (coding for chorismate mutase) is predicted to lead to a requirement for both tyrosine and phenylalanine, whereas a strain disrupted within *TYR1* (coding for prephenate dehydrogenase) is expected to require only tyrosine.

This simple picture is complicated by the complex regulatory properties of metabolic pathways, and the utilization of some pathway intermediates in secondary metabolic processes *(23,24)*. Regulation of the aromatic amino acid biosynthetic pathway by all three of its end-products is exerted through negative control over the key regulatory steps. This complex regulatory pattern can result in unintuitive nutritional phenotypes, including sensitivity to nonessential amino acids *(20)*. In some cases, disruption of a pathway may result in a new requirement for growth factors that are derived from pathway intermediates. For example, disruption of aromatic amino acid biosynthesis in *Saccharomyces* results in a requirement for both p-amino benzoic and *p*-hydroxybenzoic acids *(23)*. Disruption of biosynthetic pathways may also lead to accumulation of metabolic intermediates that are deleterious to the cell. Because of these complications, a basic characterization of the nutritional requirements of the auxotrophic strain is an essential part of its application in labeling studies. The basic methods involved in both characterization of the auxotrophic stain and the expression of isotopically labeled protein will be covered in this chapter.

2. Materials

2.1. P. pastoris *Strains and Expression Vectors*

The strains currently available for aromatic amino acid labeling include *Pichia pastoris ura3Δ1 aro1::URA3* (tyrosine, phenylalanine, tryptophan triple auxotroph), *Pichia pastoris ura3Δ1 aro7::URA3* (tyrosine, phenylalanine double auxotroph), and *Pichia pastoris ura3Δ1 tyr1::URA3* (tyrosine auxotroph). The pPICZBαGAOX expression vector was constructed by splicing the coding sequence for *Dactylium dendroides* galactose oxidase cDNA to the sequence for the α-mating factor signal peptide coding sequence in the pPICZBα vector (Invitrogen).

2.2. Maintenance

1. Phosphate-buffered SD medium: 0.17% YNB w/o amino acids or ammonium sulfate, 0.5% ammonium sulfate, 2% glucose, 100 mM sodium phosphate (pH 6.7), and specific amino acid supplementation (100 mg/L L-tyrosine) (*see* **Notes 1** and **2**).
2. Minimal agar: phosphate-buffered SD medium, 2% Difco Bacto-agar, and any required specific amino acid supplements.
3. Autoclave-sterilized glycerol.
4. 1.2 mL externally threaded sterile cryogenic vial (Nalge Co., Rochester, NY).

2.3. Nutritional Requirements

1. Phosphate-buffered SD medium: 0.17% YNB w/o amino acids or ammonium sulfate, 0.5% ammonium sulfate, 2% glucose, and 100 mM sodium phosphate (pH 6.7).
2. Filter-sterilized stock solutions of 10X amino acid supplements.
3. 14-mL Falcon polypropylene snap top tubes (Becton Dickinson).

2.4. Dose–Response Curves for Amino Acid Supplementation

1. Phosphate-buffered SD medium: 0.17% YNB w/o amino acids or ammonium sulfate, 0.5% ammonium sulfate, 2% glucose, and 100 mM sodium phosphate (pH 6.7).
2. Filter-sterilized stock solution of 10× amino acid supplement (*see* **Note 3**).
3. 125-mL baffled flasks (*see* **Note 4**).

2.5. Transformation

1. pPICZBαGAOX expression vector linearized by *Pme* I digestion.
2. Phosphate-buffered SD medium: 0.17% YNB w/o amino acids or ammonium sulfate, 0.5% ammonium sulfate, 2% glucose, 100 mM sodium phosphate (pH 6.7), and specific amino acid supplements.
3. 1 M sorbitol solution, cold.

4. SD medium (with required amino acid supplement and 2% glucose) containing 1 *M* sorbitol.
5. Zeocin selection agar: phosphate-buffered SD agar, 2% glucose, amino acid supplement, and 1 g/L Zeocin.

2.6. Recombinant Protein Production in Shake Flask Cultures

1. 2X Phosphate-buffered SD medium: 0.34% YNB w/o amino acids or ammonium sulfate, 1% ammonium sulfate, 100 m*M* sodium phosphate (pH 6.7), and 2% glucose or glycerol.
2. Filter-sterilized stock solution of isotopically labeled amino acid (e.g., D,L-4 hydroxyphenyl-d_4-alanine-2,3,3,-d_3 [D,L-d_7-tyrosine]; CDN Isotopes).
3. Filter-sterilized methanol (*see* **Note 5**).
4. Baffled shake flasks (*see* **Note 4**).

2.7. Protein Characterization

The instrumentation required for protein characterization will depend on the method used to determine the extent of label incorporation (mass spectrometry, NMR, or other spectroscopic approaches). For galactose oxidase *(21)*, EPR spectroscopy can be conveniently used to quantitatively determine the extent of tyrosine labeling in the free radical-containing apoprotein *(27)*.

3. Methods

3.1. Strain Construction

Standard methods were used to prepare the auxotroph strains from a *Pichia pastoris ura3*Δ1 parent strain by targeted gene disruption *(20)* and each was verified by PCR analysis of genomic DNA.

3.2. Strain Maintenance

Fresh cultures are prepared by plating frozen stock onto SD + glucose agar containing the required amino acid and growing at 30°C for 24 to 48 h. (*see* **Note 6**). Zeocin is not required unless reselection of the expression strain is desired. Media should be freshly prepared, because aromatic amino acid supplements are subject to autooxidation. Frozen stock is prepared by suspending cells from a fresh agar plate in 0.7 mL phosphate-buffered SD medium and adding sterile glycerol to make a 10% solution in sterile 1.2-mL cryogenic vials. After vortexing to thoroughly mix the solution, the stock culture is stored frozen at −80°C.

3.3. Characterization of Nutritional Requirements

1. Inoculate 10 mL of phosphate-buffered SD + 2% glucose without the required amino acid in a 125-mL baffled flask at moderate cell density and incubate overnight to produce an amino acid-limited starved culture (*see* **Note 7**).

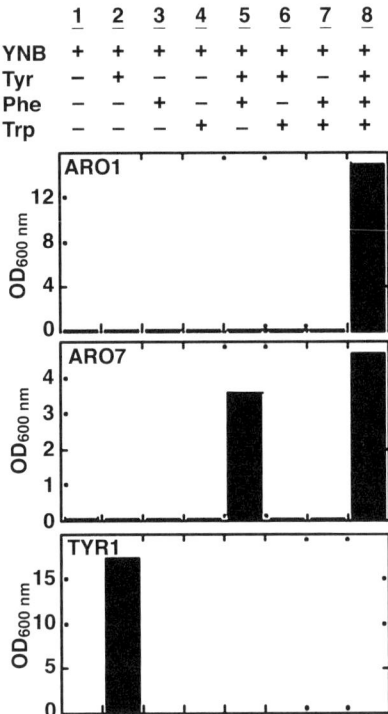

Fig. 2. Nutritional requirements of auxotrophic *Pichia pastoris* strains. *P. pastoris ura3Δ1 aro1::URA3*, *ura3Δ1 aro7::URA3*, and *ura3Δ1 tyr1::URA3* starved cultures were inoculated into 10 mL of media (1–8) to define aromatic amino acid requirement for growth, and the cell density (OD_{600}) was measured after 48 h at 30°C. The media contain YNB + ammonium sulfate + 2% glucose in 100 m*M* sodium phosphate buffer (pH 6.8), supplemented with aromatic amino acids (150 mg/L) as indicated above the bar graphs. Tryptophan and phenylalanine are growth-inhibitory for the *TYR1* strain *(20)*.

2. Inoculate 2 mL of fresh media (phosphate-buffered SD + 2% glucose + nutritional supplements) in 14-mL Falcon tubes with an aliquot of the starved culture to give a uniform cell density ($OD_{600} \cong 0.05$)
3. Shake at 30°C for 24 to 48 h.
4. Measure the final cell density (OD_{600}). Cultures that are not growth-limited will approach saturation density at this point, typically reaching $OD_{600} > 10$ (*see* **Fig. 2**).

3.4. Determining Minimal Amino Acid Requirement

1. Inoculate 10 mL of phosphate-buffered SD + 2% glucose without the required amino acid in a 125-mL baffled flask at moderate cell density and incubate overnight to produce an amino acid-limited starved culture (*see* **Note 7**).

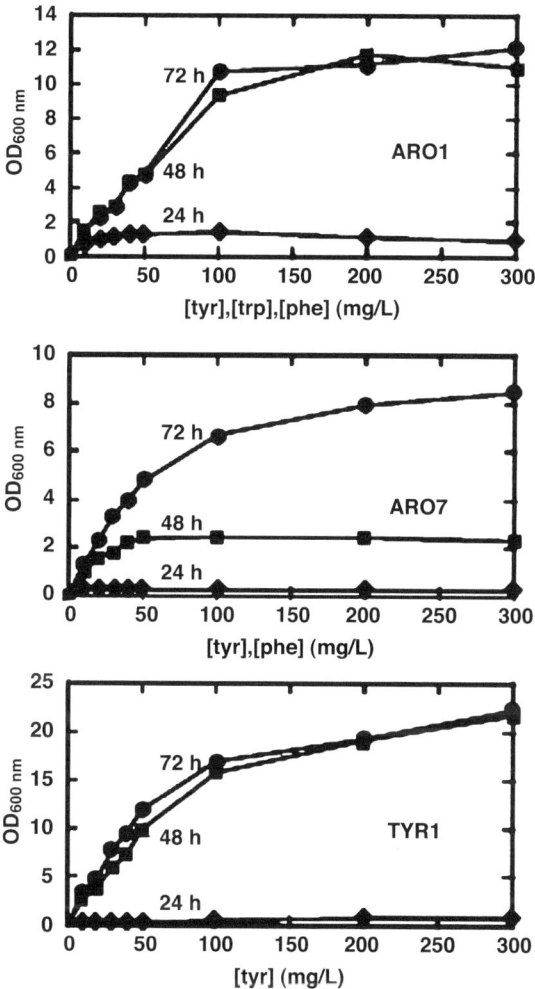

Fig. 3. Growth curves observed for *Pichia pastoris* aromatic amino acid auxotroph strains. Auxotrophic strains of *P. pastoris* were starved and inoculated into selective media containing varying amounts of the required aromatic amino acids, and the cell density (OD_{600}) was monitored after 24, 48, and 72 h at 30°C *(20)*.

2. Inoculate 10 mL of fresh media (phosphate-buffered SD + 2% glucose + nutritional requirements) in 125-mL baffled flasks with an aliquot of the starved culture to give a low, uniform cell density ($OD_{600} \cong 0.05$) (*see* **Note 8**).

3. Shake at 30°C, measuring the cell density (OD_{600}) at 24, 48, and 72 h. The resulting dose-response curves (*see* **Fig. 3**) may be used to determine the minimum amount of labeled amino acid required to avoid growth limitation.

3.5. Transformation

The procedure for preparing expression constructs using auxotrophic strains of *Pichia pastoris* closely parallels standard methods used for the wild type strain *(22,23)*.

1. Prepare electrocompetent cells from 500 mL of culture grown to $OD_{600} \cong 1$ in phosphate-buffered SD medium + 2% glucose + required amino acids (*see* **Note 9**).
2. Transform the cells with 4 µg of *Pme* I-linearized pPICZαGAOX expression vector in sterile deionized water by electroporation (0.2 cm path cell, 2500 V).
3. Immediately add 1 mL cold 1 *M* sorbitol and transfer to 14-mL Falcon tubes. Incubate at 30°C with shaking for 1 h.
4. Add 1 mL SD medium containing 1 *M* sorbitol and continue shaking for 1 h.
5. Plate an aliquot (100 µL) of the transformation mixture onto freshly prepared SD + glucose + amino acid + Zeocin agar and spread uniformly with a cell spreader.
6. Incubate at 30°C for at least 2 d to allow colonies to emerge. Pick the largest (hyper-resistant) colonies to fresh agar plates to use as stock clones for expression screening.

3.6. Recombinant Protein Production in Shake Flasks

1. Inoculate 20 mL of phosphate-buffered SD + 2% glucose + amino acid supplement (e.g., 150 mg/L unlabeled D,L-tyrosine) in a 125-mL baffled flask using freshly grown expression strain culture. Grow overnight at 30°C with shaking (*see* **Note 10**).
2. Centrifuge in sterile Oak Ridge tubes. (We use a Sorvall RC-5B centrifuge with SA600 rotor spun for 10 min at 5000 rpm.)
3. Resuspend in 1 L 2X phosphate-buffered SD + 2% glucose supplemented with the labeled amino acid (e.g., 150 mg/L D,L-d_7-tyrosine) + 1 mL PTM4 + 2 mL 0.02% biotin in a 2 L baffled flask (*see* **Notes 10** and **11**).
4. Grow at 30°C with shaking for approx 48 h, measuring OD_{600} to monitor cell density.
5. When the culture approaches stationary phase, centrifuge in sterile centrifuge bottles and resuspend the cells in 200 mL 2X phosphate-buffered SD + PTM4 trace metals solution + 2 mL 0.02% biotin + 20 µ*M* $CuSO_4$ (for galactose oxidase expression) + labeled amino acid (e.g., 150 mg/L D,L-d_7-tyrosine) in a 2-L baffled flask. Induce with 0.05% methanol (*see* **Note 12**).
6. Grow for 4 d at 30°C with shaking. Supplement every 24 h by adding sterile methanol to 0.05% of culture volume (e.g., 1 mL/200 mL).

3.7. Quantitative Characterization of the Labeled Protein

The characterization of the labeled protein can use any technique that is sensitive to the presence of the isotope in the sample. Mass spectrometry will be most widely applicable, taking advantage of the mass perturbation of the isotope. The extent of labeling may be accurately determined by mass spectrometry following proteolytic digestion, using high-resolution mass analysis to determine

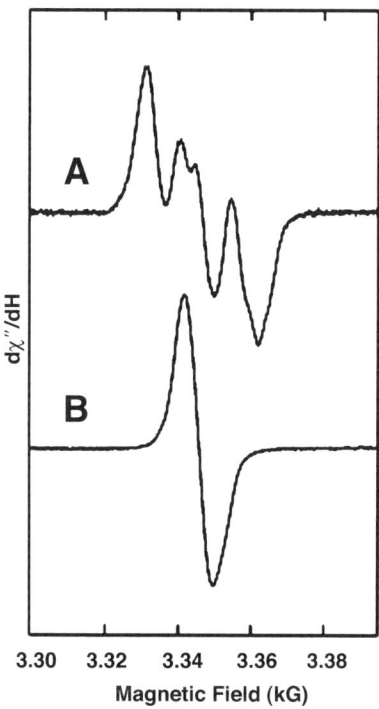

Fig. 4. EPR Spectra for Labeled Galactose Oxidase. Galactose oxidase apoprotein (15 mg/mL in 50 mM KHPO$_4$ [pH 7.0]) was treated with potassium ferricyanide and desalted. Protein containing (**A**) unlabeled L-tyrosine, (**B**) d_7-L- tyrosine. Spectra were recorded at 100 K using a Bruker E500 EPR spectrometer. Instrumental parameters: frequency, 9.4 GHz; power, 0.2 mW, modulation amplitude, 1 G; center field, 3345 G; sweep width, 100 G. Deconvolution of spectrum (**B**) and double integration of the spectra indicates that <1% of unlabeled tyrosine is present in (**B**).

the proportions of labeled and unlabeled peptide fragments containing the required amino acid.

NMR spectroscopy may also be used to quantitate the labeling of the protein, particularly when the isotope incorporation is aimed at NMR applications. Other approaches may also be useful. For example, galactose oxidase forms a stable protein-free radical on a modified tyrosine side chain *(22)*, allowing EPR spectroscopy to be used to determine the extent of incorporation of labeled tyrosine (*see* **Fig. 4**).

Whereas all three aromatic auxotroph strains efficiently incorporate isotopic label (**Table 1**), the yield of protein differs markedly among these strains (*see* **Note 13**). The highest yield of recombinant protein is found with the single auxotroph *TYR1* strain. In general, greater productivity in recombinant protein production may be expected from a more robust strain.

Table 1
**Protein Expression and Labeling Efficiency for *Pichia pastoris*
Aromatic Auxotroph Strains**

Strain	Protein yield (mg/L)[a]	Labeling efficiency (%)[b]
ARO1	7	>99
ARO7	50	>99
TYR1	120	>99

[a]Amount of galactose oxidase induced by 0.5% methanol for 4 d and purified from culture medium by ion exchange chromatography as previously described *(17)*; per liter of expression medium.

[b]Determined by quantitative analysis of the EPR spectrum of the tyrosine free radical in oxidized apo-galactose oxidase prepared from cultures supplemented with β,β-d_2-tyrosine.

4. Notes

1. *Pichia* aromatic auxotrophs strains are unable to grow on rich yeast medium (YPD).
2. A stock solution of $1M$ sodium phosphate (pH 6.7) may be sterilized by autoclaving and added to the other components after sterilization. Media containing both phosphate buffer and the carbon source (glycerol or glucose) should not be autoclaved, but may be filter-sterilized.
3. Growth curves should be performed using amino acids having the same enantiomeric composition (L- or D,L-) as that used for expression labeling, even though unlabeled amino acid is used in these experiments.
4. Baffled flasks with straight-neck cap closures are preferred because they permit efficient gas exchange and aeration of the culture.
5. Methanol is conveniently filtered using Millipore Steriflip GP units with polyether sulfone (PES) membranes.
6. The formation of reactive metabolic intermediates in the strains containing broken biosynthetic pathways may result in rapid loss of viability as a result of aging or photochemical sensitivity. All three aromatic auxotroph strains turn red under room light, most likely as a result of photochemical transformations of unutilized aromatic pathway intermediates. Cultures should therefore be freshly prepared and extended exposure to room light should be avoided.
7. Starving the culture by extending growth in defined medium containing glycerol or glucose (but without the required amino acid supplement) depletes endogenous pools of unlabeled amino acid. The cell density will typically double in this phase.
8. D,L-mixtures may also be used. In some cases, yeast may be able to utilize both isomers as a result of metabolic interconversion by racemases or amino transferases *(22)*, permitting the (usually less expensive) racemic mixture of labeled amino acid to be used for isotopic labeling. Note that both the α-hydrogen and α-amino group would be expected to exchange in this process. In other cases, however, only the L-amino acid may be utilized by the cells, and the D-isomer may even be toxic. For example, D-alanine is toxic to *Saccharomyces cerevisiae* *(30)*.

9. The doubling time of auxotrophic cultures is significantly longer than that typical of the wild-type strain, even when supplemented with the required amino acids at optimal levels. For example, the doubling time for the *P. pastoris* tyrosine auxotroph (grown in phosphate-buffered SD + 2% glucose + 150 mg/L L-tyrosine) is approx 4.4 h, compared with 1.5 h for the wild-type strain grown in YPD.

10. The amount of required amino acid needed for optimal growth in high-density fermentation conditions *(31)* is generally much greater than that required for shake cultures. For example, *P. pastoris HIS4* strain (GS115) requires supplementation with 10 g histidine/L/d to reach optimal growth during high-density fermentation, whereas an initial 40 mg/L histidine supplementation is sufficient for shake culture.

11. As an extension of these methods, the combination of uniform labeling techniques for deuterium incorporation *(9–13)* together with residue-specific labeling with protonated amino acids could be used to introduce specific protonation in a deuterium background.

12. Richer medium may be required to stabilize some secretory proteins. It is possible to use standard yeast drop-out medium (containing all except the required amino acid) supplemented with the labeled required amino acid and PTM4 trace metals solution + biotin in place of 2X SD medium during the induction phase.

13. The labeling efficiency is much lower with wild-type *P. pastoris*. Supplementation of the wild-type expression strain with labeled tyrosine during the methanol induction phase results in approx 30% label incorporation into galactose oxidase. When labeled tyrosine is present in both growth and induction phases, up to 70% incorporation may be achieved with the wild type expression host.

Acknowledgments

The author thanks Dr. J.M. Cregg (Keck Graduate Institute, Claremont, CA) for providing partial sequences of *P. pastoris ARO1*, *ARO7* and *TYR1* genes. This work was supported by the National Institutes of Health (GM 46749).

References

1. Kennedy, M. A., Montelione, G. T., Arrowsmith, C. H., and Markley, J. L. (2002) Role for NMR in structural genomics. *J. Struct. Funct. Genomics* **2**, 155–169.

2. Montelione, G. T., Zheng, D., Huang, Y. J., Gunsalus, K. C., and Szyperski, T. (2000) Protein NMR spectroscopy in structural genomics. *Nat. Struct. Biol.* **7**, 982–985.

3. Veenstra, T. D., Martinovic, S., Anderson, G. A., Pasa-Tolic, L., and Smith, R. D. (2000) Proteome analysis using selective incorporation of isotopically labeled amino acids. *J. Am. Soc. Mass Spectrom.* **11**, 78–82.

4. Gardner, K. H. and Kay, L. E. (1998) The use of ^{2}H, ^{13}C, ^{15}N multidimensional NMR to study the structures and dynamics of proteins. *Annu. Rev. Biophys. Biomol. Struct.* **27**, 357–406.

5. Kushner, D. J., Baker, A., and Dunstall, T. G. (1999) Pharmacological uses and perspectives of heavy water and deuterated compounds. *Can. J. Physiol. Pharmacol.* **77**, 79–88.

6. Cereghino, G. P., Cereghino, J. L., Ilgen, C., and Cregg, J. M. (2002) Production of recombinant proteins in fermenter cultures of the yeast *Pichia pastoris. Curr Opin Biotechnol.* **13,** 329–332.

7. Cereghino, J. L. and Cregg, J. M. (2000) Heterologous protein expression in the methylotrophic yeast *Pichia pastoris. FEMS Microbiol. Rev.* **24,** 45–66.

8. Wood, M. J. and Komives, E. A. (1999) Production of large quantities of isotopically labeled protein in *Pichia pastoris* by fermentation. *J. Biomolecular NMR* **13,** 149–159.

9. de Lamotte, F., Boze, H., Blanchard, C., et al. (2001) NMR monitoring of accumulation and folding of ^{15}N-labeled protein overexpressed in *Pichia pastoris. Prot. Expr. Purif.* **22,** 318–324.

10. Tomida, M., Kimura, M., Kuwata, K., Tomoya, H., Okano, Y., and Era, S. (2003) Development of a high-level expression system for deuterium-labeled human serum albumin. *Jap. J. Physiol.* **53,** 65–69.

11. Rodriguez, E. and Krishna, N. K. (2001) An economical method for ^{15}N/^{13}C isotopic labeling of proteins expressed in *Pichia pastoris. J. Biochem. (Tokyo)* **130,** 19–22.

12. Solà, A., Maaheimo, H., Ylönen, K., Ferrer, P., and Szyperski, T. (2004) Amino acid biosynthesis and metabolic flux profiling of *Pichia pastoris. Eur. J. Biochem.* **271,** 2462–2470.

13. Pickford, A. R. and O'Leary, J. M. (2004) Isotopic labeling of recombinant proteins from the methylotrophic yeast *Pichia pastoris. Methods Mol. Biol.* **278,** 17–33.

14. Replogle, K., Hovland, L., and Rivier, D. H. (1999) Designer deletion and prototrophic strains derived from *Saccharomyces cerevisiae* W303-1a. *Yeast* **15,** 1141–1149.

15. von Borstel, R. C., Savage, E. A., Wang, Q., et al. (1998) Topical reversion at the *HIS1* locus of *Saccharomyces cerevisiae*. A tale of three mutants. *Genetics* **148,** 1647–1654.

16. Wloch, D. M., Szafraniec, K., Borts, R. H., and Korona, R. (2001) Direct estimate of the mutation rate and the distribution of fitness effects in the yeast *Saccharomyces cerevisiae. Genetics* **159,** 441–452.

17. Crane, D. I. and Gould, S. J. (1994) The *Pichia pastoris HIS4* gene: nucleotide sequence, creation of a non-reverting *his4* deletion mutant, and development of *HIS4*-based replicating and integrating plasmids. *Curr. Genet.* **26,** 443–450.

18. Lin Cereghino, G. P., Lin Cereghino, J., Sunga, A. J., et al. (2001) New selectable marker/auxotrophic host strain combinations for molecular genetic manipulation of *Pichia pastoris. Gene* **263,** 159–169.

19. Li, D.-Y., Ji, X.-S., Yu, J., Chen, M.-J., and Yuan, Z.-Y. (2001) PCR based cloning and sequence analysis of the *Pichia pastoris* cystathionine β-synthase gene. *ACTA Biochim. Biophys. Sinica* **33,** 600–606.

20. Whittaker, M. M. and Whittaker, J. W. (2005) Construction and characterization of *Pichia pastoris* strains for labeling aromatic amino acids in recombinant proteins. *Prot. Expr. Purif.* **41,** 266–274.

21. Whittaker, J. W. (2002) Galactose oxidase. *Adv. Protein Chem.* **60,** 1–49.
22. Whittaker, M. M. and Whittaker, J. W. (2000) Expression of recombinant galactose oxidase. *Protein Expr. Purif.* **20,** 105–111.
23. Braus, G. H. (1991) Aromatic amino acid biosynthesis in the yeast *Saccharomyces cerevisiae*: a model system for the regulation of a eukaryotic biosynthetic pathway, *Microbiol. Revs.* **55,** 349–370.
24. Jensen, R. and Fischer, R. (1987) The postprephenate biochemical pathways to phenylalanine and tyrosine: an overview. *Meth. Enzymol.* **142,** 472–478.
25. Sousa, S., McLaughlin, M. M., Pereira, S. A., et al. (2002) The *ARO4* gene of *Candida albicans* encodes a tyrosine-sensitive DAHP synthase: evolution, functional conservation and phenotype of Aro3p-, Aro4p-deficient mutants. *Microbiology* **148,** 1291–1303.
26. Krappmann, S., Pries, R., Gellissen, G., Hiller, M., and Braus, G. H. (2000) *HARO7* encodes chorismate mutase of the methyltrophic yeast *Hansenula polymorpha* and is derepressed upon methanol utilization. *J. Bact.* **182,** 4188–4197.
27. Whittaker, M. M. and Whittaker, J. W. (1990) A tyrosine-derived free radical in apogalactose oxidase. *J. Biol. Chem.* **265,** 9610–9613.
28. Rose, M. D., Winston, F., and Hieter, P. (1990) *Methods in yeast genetics. A laboratory course manual*, *1st edition*. Cold Spring Harbor Press, Cold Spring Harbor, NY.
29. Cregg, J. M. and Russell, K. A. (1998) Transformation. *Methods Molec. Biol.* **103,** 27–39.
30. Uo, T., Yoshimura, T., Tanaka, N., Takegawa, K., and Esaki, N. (2001) Functional characterization of alanine racemase from *Schizosaccharomyces pombe*: a eucaryotic counterpart to bacterial alanine racemase. *J. Bacteriol.* **183,** 2226–2233.
31. Stratton, J., Chiruvolu, V., and Meagher, M. (1998) High cell-density fermentation, *Methods Mol. Biol.* **103,** 107–120.

14

Classical Genetics

Ilya Tolstorukov and James M. Cregg

Abstract

A significant advantage of *Pichia pastoris* as an experimental system is the ability to readily bring to bear both classical and molecular genetic approaches to a research problem. Although the advent of yeast molecular genetics has introduced new and exciting capabilities, classical genetics remains the approach of choice in many instances. These include the generation of mutations in previously unidentified genes (mutagenesis), the removal of unwanted secondary mutations (backcrossing), the assignment of mutations to specific genes (complementation analysis), and the construction of strains with new combinations of mutant alleles. This chapter describes these genetic manipulation methods for *P. pastoris*. In addition, certain yeast genes are essential for survival of the organism. However, determining whether a newly cloned gene is essential or not can be difficult with *P. pastoris*. In this chapter, we also describe a series of experiments to investigate the potential essential nature of a cloned gene in this yeast.

Key Words: *Pichia pastoris*; molecular and classical genetics; backcrossing; complementation analysis; strain construction.

1. Introduction

To comprehend the genetic strategies employed with *P. pastoris*, it is first necessary to understand basic features of the life cycle of this yeast *(1,2)*. *P. pastoris* is an ascomycetous budding yeast that most commonly exists in a vegetative haploid state (*see* **Fig. 1**). Upon nitrogen limitation, mating occurs and diploid cells are formed. Because cells of the same strain can readily mate with each other, *P. pastoris* is by definition homothallic. Of homothallic yeasts analyzed to date, all have been found to operate using a mating type switching mechanism. Although it has not been identified in *P. pastoris*, it is likely as the mechanism has been demonstrated in the related methylotrophic yeast *Pichia methanolica* (a.k.a. *P. pinus*) *(3,4)*. After mating, the resulting diploid products

From: *Methods in Molecular Biology, vol. 389:* Pichia *Protocols, Second Edition*
Edited by: J. M. Cregg © Humana Press Inc., Totowa, NJ

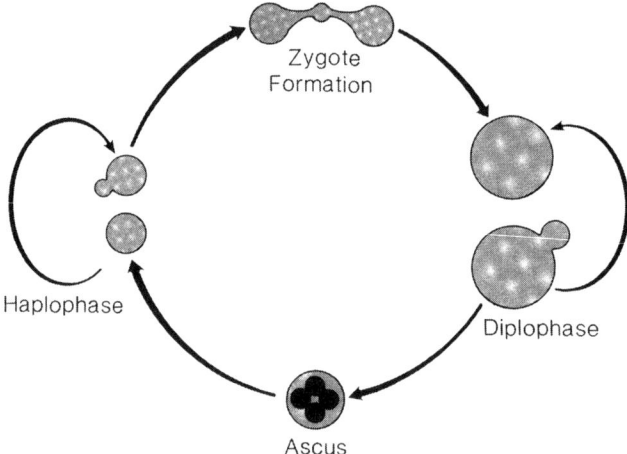

Fig. 1. Diagram of the *P. pastoris* life cycle. Reprinted with permission from **ref. *14***.

can be maintained in that state by shifting them to a standard vegetative growth medium or allowed to proceed through meiosis and to the production of asci which typically contain four haploid spores.

The key feature of the *P. pastoris* life cycle that permits genetic manipulation is its physiological regulation of mating. *P. pastoris* is most stable in its vegetative haploid state, a great advantage in the isolation and phenotypic characterization of mutants. In wild-type homothallic strains of *Saccharomyces cerevisiae*, the reverse is true: haploid cells are unstable and rapidly mate to form diploids *(5)*. To cross *P. pastoris*, selected pairs of complementarily marked parental strains are mixed and subjected to nitrogen limitation for a time period sufficient to initiate mating. The strains are then shifted to a nonlimiting medium supplemented with a combination of nutrients that select for growth of hybrid diploid strains and against the growth of the haploid parental strains and self-mated diploid strains. To initiate meiosis and sporulation, diploid strains are simply returned to a nitrogen-limited medium.

2. Materials

2.1. Strains

All *P. pastoris* strains are derivatives of the wild-type strain NRRL-11430 (Northern Regional Research Laboratories, Peoria, IL). Auxotrophically marked strains are convenient for selection of diploid strains; a representative selection of such strains is listed in **Table 1**. The identity of the biosynthetic genes affected in these strains is known for several of the mutant groups: *his4*, histidinol dehydrogenase; *arg4*, argininosuccinate lyase; *ade1*, PR-aminoimidazolesuccinocarboxamide

Table 1
Auxotrophic Mutants of *P. pastoris*

Representative strain	Genotype
JC233	*his1*
JC234	*his2*
GS115	*his4*
JC247	*arg1*
JC248	*arg2*
GS190	*arg4*
JC235	*lys1 his4*
JC236	*lys2 his4*
JC237	*lys3*
JC251	*pro1*
JC252	*pro2*
JC239	*met1*
JC240	*met2**
JC241	*met3**
JC242	*met4**
JC220	*ade1*
JC221	*ade2*
JC222	*ade3*
JC223	*ade4*
JC224	*ade5*
JC225	*ade6*
JC226	*ade7 his4*
JC254	*ura3*
JC225	*ura5*

*Mutants in these groups will grow when supplemented with either methionine or cysteine.

synthase; *ura3*, orotidine-5′-phosphate decarboxylase *(6,7)*; *ura5*, orotate phosphoribosyltransferase *(8)*; and *met2*, homoserine-*O*-transacetylase *(9)*. Typically, *P. pastoris* cultures are grown at 30°C with shaking at ~200 rpm.

2.2. Media and Solutions

1. YPD medium: 1% *(w/v)* yeast extract, 2% *(w/v)* peptone, and 0.4% *(w/v)* glucose (dextrose).
2. YNB medium: 0.67% *(w/v)* yeast nitrogen base without amino acids or ammonium sulfate, 0.5% *(w/v)* ammonium sulfate, and supplemented with either 0.4% glucose (YNB-glucose) or 0.5% methanol (YNB-methanol). Amino acids and nucleotides are added to 50 μg/mL as required for auxotrophic strains.

3. 5-FOA medium (for uracil-requiring mutants): YNB-glucose medium supplemented with 50 µg/mL uracil and 750 µg/mL 5-fluoroorotic acid (PCR, Inc., Gainesville, FL).
4. For plates, agar is added to 2% *(w/v)* final concentration.
5. 100% glycerol.
6. Diethyl ether.

2.3. Reagents for Mutagenesis

1. 1 mL of a 10-mg/mL solution of *N*-methyl-*N'*-nitro-*N*-nitrosoguanidine (NTG) (Sigma Chemical, St. Louis, MO) in acetone, stored frozen at −20°C (*see* **Note 1**).
2. 1 L Mutagenesis buffer: 50 mM KPO4 (pH 7.0).
3. 2 L 10% Na thiosulfate.

3. Methods
3.1. Long-Term Strain Storage

Viable *P. pastoris* strains are readily stored frozen for long periods (>20 yr). For each strain to be stored, pick a single fresh colony from a plate containing a selective medium and inoculate the colony into a sterile tube containing 2 mL of YPD medium. After overnight incubation with shaking, transfer 1.2 mL of the culture to a sterile 2.0-mL cryovial containing 0.6 mL of 100% glycerol (~30% final concentration) (*see* **Note 2**). Mix the culture and glycerol thoroughly and freeze at −70°C.

To resurrect a stored strain, remove the cryovial from the freezer, immediately plunge a hot sterile inoculation loop (or a sterile toothpick) into the frozen culture, transfer a small portion of the culture to an agar plate containing an appropriate medium, and immediately return the culture to the freezer (*see* **Note 3**).

3.2. NTG Mutagenesis

1. Inoculate the strain to be mutagenized into a 10-mL preculture of YPD and incubate at 30°C with shaking overnight (*see* **Note 4**).
2. On the next morning, dilute the preculture with fresh YPD and maintain it in logarithmic growth phase (OD_{600} < 1.0) throughout the day.
3. In late afternoon, use a portion of the preculture to inoculate a 500-mL culture of YPD medium in a baffled Fernbach culture flask (or alternatively two 250-mL cultures in 1-L baffled culture flasks) to an OD_{600} of approx 0.005 and incubate with shaking overnight (*see* **Note 5**).
4. On the following morning, harvest the culture at an OD_{600} of approx 1.0 by centrifugation at 5000g at 4°C and suspend the culture in cold sterile mutagenesis buffer to obtain final OD_{600} ≈ 1.
5. Transfer the suspension into four 250-mL sterile plastic centrifuge bottles.
6. Wash the cells once more by centrifugation and resuspend in the buffer to obtain a final OD_{600} ≈ 2.

7. Add aliquots of 100 µL, 200 µL, and 400 µL of the NTG solution to each of three cultures and hold the fourth as an untreated control. Incubate the cultures for 1 hour at 30°C and stop the mutagenesis by adding 50 mL of a 10% Na thiosulfate solution to each culture.

8. Wash each mutagenized culture by centrifugation once with 100 mL of mutagenesis buffer, once with 100 mL of YPD medium, then resuspend each in 150 mL of YPD medium.

9. Remove 100 µL samples of each culture, prepare 100- and 10,000-fold serial dilutions of each and spread 100 µL aliquots of the dilutions on YPD plates to determine the percentage of cells that have survived mutagenesis in each culture. Optimal survival rates are between 2 and 20% of the untreated control culture.

10. Transfer each mutagenized culture to a 500-mL shake flask and allowed to recover for 4 hours at 30°C with shaking.

11. Concentrate the final cultures by centrifugation, and resuspended in 15 mL of YPD medium containing 30% glycerol.

12. Place aliquots of 0.5 mL of NTG-treated samples into sterile microcentrifuge tubes or cryovials and stored frozen at –70°C for future use.

13. To estimate viability of the mutagenized cells, thaw a tube of cells, serially dilute in sterile water, and spread on non-selective medium such as YPD or YNB-glucose plates.

14. To screen for auxotrophic mutants, replica plating onto sets of plates containing appropriate diagnostic media. Use the estimate from **step 13** to generate plates with approx 100 to 500 colonies formed by single cells.

3.3. Selection for Uracil Auxotrophs Using 5-Fluoroorotic Acid

5-Fluoroorotic acid (5-FOA), a uracil biosynthetic pathway analogue, is metabolized to yield a toxic compound by certain enzymes in the pathway *(6)*. As a result, organisms that are prototrophic for uracil synthesis (Ura+) are sensitive to 5-FOA whereas certain Ura⁻ auxotrophs cannot metabolize the drug and thus, are resistant to it. Selection for 5-FOA resistant strains of *P. pastoris* is a highly effective means of isolating Ura⁻ mutants affected in either of two Ura pathway genes. One of these genes, *URA3*, encodes orotidine-5'-phosphate decarboxylase. The other is the homolog of the *S. cerevisiae* orotidine-5'-phosphate pyrophosphorylase gene (*URA5*) as *ura5* mutants represent the other complementation group selected by 5-FOA in this yeast.

1. To select for Ura⁻ strains of *P. pastoris*, spread approx 2 OD$_{600}$ units (~5 × 10^7 cells) on a 5-FOA plate. Resistant colonies will appear after approx 1 wk at 30°C.

2. Test the 5-FOA-resistant colonies for Ura phenotype by streaking them onto each of two YNB-glucose plates, one with and one without uracil. The highest frequency of Ura⁻ mutants are found in mutagenized cultures like those described above. However, Ura⁻ stains often exist at a low but significant frequency within unmutagenized cell populations as well and can be readily selected by simply suspending cells from an YPD plate in sterile water and spreading them onto a 5-FOA plate.

3. If it is necessary to determine which *URA* gene is defective in new Ura⁻ strains, the strains can either be subjected to complementation testing against the known *ura* mutants (**Table 1**) or transformed with a vector containing the *P. pastoris* *URA3* gene.

3.4. Standard Genetic Crossing Procedure

3.4.1. Mating and Selection of Diploids

The mating and selection of diploid strains constitutes the core of complementation analysis and is the first step in strain construction and backcrossing. Because *P. pastoris* is functionally homothallic, the mating type of strains is not a consideration in planning a genetic cross. However, because cells of the same strain will also mate and the mating efficiency between *P. pastoris* cells is low, it is essential that strains to be crossed contain complementary markers that allow for the selective growth of crossed diploids and against the growth of self-mated diploids and parental strains. Auxotrophic markers are generally most convenient for this purpose, but mutations in any gene that affect the growth phenotype of *P. pastoris* such as genes required for utilization of a specific carbon source (e.g., methanol or ethanol) or nitrogen source (e.g., methylamine) can be used as well.

1. To begin a mating experiment, select a fresh colony (no more than 1-wk old) of each strain to be mated from a YPD plate using an inoculation loop and streak across the length of each of two YPD plates (*see* **Fig. 2A**).
2. After overnight incubation, transfer the cell streaks from both plates onto a sterile velvet such that the streaks from one plate are perpendicular to those on the other.
3. Transfer the cross streaks from the velvet to a mating medium plate and incubate for 2 to 3 d at room temperature to initiate mating.
4. After incubation, replica plate to an appropriate agar medium for the selection of complementing diploid cells. Diploid colonies will arise at the junctions of the streaks after approx 3 d of incubation at 30°C (*see* **Fig. 2B**). Diploid cells of *P. pastoris* are approx twice as large as haploid cells and are easily distinguished by examination under a light microscope by size or by their high efficiency of sporulation (*see* **Subheading 3.4.2**).
5. Colony-purify diploid strains by streaking at least once for single colonies on diploid selection medium (*see* **Note 6**).

3.4.2. Sporulation and Spore Analysis

Diploid *P. pastoris* strains efficiently undergo meiosis and sporulation in response to nitrogen limitation.

1. To initiate sporulation, transfer freshly grown diploid colonies from a YPD or YNB-glucose plate to a mating (sporulation) plate either by replica plating or with an inoculation loop. Incubate the plate for 3 to 4 d at room temperature. The sporulation in all *Pichia* species correlates with accumulation of a red pigment in

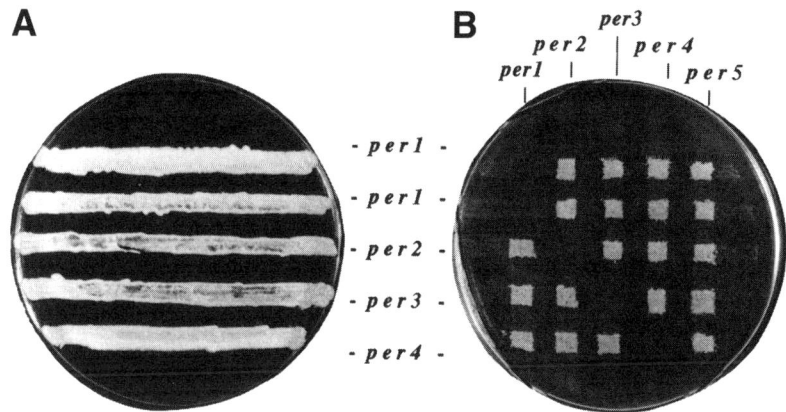

Fig. 2. Complementation analysis plates. On the left is a YPD medium plate in which five methanol-utilization-defective strains have been streaked in preparation for complementation testing. On the right is a YNB methanol medium plate on which complementing diploid strains have selectively grown.

the ascus. Sporulated diploid samples are easily distinguished by their tan color relative to the white color of haploid cultures or by a high number of asci in the cell culture as observed by normal or phase-contrast microscopy. Because *P. pastoris* spores are small and adhere to one another, tetrad dissection via micromanipulation is difficult. Therefore, spore products are analyzed using a Random Spore Analysis (RSA) procedure.

2. For RSA, transfer an inoculation loop full of sporulated material to a 1.5-mL microcentrifuge tube containing 0.7 mL of sterile water and vortex the mixture.

3. In a fume hood, add 0.7 mL of diethyl ether to the spore preparation, and vortex thoroughly for approx 5 min at room temperature. The ether treatment selectively kills vegetative cells remaining in spore preparations.

4. In the hood, centrifuge the cells for 2 min, remove the ether (the upper phase) and resuspend the pellet in the remaining water. Serially dilute samples of the preparation (the bottom aqueous phase), and spread 10 and 100 µL of the suspension onto a nonselective medium (e.g., YPD).

5. After incubation, replica plate from the plates that contain in the range of 100 to 1000 colonies onto a set of master YPD plates and than replicate each onto a series of plates containing suitable diagnostic media. For example, to analyze the spore products resulting from a cross of GS115 (*his4*) and GS190 (*arg4*), appropriate diagnostic media would be YNB-glucose supplemented with:

a. No amino acids.
b. Arginine.
c. Histidine.
d. Arginine and histidine.

PPF1 X KM7121 MATING PRODUCTS
NON-PARENTAL
SPORE PRODUCTS

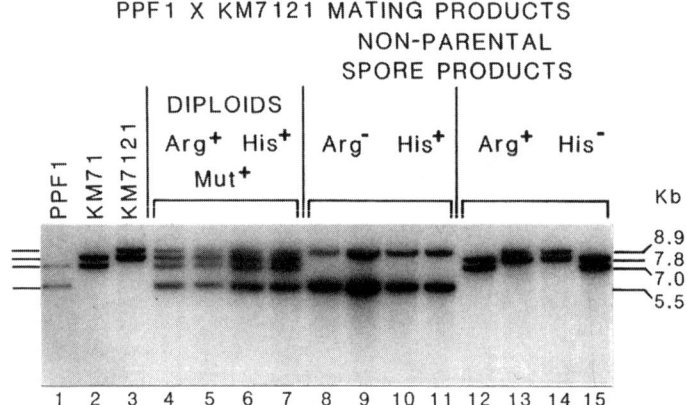

Fig. 3. Southern blot of selected strains resulting from the cross of KM7121 and PPF1. EcoRI digested genomic DNA samples from the following strains are shown: *lane 1:* PPF1 (*arg4 his4 AOX1 AOX2*); *lane 2:* KM7121 (*arg4 his4 aox1Δ::SARG4 aox2Δ::PHIS4*); *lane 3:* MC100 (diploid); *lane 4:* MC100-1 (*arg4 his4 AOX1 aox2Δ::PHIS4*); *lane 5:* MC100-2 (*arg4 his4 aox1Δ::SARG4 AOX2*); *lane 6:* MC100-3 (*arg4 his4 aox1Δ::SARG4 aox2Δ::Phis4*).

6. Compare or score the phenotype of individual colonies on each of the diagnostic plates and identify ones with the desired phenotype(s). This can also be done at the DNA level as shown in **Fig. 3**.

7. Select several colonies that appear to have the appropriate phenotype and streak for single colonies onto a nonselective medium plate and retest a single colony from each streak on the same set of diagnostic medium-containing plates. This step is important because *P. pastoris* spores adhere tightly to one another and colonies resulting from spore germination frequently contain cells derived from more than one spore. Another consequence of spore clumping is that markers appear not to segregate 1:1 but to be biased toward the dominant or wild-type phenotype. For example, in the GS115 (*his4*) × GS190 (*arg4*) cross described above, more His⁺Arg⁺ spore products will be apparent than the 25% expected in the population and His⁻Arg⁻ spore products will appear to be underrepresented.

3.5. Analysis of Essential Genes

Because the spore viability of *P. pastoris* cultures is usually not high enough to analyze the behavior of genetic markers by tetrad analysis, there is a need for alternative genetic approaches to investigate certain questions in this yeast. One set of questions that frequently arises is whether a cloned gene is essential for the survival of the organism (an essential gene) and whether that cloned essential gene is allelic with regard to a specific mutation in a gene.

Typically, the first hint an investigator receives that a cloned gene is essential is an inability to construct a gene replacement or knock-out strain in that gene despite repeated attempts with multiple gene replacement constructs.

3.5.1. Analysis of Diploid Strains With a Deleted Allele of an Essential Gene

More substantial evidence that the cloned gene in question is essential may be obtained by constructing a gene replacement allele in a diploid strain of *P. pastoris*, sporulating that strain and examining the spore products *(14)*. To clarify how this is done, an example from our group is described as follows.

A *P. pastoris* strain was isolated with a mutation that conferred an inability to grow on gluconeogenic carbon sources, including both methanol and ethanol. We termed the mutant gene in this strain *mxr2*, and the phenotype of the *mxr2* mutant as Mxr2. We then cloned a DNA fragment that complemented the *mxr2* mutation by transforming the strain with a genomic DNA library and asking for *P. pastoris* transformants that restored growth on ethanol and methanol (Mxr2⁺ phenotype). Sequencing of this DNA fragment revealed a number of ORFs, one of which potentially encoded a protein with strong similarity to the *S. cerevisiae GAL83* gene product (Gal83p).

We suspected that because the *P. pastoris* gene (*PpGAL83*) complemented the mutant *mxr2* strain, the chromosomal *MXR2* and cloned *PpGAL83* were the same gene; however, we did not have formal genetic evidence that this was true. It could be that the cloned *PpGAL83* gene complemented the *mxr2* mutation by suppression. Thus, we continued to refer to the mutant gene as *mxr2* and the cloned gene as *PpGAL83* until we had formal genetic evidence that they were equivalent. We proceeded to use the cloned *PpGAL83* fragment to construct gene replacement alleles of the gene. However, several constructs using either the *ARG4* or *ZEO*^R genes as markers failed to yield a gene replacement strain in haploid *P. pastoris*.

Next, we constructed a His⁻ Arg⁻ diploid strain from *ade1 his4 arg4* and *his4 arg4 met1* haploid strains and transformed the gene replacement fragments into this diploid strain. Transformants were selected by marker phenotype (Arg⁺ or Zeo^R), single colony purified once and genomic DNA from samples of several strains analyzed by polymerase chain reaction (PCR) to distinguish between random and targeted integration. For this purpose, specific sets of primers were used that could amplify only the disrupted chromosomal copy of *PpGAL83* (gene replacement) but neither the wild-type genomic DNA nor the transforming DNA. As positive control, primers specific for amplifying a portion of the nondisrupted *PpGAL83* gene were used.

Results suggested that several transformants were truly heterozygotic carrying both a wild-type and a disrupted *Pp-gal83* allele. Both types of diploid transformants

(with correct gene replacement and with nonhomologous integration) were sporulated and subjected to random spore analysis. It was expected that transformants
with randomly integrated fragments should produce substantial amounts (up to
25%) of spore progeny carrying the selective marker used for gene disruption
(either *ARG4* or *Zeo^R*). However, RSA found none of the spore products from the
potential disrupted diploids revealed a meiotic product with the selective marker;
this suggested that they each contained a disruption of the *PpGAL83* locus and
that this disruption led to a lethal phenotype (i.e., no progeny with an Arg$^+$ or
Zeo^R phenotype). However, we were not able to formally determine which ORF,
PpGAL83 or another ORF was responsible for the lethality.

3.5.2. Rescue of a Gene-Replacement Allele

To further demonstrate the essential nature of the *PpGAL83* locus, a gene
rescue approach was used. The concept of this approach is to introduce an additional copy of the *PpGAL83* DNA fragment into the *P. pastoris* genome at a site
unlinked to *PpGAL83* locus. This fragment rescues the lethal phenotype of the
gene-replacement *PpGAL83* allele at the normal *PpGAL83* chromosomal locus.
For easy segregation of the two *PpGAL83* gene copies, the second copy is integrated at an ectopic location in the genome (i.e., one not linked to the normal
PpGAL83 locus).

The steps in the construction and verification of this strain were as follows:

1. Make competent cells of a gene replacement diploid candidate obtained by transformation of the *ARG4* containing cassette and supposedly having *Pp-gal83::ARG4*
   disruption (designated D1). The phenotype of the diploid transformant D1 ws His$^-$
   and the genotype:

   ade1 his4 arg4 MET1 [gal83::ARG4]
   ADE1 his4 arg4 met1 GAL83

2. Digest the pYM8-*PpGAL83* vector (containing a *HIS4* marker gene) with an
   enzyme making a single cut outside of the inserted chromosomal fragment with
   the *PpGAL83* gene to increase the probability of integration of the vector into a
   site ectopic to the *PpGAL83* locus. Transform the linearized pYM8[*HIS4*]-
   PpGAL83 vector into competent cells of the D1 strain described above and select
   for His$^+$ transformants.

3. The transformants must be checked to be sure they are Mxr2$^+$ and analyzed by
   PCR for the presence of the two wild-type and disrupted copies of *PpGAL83* loci
   by using a primer set specific for each DNA sequence. The genotype of the
   selected His$^+$ transformants (designated D2) is:

   ade1 his4 arg4 MET1 [gal83::ARG4] pYM8[HIS4]-GAL83
   ADE1 his4 arg4 met1 GAL83 unknown GENE

4. To demonstrate that the integrated cassette of pYM8-*PpGAL83* has an ectopic
   (independent) location relative to the *PpGAL83* locus, and to obtain evidence that

the *gal83::ARG4* disruption is lethal, and can be rescued by the cloned *PpGAL83* fragment, the diploid D2 was subjected to a random spore analysis in which, unlike D1, the Arg$^+$ haploid segregants, which inherit both the *Ppgal83::ARG4* and pYM8-*PpGAL83* cassettes (designated as h2) survive.

3.5.3. Analysis of Allelic Relationship of an Essential Gene

Once haploid segregants (h2) have been isolated and their genotype [*Ppgal83::ARG4*] [pYM8-*PpGAL83*] verified by PCR, they can be used in further genetic analysis with an induced *mxr2* mutant in order to demonstrate that the *PpGAL83* and *MXR2* genes are the same. Two potential genotypes of *Ppgal83::ARG4* are possible:

1. If *MXR2* = *PpGAL83,* the genotype of a haploid segregant h2 will be: *his4* [*HIS4-PpGAL83*] *arg4 gal83::ARG4.*
2. If *MXR2* and *PpGAL83* are not the same gene, the genotype of the strain would be: *his4* [*HIS4-PpGAL83*] *arg4 gal83::ARG4 MXR2.*

The difference between these two genotypes is either the (1) absence or (2) presence of the *MXR2* gene at its native chromosomal locus. In order to analyze the *GAL83*-knockout strain, an auxotrophic segregant h2 (Ade$^-$, Met$^-$ or Ade$^-$ Met$^-$) was crossed to a *mxr2* mutant strain (e.g., *mxr2 his4 arg4*) to obtain an Mxr2$^+$ diploid D3 (h2 x *mxr2*). The meiotic segregation of the D3 diploid was compared with that obtained with the D2 diploid:

1. If the *MXR2* = *GAL83*:

   then D2-1: *ade1 his4 arg4 MET1* [*gal83::ARG4*] pYM8[*HIS4*]-*GAL83*
   ADE1 his4 arg4 met1 GAL83=MXR2 unknown GENE
   or D3-1: *ade1 his4 arg4 MET1* [*gal83::ARG4*] pYM8[*HIS4*]-*GAL83*
   ADE1 his4 arg4 met1 gal83=mxr2 unknown GENE

2. If *MXR2* and *GAL83* are different genes:

   then D2-2: *ade1 his4 arg4 MET1* [*gal83::ARG4*] pYM8[*HIS4*]-*GAL83 MXR2*
   ADE1 his4 arg4 met1 GAL83 unknown GENE MXR2
   or D3-2: *ade1 his4 arg4 MET1* [*gal83::ARG4*] pYM8[*HIS4*]-*GAL83 MXR2*
   ADE1 his4 arg4 met1 GAL83 unknown GENE mxr2

Comparing the genotypes, it can be seen that if the disruption of the *PpGAL83* locus is lethal, neither of the two diploid lines D2 nor D3 can produce Arg$^+$ His$^-$ meiotic segregants (*his4 arg4* [*gal83::ARG4*]), because the [*gal83::ARG4*] and [*HIS4-GAL83*] cassettes are the only genetic elements in the strains providing Arg$^+$ and His$^+$ phenotypes, respectively. It is clear that all Arg$^+$ His$^-$ spores must inherit the disrupted locus [*gal83::ARG4*] without the rescuing cassette [*HIS4-GAL83*]. Moreover, the segregation of D3-1 (*gal83* = *mxr2*) and D3-2 (*GAL83* and *MXR2* at different loci) will be different. D3-1 will be unable to produce either viable segregants His$^-$ Mxr$^+$ or Arg$^+$ Mxr$^-$, because the only wild-type *GAL83 (= MXR2)* gene present in the diploid must cosegregate with the *HIS4* marker, and the *mxr2* mutation cannot co-segregate with the *ARG4* gene. However, D3-2, which contains one allele of the wild-type *MXR2* gene, is able to form His⁻ Arg⁻ Mxr⁺ segregants that carry none of the transformed sequences (*his4 arg4 GAL83 MXR2*), but is still unable to produce Arg⁺ Mxr⁻ spores because the disrupted *gal83::ARG4* locus requires the presence of the rescuing cassette *HIS4-GAL83* which complements the Mxr⁻ phenotype.

Note that **Subheading 3.5.** is intended only as an example of a more complicated genetic selection, a reminder of how classical genetic selection techniques may be used when contemporary one are inadequate.

4. Notes

1. NTG is a powerful mutagen and a potent carcinogen. Therefore, great care should be exercised in handling this hazardous compound. Gloves, eye protection, and lab coat should be warn when working with NTG. The compound is most dangerous as a dry powder, and therefore a particle filter mask should be worn when weighing out the powder. All materials that come in contact with NTG should be soaked overnight in a 10% solution of Na thiosulfate prior to disposal or washing.
2. 100% glycerol stock is extremely viscous. One may instead use a stock solution of 50% (w/v) glycerol with water that has been sterile autoclaved, so long as the final concentration is ~30% glycerol for the frozen cells.
3. Freezing kills approx 90% of the cells. However, once frozen the remaining cells maintain their viability. Thus, it is critical not to allow the frozen culture to thaw because approx 90% of the remaining viable cells will be killed with each round of freezing.
4. This mutagenesis procedure is a modified version of that described by Gleeson and Sudbury *(10)* and Liu et al. *(2)*. Alternative methods such as with ethylmethane sulfonate or ultraviolet light may also be effective with *P. pastoris* and are described in Rose et al. *(11)*, Spencer and Spencer *(12)*, and Sherman *(13)*.
5. The starting density of this culture can be adjusted to compensate for changes in the length of the incubation period. Adjust the density assuming that *P. pastoris* has a generation time of between 90 and 120 min at 30°C in YPD medium.
6. *P. pastoris* diploid strains are unstable relative to haploid strains and will sporulate if subjected to the slightest stress (e.g., 1 or 2 wk on a YPD plate at room temperature). Thus, to maintain diploid strains, either transfer frequently to fresh selective plates or store frozen at −70°C. When working with these strains, check under the microscope frequently to be sure that strains are still diploid.

References

1. Cregg, J. M. (1987) Genetics of methylotrophic yeasts, in *Proceedings of the Fifth International Symposium on Microbial Growth on C1 Compounds*, (Duine, J. A. and Verseveld, H. W., eds.), Martinus Nijhoff Publishers, Dordrecht, The Netherlands, pp. 158–167.

2. Liu, H., Tan, X., Veenhuis, M., McCollum, D., and Cregg, J. M. (1992) An efficient screen for peroxisome-deficient mutants of *Pichia pastoris*. *J. Bacteriol.* **174,** 4943–4951.

3. Tolstorukov, I. I. and Benevolenskii, S. V. (1978) Study of the mechanism of mating and self-diploidization in haploid yeasts *Pichia pinus*: I. Bipolarity of mating. *Genetika* **14,** 519–526.

4. Tolstorukov, I. I. and Benevolenskii, S. V. (1980) Study of the mechanism of mating and self-diploidization in haploid yeasts *Pichia pinus*: II. Mutations in the mating type locus. *Genetika* **16,** 1335–1341.

5. Hicks, J. B. and Herskowitz, I. (1976) Interconversion of yeast mating types. I: Direct observation of the action of the homothalism (*HO*) gene. *Genetics* **83,** 245–258.

6. Boeke, J. D., Trueheart, J., Natsoulis, G., and Fink, G. R. (1987) 5-Fluoroorotic acid as a selective agent in yeast molecular genetics. *Methods Enzymol.* **154,** 164–175.

7. Krooth, R. S., Hsiao, W.-L., and Potvin, B. W. (1979) Resistance to 5-fluoroorotic acid and pyrimidine auxotrophy: a new bidirectional selective system for mammalian cells. *Som. Cell Genet.* **5,** 551–569.

8. Nett, J. H. and Gerngross, T. U. (2003) Cloning and disruption of the *PpURA5* gene and construction of a set of integration vectors for the stable genetic modification of *Pichia pastoris*. *Yeast* **20,** 1279–1290.

9. Thor, D., Xiong, S., Orazem, C. C., Kwan, A. C., Cregg, J. M., Lin-Cereghino, J., Lin-Cereghino, G. P. (2005) Cloning and characterization of the *Pichia pastoris MET2* gene as a selectable marker. *FEMS Yeast Res.* **5,** 935–942.

10. Gleeson, M. A. and Sudbery, P. E. (1988) Genetic analysis in the methylotrophic yeast *Hansenula polymorpha*. *Yeast* **4,** 293–303.

11. Rose, M. D., Winston, F., and Hieter, P. (1988) *Methods in Yeast Genetics Laboratory Manual*. Cold Spring Harbor Laboratory, Cold Spring Harbor, N. Y.

12. Spencer, J. F. T. and Spencer, D. M. (1988) Yeast genetics, in *Yeast: A Practical Approach* (Campbell, I. and Dufus, J. H. eds.) IRL Press, Oxford, England, pp. 65–106.

13. Sherman, F. (1991) Getting started with yeast. *Methods Enzymol.* **194,** 3–21.

14. Tolstorukov, I. I., Dutova, T. A., Benevolensky, S. V., and Soom, Ya. O. (1977) Hybridization and genetic analysis of methanol-utilizing yeast *Pichia pinus*. *Genetika* **13,** 322–329.

15

Identification of Pexophagy Genes by Restriction Enzyme-Mediated Integration

Laura A. Schroder, Benjamin S. Glick, and William A. Dunn, Jr.

Abstract

Pichia pastoris has proven to be a valuable model for examining the molecular events of the selective degradation of peroxisomes by a process called pexophagy. We have developed a protocol to rapidly identify genes essential for glucose-induced pexophagy. This method utilizes the random integration of a Zeocin resistance cassette vector into the genomic DNA thereby disrupting gene expression. Transformed yeast are selected by growth on Zeocin and pexophagy mutants identified by their inability to degrade the peroxisomal enzyme alcohol oxidase. The Zeocin vector along with flanking genomic DNA is then isolated from the mutants and the disrupted genes identified by sequencing. We have been able to isolate 59 mutants and identify 8 unique genes. The identification of 24 genes in *P. pastoris* and 7 genes in *H. polymorpha* have been reported using this approach which has been referred to as restriction-mediated integration.

Key Words: *Pichia pastoris*; random DNA integration; insertional mutagenesis; REMI; RALF; pexophagy; autophagy; peroxisomes; alcohol oxidase. (REMI) and random integration of linear DNA fragments (RALF).

1. Introduction

The methylotropic yeast, *Pichia pastoris*, can grow on methanol by synthesizing peroxisomal (e.g., alcohol oxidase and dihydroxyacetone synthase) and cytosolic (e.g., formate dehydrogenase) enzymes necessary to metabolize and assimilate this carbon source. Upon adapting these cells from growth on methanol to a medium containing glucose, the peroxisomes are rapidly and selectively degraded by a process called micropexophagy *(1–4)*. During glucose adaptation, clusters of peroxisomes are surrounded by arm-like extensions from the vacuole. Upon homotypic membrane fusion of the arms, the peroxisome cluster is incorporated into the vacuolar lumen where it is subsequently degraded.

From: *Methods in Molecular Biology, vol. 389:* Pichia *Protocols, Second Edition*
Edited by: J. M. Cregg © Humana Press Inc., Totowa, NJ

The highly regulated degradation of these large peroxisomes combined with classical yeast genetics makes *P. pastoris* an ideal model to characterize the molecular events of pexophagy. In order to identify those genes essential for pexophagy, we have developed a direct colony assay by which to screen thousands of nitrosoguanidine-generated mutants that were unable to degrade peroxisomal alcohol oxidase. Once the mutant was verified, we identified the mutated gene by complementation with a genomic library. This approach was quite time consuming and false positives were inevitable. Therefore, we set out to design a new approach to identify those genes essential for glucose-induced pexophagy.

We have utilized restriction enzyme-mediated integration (REMI) to facilitate a random disruption of genes in the *P. pastoris* genome by the incorporation of a Zeocin-resistance gene. Transformation of *P. pastoris* with a pREMI-Z vector (linearized with *Bam*HI) was done in the presence of *Bam*HI or *Dpn*II that would randomly cleave the genomic DNA leaving four base overhangs compatible with the linearized vector. Transformed cells were selected by growth on Zeocin plates. Those transformed cells unable to degrade peroxisomes during glucose adaptation were identified by direct colony assays and the interrupted genes (*GSA*, glucose-induced selective autophagy) sequenced. REMI mutagenesis provides a novel approach to identifying key genes involved in autophagic processes and straightforward assays estimate the impact of the interrupted gene on pexophagy.

2. Materials

2.1. Pichia pastoris *and* Escherichia coli *Strains*

1. GS115 (*his4*).
2. PPF1 (*his4 arg4*).
3. DH5α.

2.2. *Culture Media*

1. YPD: 1% Bacto yeast extract, 2% Bacto peptone, and 2% dextrose.
2. YND: 0.67% yeast nitrogen base, 0.4 mg/L biotin, and 2% glucose.
3. YNDH: 0.67% yeast nitrogen base, 0.4 mg/L biotin, 40 µg/L histidine, and 2% glucose.
4. YPD+Z: 1% Bacto yeast extract, 2% Bacto peptone, and 2% dextrose plus 100 µg/mL Zeocin.
5. YNM: 0.67% yeast nitrogen base, 0.4 mg/L biotin, and 0.5% methanol.
6. YNMH: 0.67% yeast nitrogen base, 0.4 mg/L biotin, 0.5% methanol, and 40 µg/L histidine.
7. LB+Z: 0.5% Bacto yeast extract, 1% Bacto tryptone, and 0.5% NaCl plus 25 µg/mL Zeocin.
8. YETM: 0.5% Bacto yeast extract, 1% Bacto tryptone, and 5% $MgSO_4 \cdot 7H_2O$. Adjust pH to 7.5 with KOH.

2.3. Vector

1. pREMI-Z (NCBI accession number AF282723).

2.4. Transformation of P. pastoris

1. YPD.
2. $1M$ NaHEPES (pH 8.0).
3. $1M$ Dithiothreitol (DTT).
4. $1M$ Sorbitol.

2.5. Qualitative Assessment of Alcohol Oxidase Degradation by Direct Colony Assay

1. P5 Filter Paper, 9-cm circle.
2. NitroBind, nitrocellulose 0.45 μm, 85-mm circle.
3. AOX Detection Solution: 3.4 U/mL horseradish peroxidase, 0.56 mg/mL 2,2′-azinobis (3- ethylbenzthazoline-6-sulfonic acid), 33 mM potassium phosphate buffer, (pH 7.5), and 0.13% methanol.
4. Liquid nitrogen.

2.6. Quantitative Assessment of AOX Degradation by Liquid Medium Assay

1. Cell Lysis Buffer: 20 mM Tris (pH 7.5), 50 mM NaCl, 1 mM ethylene diamine tetraacetic acid (EDTA), 1 mM phenylmethylsulfonyl fluoride, 1 μg/mL pepstatin A, and 0.5 μg/mL leupeptin.
2. 425 to 600 μm glass beads (Sigma; cat. no. G-8772).
3. AOX Assay Solution: 3.4 U/mL horseradish peroxidase (HRP), 0.56 mg/mL 2,2′-azinobis (3- ethylbenzthazoline-6-sulfonic acid), and 33 mM potassium phosphate buffer (pH 7.5).
4. Methanol.
5. $4N$ HCl.

2.7. Isolation of Yeast Genomic DNA

1. DNA Extraction Buffer: 10 mM Tris (pH 8.0), 2% Triton X-100, 1% sodium dodecyl sulfate (SDS), 100 mM NaCl, and 1 mM Na$_2$EDTA.
2. Phenol:chloroform:isoamyl alcohol (25:24:1 *[v/v]*).
3. Tris-EDTA (10 mM Tris [pH 7.5], 1 mM Na$_2$EDTA).
4. 10 mg/mL RNase A.
5. Chloroform:isoamyl alcohol (24:1 *[v/v]*).
6. $3M$ NaOAc.
7. 425 to 600 μm glass beads (Sigma; cat. no. G-8772).
8. Ethanol (ice cold).
9. QIAprep Spin Miniprep Kit, (Qiagen, Valencia, CA; cat. no. 27106).

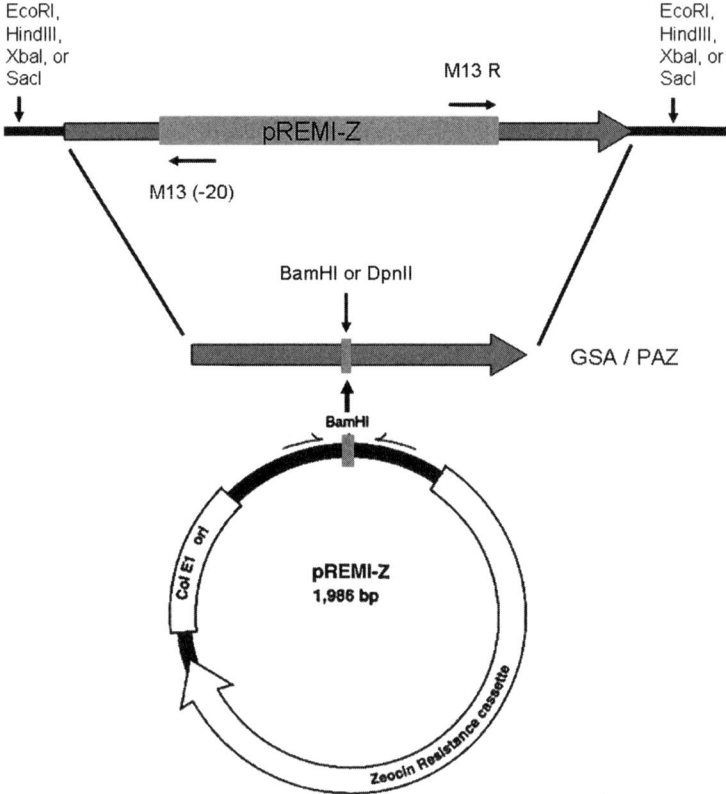

Fig. 1. Insertional mutagenesis by restriction enzyme-mediated integration of linearized pREMI-Z. GS115 cells are transformed by the insertion of linearized pREMI-Z (NCBI accession number AF282723) into the *Bam*HI or *Dpn*II site of a putative GSA gene. Zeocin-resistant cells that are defective in pexophagy are isolated and the GSA/PAZ gene identified by excising the pREMI-Z and flanking genomic DNA by restriction digest.

3. Methods

We have developed a protocol to rapidly identify genes essential for glucose-induced micropexophagy (*see* **Fig. 1**). This method utilizes a random integration assisted by restriction enzymes of a Zeocin-resistance cassette vector into the genomic DNA thereby disrupting gene expression. The Zeocin-resistant yeast defective in peroxisome degradation are then identified by direct colony assays. Afterward, the site of insertion of the REMI-Z is determined by first digesting the genomic DNA with restriction enzymes and then isolating the pREMI-Z with flanking genomic DNA by amplifying in *E. coli*. Once this vector is isolated, the flanking genomic DNA can be sequenced and the gene identified by BLAST searches of the NCBI databases. The methods described below outline: (i) vector construction, (ii) yeast transformation with the pREMI-Z vector, (iii) identification and isolation of pexophagy mutants, (iv) isolation of the pREMI-Z vector with flanking genomic DNA, and (v) identification of mutated genes.

3.1. Vector Construction

A pREMI-Z vector containing the Zeocin-resistance gene behind the *S. cerevisiae* TEF and bacterial EM7 promoters to facilitate expression in yeast and bacteria (*E. coli*) was constructed. This includes (a) construction of vector and (b) amplification of vector.

Two 68-base oligos were annealed to yield the adapter that contains multiple stop codons in all 6 frames and the M13 reverse and M13 forward (–20) sequencing primers pointed in towards the central *Bam*HI site:

$$\overrightarrow{}$$
GATCGGAAACAGCTATGACCATGTCAGTCAGTCA**GGATCC**
TAGCTAGCTAGACTGGCCGTCGTTTTAC

CCTTTGTCGATACTGGTACAGTCAGTCAGT**CCTAGG**ATCGAT
CGATCTGACCGGCAGCAAAATGCTAG
$$\overleftarrow{}$$

The parent vector pPICZ-A (Invitrogen, San Diego) was cut at unique *Bam*HI and *Bgl*II sites. The above adapter was then ligated into the pPICZ-A fragment at the *Bam*HI and *Bgl*II sites to create pREMI-Z (*see* **Fig. 1**). The vector was amplified by transforming Rb-competent DH5α and growing the transformants on LB+Z.

3.2. Yeast Transformation

The next step in this process involves the transformation of yeast and genomic integration of the pREMI-Z vector. This section includes (a) preparation of electro competent yeast, (b) transformation of yeast by electroporation, (c) selection, and (d) verification of vector integration.

3.2.1. Electrocompetent Yeast

1. GS115 or PPF1 cells are grown in 250 mL YPD to an OD_{600} of 1.0 –1.5.
2. Cells are harvested at 4°C in sterile 250 mL Nalgene centrifuge bottles at 4000*g* for 20 min.
3. Cells are then resuspended in 50 mL YPD containing 1 mL 1*M* HEPES (pH 8.0), 1.25 mL of 1 *M* DTT is added dropwise while swirling the cells and the cells are incubated at 30°C for 15 min with gentle shaking.
4. Bring to 250 mL with sterile cold water and harvest the cells as above.
5. Wash cells two times with 250 mL of sterile cold water.

6. Resuspend cells in 25 mL sterile cold 1 *M* sorbitol, transfer to 50-mL conical tube, and harvest cells at 2000*g* for 10 min.
7. Resuspend cells into 0.5 mL sterile cold 1 *M* sorbitol.
8. Electro-competent *P. pastoris* are more effective when used immediately, but can be frozen at –80°C for later use.

3.2.2. Transformation of Yeast by Electroporation

1. pREMI-Z is linearized by digestion with *Bam*HI.
2. 50 µL of competent yeast cells are mixed with 1 to 2 µg of linearized pREMI-Z and 0.5 to 1.0 units of *Bam*HI or *Dpn*II.
3. The solution is then transferred to a 0.2-cm gap cuvette for electroporation (1.5 kV, 25 µF, 400 Ω).
4. Immediately after electroporation the cells are suspended in 1 mL of cold 1*M* sorbitol.

3.2.3. Selection of Transformants

1. The entire sample of electroporated cells is transferred to multiple YPD+Z plates (200–250 µL/plate).
2. Plates are then incubated for 2 to 4 d at 30°C until colonies appear.

3.2.4. Verifying Vector Integration

1. Pexophagy mutants are grown to stationary growth in 2 mL of YPD.
2. Cells are lysed and genomic DNA isolated as described in **Subheading 3.4.1.**
3. 10 µL of genomic DNA is digested with *Eco*RI.
4. The genomic DNA fragments are then separated on an agarose gel and transferred to nylon membranes.
5. The DNA probe was made from *Bam*HI-linearized pREMI-Z following instructions included in the North2South Biotin Random Primer Kit (Pierce, Rockford, IL).
6. pREMI-Z vectors containing DNA fragments on the nylon membranes were detected following instruction included in the North2South Chemiluminescent Hybridization and Detection Kit (Pierce, Rockford, IL).
7. The visualized bands demonstrate pREMI-Z insertions (*see* **Fig. 2**). Virtually all the Zeocin-resistant clones have a single band suggesting a single insertion site. However, it appears that some clones have pREMI-Z inserted into two genes (*see* **Fig. 2**, *R23*).

3.3. Identification and Isolation of Glucose-Induced Pexophagy Mutants

After isolating hundreds to thousands of Zeocin-resistant clones, the next step is to screen for pexophagy mutants. This is done by direct colony assay to identify those clones that do not degrade AOX when adapted from methanol to glucose. Those pexophagy mutants identified by direct colony assay are then isolated and verified by a liquid medium assay. This section includes (a) identification of pexophagy mutants by direct colony assay and (b) quantification of the pexophagy defect by liquid medium assay.

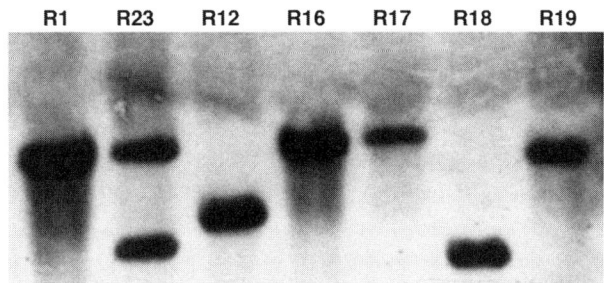

Fig. 2. Insertion of pREMI-Z into the genomic DNA of REMI mutants as visualized on Southern blots. Genomic DNA isolated from REMI mutants was digested with *Eco*RI, the resulting DNA fragments separated on agarose gels, and those DNA fragments containing pREMI detected on Southern blots using biotinylated probes prepared from pREMI-Z.

3.3.1. Direct Colony Assay

1. Colonies are grown on YPD+Z selection plates overnight 2 to 3 d at 30°C.
2. Colonies are replica-plated from the YPD+Z plates to YNM plates (with appropriate amino acid supplements) and grown for 3 to 4 d at 30°C.
3. Colonies are then replica-plated from the YNM plates to nitrocellulose which is placed onto YND plates, colonies face up, and grown for 16 to 18 h at 30°C.
4. The colonies, on nitrocellulose, are removed from the YND plates and the cells lysed by freezing in liquid nitrogen for 20 s, taking care not to break the paper into shards.
5. The frozen nitrocellulose is carefully placed on Whatman paper soaked with AOX Detection Solution (*see* **Subheading 2.5.**, **item 3**).
6. The nitrocellulose is incubated for 60 to 90 min at room temperature.
7. AOX activity is visualized with the appearance of periwinkle (purple) dots (*see* **Fig. 3**). When grown on YNM plates, all colonies have AOX activity. However, only those pexophagy mutants unable to degrade AOX will have AOX activity when transferred from YNM to YND.

3.3.2. AOX Degradation by Liquid Medium Assay

1. Pexophagy mutants identified by direct colony assay are isolated either from the original YPD-Z or the YNM plates and grown in YPD medium.
2. Cells (0.6 mL of saturated YPD culture) are grown in 20 mL YNM with appropriate amino acid supplements.
3. After 36 to 38 h growth on YNM, 0.4 g glucose is added.
4. Two mL aliquots of cells at 0 and 6 h of glucose adaptation are harvested by centrifugation.
5. The pellets are resuspended in 1 mL of cell lysis buffer (*see* **Subheading 2.6.**, **item 1**).
6. The cells are then lysed by vortexing in the presence of 0.5 g of 425 to 600 μm glass beads.

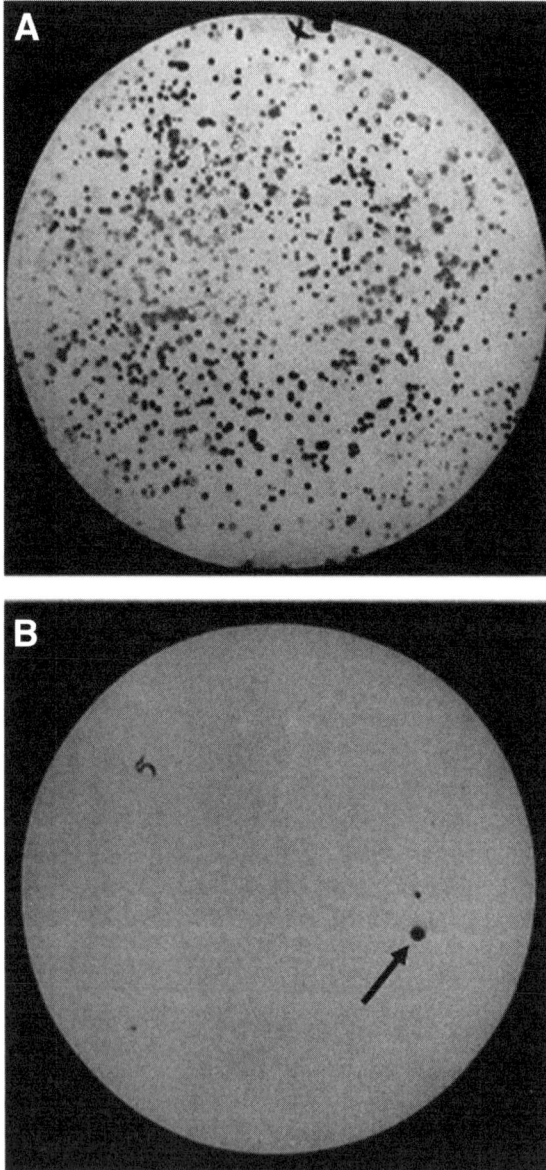

Fig. 3. The direct colony assay of AOX activity in REMI-mutated cells grown on (**A**) YNM or (**B**) adapted from YNM to YND. The arrow indicates a colony defective in degrading AOX.

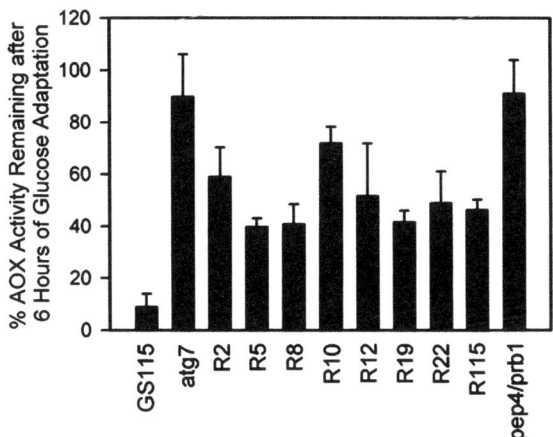

Fig. 4. Glucose-induced degradation of AOX in REMI mutants. Parental GS115, autophagy-defective *atg7* mutants, vacuole-defective *pep4/prb1* mutants, and REMI mutants (R2, R5, R8, R10, R12, R19, R22, and R115), were grown in YNM medium and then switched to YND medium. At 0 h and 6 h, the cells were lysed and AOX activity determined. Only 10% of the AOX remained at 6 h of glucose adaptation in GS115, whereas 90% was present in *atg7* and *pep4/prb1* mutants. The pexophagy defect in the REMI mutants was not as severe as that observed in the *atg7* mutants with 40 to 70% of the AOX activity remaining.

7. The glass beads and cellular debris are removed by centrifugation.
8. Alcohol oxidase is measured by adding 50 μL of this extract to 3 mL of AOX Assay Solution (*see* **Subheading 2.6.,** **item 3**).
9. The reaction is started by adding 10 μL methanol and continued at room temperature for 15 to 30 min until the appearance of a teal color.
10. The reaction is stopped by addition of 200 μL 4*N* HCl.
11. The developed teal color is measured at 410 nm.

A representation of AOX degradation measured by liquid medium can be seen in **Fig. 4**. Parental GS115 cells degrade over 90% of the AOX after six hours of glucose adaptation compared with 10% in *atg7* and *pep4/prb1* mutants. In comparison, 30 to 60% of the AOX is degraded by eight unique REMI mutants.

3.4. Identification of the Disrupted Gene Caused by the Insertion of pREMI-Z

After isolating the pexophagy mutants and verifying on Southerns (*see* **Subheading 3.2.4.**) that the pREMI-Z inserted into a single site, the next step is to identify the site of insertion and the disrupted GSA/PAX gene. This is done by isolating the genomic DNA, digesting the DNA with selected restriction enzymes, ligating the pREMI-Z and its flanking DNA into a circular vector, and amplifying the vector in *E. coli* for DNA sequencing. This section will cover (a) the isolation of gDNA in yeast cells, (b) the isolation of the pREMI-Z vector, and (c) sequencing the genomic DNA that flanks the pREMI-Z.

3.4.1. Genomic DNA Isolation

1. Each *P. pastoris* mutant strain is grown to stationary phase in 1 mL YPD at 30°C.
2. Harvest the cells by centrifugation (14,000*g* for 2 min).
3. To the cell pellets add 0.2 mL of DNA extraction buffer (*see* **Subheading 2.7., item 1**), 0.2 mL of phenol:chloroform:isoamyl alcohol (25:24:1), and 0.3 g 425 to 600 μm glass beads.
4. Vortex on high setting for 3 to 4 min.
5. Add 0.2 mL of Tris-EDTA (pH 8.0).
6. Microfuge at 14,000*g* for 2 min.
7. Transfer the top, aqueous layer to a fresh 1.5-mL microfuge tube and add a 2:1 volume of 100% ethanol (EtOH) (about 1 mL) and gently mix by inversion.
8. Microfuge at 14,000*g* for 2 min and decant supernatant.
9. Resuspend the pellet in 0.4 mL Tris-EDTA (pH 8.0) and 3 μL RNase A (10 mg/mL).
10. Incubate for 5 mins at 37°C.
11. Add 10 μL 4*M* ammonium acetate and 1 mL 100% cold ethanol and gently mix by inversion.
12. Microfuge at 14,000*g* for 2 min and decant supernatant.
13. Air-dry the pellet (approx 20 min; do not allow the pellet to dry completely).
14. Resuspend the pellet in 50 μL Tris-EDTA (pH 8.0), or in 50 μL of water.

3.4.2. Amplification of pREMI-Z Isolated From Pexophagy Mutants

1. Genomic DNA (5-10 μL) is digested in a total volume of 20 μL with *Eco*RI, *Hind*III, *Xba*I, *Sac*I, or *Bgl*II (enzymes that do not cut pREMI-Z) for 20 to 24 h at 37°C.
2. Add 180 μL ddH$_2$O and 200 μL phenol/chloroform/isoamyl alcohol (25:24:1).
3. Vortex the solution for 30 s, then microfuge at 14,000*g* for 2 min.
4. Remove the majority of the aqueous top phase and discard, avoiding the proteins at the aqueous:phenol interface.
5. To the aqueous phase, add half a volume (200 μL) of chloroform/isoamyl alcohol.
6. Again vortex the solution for 30 s, then microfuge at 14,000*g* for 2 min.
7. Remove 180 μL of the top phase (aqueous layer) and transfer to a clean tube.
8. Add 20 μL of 3 *M* sodium acetate (NaOAc) and 400 μL 100% ethanol.
9. Place the vial for 20 min in the –80°C freezer or in dry ice.
10. Microfuge at 14,000*g* for 2 min and decant supernatant.
11. Wash the pellet with 0.5 mL cold 70% ethanol.
12. Microfuge at 14,000*g* for 2 min and decant supernatant.

13. Air-dry the pellet (approx 20 min; do not allow the pellet to dry completely).
14. Resuspend the pellet in 20 μL water.
15. Incubate 5 μL of digested genomic DNA in a total volume of 10 μL with T4 DNA ligase.
16. Incubate for 20 to 24 h at 16°C.
17. Add 5 μL of the ligation reaction to 100 μL Rb-competent DH5α cells.
18. Place the solution on ice for 2 min, then 42°C for 90 s, then ice for 2 min.
19. Add 0.5 mL YETM and incubate for 1 h at 37°C with gentle shaking.
20. Plate onto LB+Z.
21. Colonies are picked and grown in liquid LB+Z.
22. The pREMI-Z vector with flanking DNA is isolated using a QIAprep miniprep kit (*see* **Subheading 2.7.**, **item 9**).

3.4.3. Sequencing Flanking Genomic DNA

1. The Big Dye sequencing reaction is done following routine procedures (Applied Biosystems) using M13 reverse and M13 (–20) primers.
2. Sequencing is done on an Applied Biosystems (ABI) 310 Prism Capillary Electrophoresis DNA Sequencer.
3. The disrupted gene is then identified by BLAST analysis on the NCBI and *S. cerevisiae* databases.

4. Notes

1. In some studies, the linearized pREMI-Z is dephosphorylated with calf intestinal phosphatase prior to electroporation (*4*).
2. The addition of a restriction enzyme during electroporation with the linearized pREMI-Z is believed to generate free chromosomal ends in the yeast genome for incorporation of the REMI-Z plasmid. However, the number of transformations was only increased twofold when investigated in *H. polymorpha* (*5*). Furthermore, doses of BamHI higher than 2 units had a detrimental affect on the efficiency of transformation. Interestingly, only 30% of the integrations had occurred at a *Bam*HI site and in many cases the *Bam*HI site was not conserved because of nucleotide loss from the pREMI-Z vector (*5*). However, REMI mutagenesis reported for *C. albicans* and *S. cerevisiae* do require the addition of restriction enzymes (*6*,*7*).
3. The integration of the pREMI-Z vector is stable for several generations of growth in nonselective YPD medium (*5*). A majority of the integrations occur at a single site within an open reading frame that can be easily characterized. However, there are some genomic integrations that could complicate the characterization of the mutant as well as the identification of the disrupted gene. First, the pREMI-Z could insert into more than one gene. Although we have found that this is a rare event, each disrupted gene would have to be characterized individually. Van Dijk et al. have observed only two out of forty *H. polymorpha* transformants contained insertions into two genomic loci (*5*). Second, there could be multiple insertions of the pREMI-Z vector into the same site. Although a single gene would be disrupted, the tandem insertion of pREMI-Z vectors would make it difficult to obtain

any meaningful sequence data. Studies in *H. polymorpha* have shown that single copy integration occurred in 50% of the transformants, double copy integration occurred in 25% of the transformants, and 3 or more copies of pREMI-Z occurred in 25% of the transformants *(5)*. It appears that the pREMI-Z cassette formed multimers prior to integration because it occurred independent of the addition of restriction enzymes. Finally, the insertion could cause a genomic deletion. These can be identified by sequencing the flanking DNA isolated with the pREMI-Z vector and noting that the sequence on both sides of the pREMI-Z is not continuous. Finally, intergenic insertions are also difficult to interpret.

4. The REMI protocol has been used successfully in both *P. pastoris* and *H. polymorpha* to isolate genes (GSA, glucose-induced selective autophagy; PAZ, pexophagy zeocin-resistance; PDD, peroxisome degradation-deficient) essential for peroxisome degradation (**Table 1**). The quantification of AOX activity has been the screen of choice for pexophagy mutants in *P. pastoris*. A more detailed characterization of these mutants can be done by observing the vacuole morphology labeled with FM4-64 relative to the peroxisomes labeled with BFP-SKL or GFP-SKL *(1,4,8)*. However, a second screen has been used to identify pexophagy mutants in *H. polymorpha (5)*. This qualitative screen requires the microscopic observation of the prolonged presence of eGFP-SKL (eGFP protein which is C-terminally extended with a peroxisomal targeting signal of serine-lysine-leucine-COOH) in cells after the shift from YNM to YND medium. The REMI protocol has also been utilized to identify mutants defective in peroxisome biogenesis *(5)*. In general, these mutants fail to thrive when grown on YNM plates. However, a more detailed characterization of the mutant regarding protein sorting or aberrant number or size of peroxisomes can be done by microscopic visualization of eGFP-SKL *(5)*.

5. Efficient lysing of the yeast is critical for accurate measurements of AOX activity. For the direct colony assay, cells were originally lysed by digestion of the cell wall with Zymolase 20T (0.25 mg/mL). However, this procedure yielded variable results and we found that liquid nitrogen freezing and thawing was more efficient. For cells in liquid suspension, the cells are lysed by vortexing in the presence of glass beads.

6. It is possible that some genomic fragments containing the pREMI-Z vector may be too large to be amplified in *E. coli*. However, we have been able to amplify vectors that are 10 to 12 kb. If the vectors are too large, then we suggest using a different restriction enzyme to digest the genomic DNA for recovery of the pREMI-Z. In principal, any enzyme that does not cut pREMI-Z would be appropriate.

7. The Zeocin antibiotic is sensitive to high salt concentrations. Normally, the selection of yeast that had been electroporated consists of growth on minimal medium plates (0.67% yeast nitrogen base without amino acids, 2% dextrose, 1*M* sorbitol, 0.4 mg/L biotin, and 2% agar). However, we have found that Zeocin does not appear to be active in such medium. Therefore, we have utilized YPD plates with Zeocin. Furthermore, the selection of Zeocin-resistant *E. coli* is done in medium containing only 0.5% NaCl.

Table 1
Pexophagy Genes Identify by REMI

	Pichia pastoris[a]	Hansenula polymorpha[b]	Characteristics	Ref.
GSA10/				
PAZ1	PDD7	ATG1[c]	Serine/threonine-kinase that complexes with Atg11 and Vac8	1,4,13
PAZ2	HpAtg8	ATG8	Soluble protein that becomes conjugated to lipids	14,15
PAZ3		ATG16	Component of Atg12-Atg5 complex	4
PAZ4		ATG26	UDP-glucose:sterol glucosyl-transferase	4
PAZ5		GCN3	Translation initiation factor	4
GSA9/				
PAZ6	PDD18	ATG11	Coiled-coil domain found at vacuole surface	8,16
GSA11/				
PAZ7		ATG2	Peripheral membrane protein that interacts with Atg9	1,4
PAZ8		ATG4	Cysteine proteinase	4
GSA14/				
PAZ9		ATG9	Integral membrane protein associated with organelles juxtaposed to the vacuole	1,4
PAZ10		GCN1	Regulates translation elongation	4
PAZ11		GCN2	Regulates translation initiation	4

(Continued)

Table 1 *(Continued)*

Pichia pastoris[a]	*Hansenula polymorpha*[b]	Characteristics	Ref.
PAZ12	ATG7	E1-like enzyme responsible for the conjugation of Atg12 to Atg5 and Atg8 to lipids	4
GSA19/ PAZ13	VPS15	Membrane-anchored Serine/threonine-kinase with WD40 domains	1,4
GSA15/ PAZ14	PEP4	Endopeptidase	4
PAZ16	ATG24	Sorting nexin with PX domain	17
PAZ19	GCN4	Transcriptional activator	
GSA12	ATG18	WD40 protein	18
GSA20	ATG3	E2-like enzyme responsible for the conjugation of Atg8 to lipids	
PDD13	MPP1	$Zn(II)_2Cys_6$ Transcription Factor	19
PDD15	ATG21	WD40 protein	
PDD20	VAM3	Vacuolar t-SNARE	20

[a] GSA, glucose-induced selective autophagy; PAZ, pexophagy zeocin-resistance.
[b] PDD, peroxisome degradation-deficient.
[c] A unified gene nomenclature of autophagy and autophagy-related processes such as pexophagy has been standardized across species to be ATG (autophagy-related) for the gene names (*16*).

8. Restriction enzyme-mediated integration has been utilized successfully to generate mutants and identify genes essential for a number of cellular pathways in many different organisms including: *Saccharomyces cerevisiae (7)*, *Candida albicans (6)*, *Coprinus cinereus (9)*, *Lentinus edodes (10)*, *Gibberella fujikuroi (11)*, and *Aspergillus fumigatus (12)*.

Acknowledgments

This work was supported by NSF (MCB-9817002) and NIH (CA95552) grants to W.A. Dunn, Jr. and NIH (GM61156) grant to B. S. Glick.

References

1. Stromhaug, P. E., Bevan, A., and Dunn, W. A., Jr. (2001) GSA11 encodes a unique 208-kDa protein required for pexophagy and autophagy in *Pichia pastoris*. *J. Biol. Chem.* **276,** 42,422–42,435.
2. Habibzadegah-Tari, P., and Dunn, W. A., Jr. (2003) Glucose-induced pexophagy in *Pichia pastoris*, in *Autophagy*, (Klionsky, D. J., ed), Landes Bioscience. Austin, TX.
3. Sakai, Y., Koller, A., Rangell, L. K., Keller, G. A., and Subramani, S. (1998) Peroxisome degradation by microautophagy in *Pichia pastoris*: identification of specific steps and morphological intermediates. *J. Cell. Biol.* **141,** 625–636.
4. Mukaiyama, H., Oku, M., Baba, M., et al. (2002) Paz2 and 13 other PAZ gene products regulate vacuolar engulfment of peroxisomes during micropexophagy. *Genes Cells* **7,** 75–90.
5. van Dijk, R., Faber, K. N., Hammond, A. T., Glick, B. S., Veenhuis, M., and Kiel, J. A. (2001) Tagging *Hansenula polymorpha* genes by random integration of linear DNA fragments (RALF). *Mol. Genet. Genomics* **266,** 646–656.
6. Brown, D. H., Jr., Slobodkin, I. V., and Kumamoto, C. A. (1996) Stable transformation and regulated expression of an inducible reporter construct in *Candida albicans* using restriction enzyme-mediated integration. *Mol. Gen. Genet.* **251,** 75–80.
7. Schiestl, R. H. and Petes, T. D. (1991) Integration of DNA fragments by illegitimate recombination in *Saccharomyces cerevisiae*. *Proc. Natl. Acad. Sci. USA* **88,** 7585–7589.
8. Kim, J., Kamada, Y., Stromhaug, P. E., et al. (2001) Cvt9/gsa9 functions in sequestering selective cytosolic cargo destined for the vacuole. *J. Cell Biol.* **153,** 381–396.
9. Granado, J. D., Kertesz-Chaloupkova, K., Aebi, M., and Kues, U. (1997) Restriction enzyme-mediated DNA integration in *Coprinus cinereus*. *Mol. Gen. Genet.* **256,** 28–36.
10. Sato, T., Yaegashi, K., Ishii, S., Hirano, T., Kajiwara, S., Shishido, K., and Enei, H. (1998) Transformation of the edible basidiomycete *Lentinus edodes* by restriction enzyme-mediated integration of plasmid DNA. *Biosci. Biotechnol. Biochem.* **62,** 2346–2350.
11. Linnemannstons, P., Voss, T., Hedden, P., Gaskin, P., and Tudzynski, B. (1999) Deletions in the gibberellin biosynthesis gene cluster of *Gibberella fujikuroi* by restriction enzyme-mediated integration and conventional transformation-mediated mutagenesis. *Appl. Environ. Microbiol.* **65,** 2558–2564.

12. Brown, J. S., Aufauvre-Brown, A., and Holden, D. W. (1998) Insertional mutagenesis of Aspergillus fumigatus. *Mol. Gen. Genet.* **259,** 327–335.

13. Komduur, J. A., Veenhuis, M., and Kiel, J. A. (2003) The *Hansenula polymorpha* PDD7 gene is essential for macropexophagy and microautophagy. *FEMS Yeast Res.* **3,** 27–34.

14. Monastyrska, I., van der Heide, M., Krikken, A. M., Kiel, J. A., van der Klei, I. J., and Veenhuis, M. (2005) Atg8 is Essential for Macropexophagy in *Hansenula polymorpha. Traffic* **6,** 66–74.

15. Mukaiyama, H., Baba, M., Osumi, M., Aoyagi, S., Kato, N., Ohsumi, Y., and Sakai, Y. (2004) Modification of a ubiquitin-like protein Paz2 conducted micropexophagy through formation of a novel membrane structure. *Mol. Biol. Cell* **15,** 58–70.

16. Klionsky, D. J., Cregg, J. M., Dunn, W. A., Jr., et al. (2003) A unified nomenclature for yeast autophagy-related genes. *Dev. Cell* **5,** 539–545.

17. Ano, Y., Hattori, T., Oku, M., et al. (2005) A sorting nexin PpAtg24 regulates vacuolar membrane dynamics during pexophagy via binding to phosphatidylinositol-3-phosphate. *Mol. Biol. Cell* **16,** 446–457.

18. Guan, J., Stromhaug, P. E., George, M. D., et al. (2001) Cvt18/Gsa12 is required for cytoplasm-to-vacuole transport, pexophagy, and autophagy in *Saccharomyces cerevisiae* and *Pichia pastoris. Mol. Biol. Cell* **12,** 3821–3838.

19. Leao-Helder, A. N., Krikken, A. M., van der Klei, I. J., Kiel, J. A., and Veenhuis, M. (2003) Transcriptional down-regulation of peroxisome numbers affects selective peroxisome degradation in *Hansenula polymorpha. J Biol. Chem* **278,** 40,749–40,756.

20. Leao-Helder, A. N., Krikken, A. M., Gellissen, G., van der Klei, I. J., Veenhuis, M., and Kiel, J. A. (2004) Atg21p is essential for macropexophagy and microautophagy in the yeast *Hansenula polymorpha. FEBS Lett.* **577,** 491–495.

16

Characterization of Protein–Protein Interactions

Application to the Understanding of Peroxisome Biogenesis

Sebastien Leon, Ivet Suriapranata, Mingda Yan, Naganand Rayapuram, Amar Patel, and Suresh Subramani

Abstract

With the approaching completion of the *Pichia pastoris* genome, a greater emphasis will have to be placed on the proteome and the protein–protein interactions between its constituents. This chapter discusses methods that have been used for the study of such interactions among both soluble and membrane-associated proteins in peroxisome biogenesis. The procedures are equally applicable to other cellular processes.

Key Words: Protein complexes; protein interactions; yeast two-hybrid, co-immunoprecipitation; tandem affinity purification; TAP-tag; mass spectrometry; membrane protein complexes; soluble protein complexes.

1. Description of Techniques

1.1. Biochemical Methods Using Homologous Systems

Several biochemical methods, including co-immunoprecipitation and the recently described tandem affinity purification (TAP) method, are available and have been used successfully in *Pichia pastoris*. These biochemical methods allow the extraction of proteins from their in vivo context together with their interacting polypeptides. It should be noted that these procedures may yield proteins that interact directly or indirectly with the protein under study.

1.1.1. Co-Immunoprecipitation

Co-immunoprecipitation (Co-IP) allows the identification of physical interactions between proteins both in vivo and in vitro, independently of the fact that this interaction is direct or bridged by another protein (second-order interaction).

From: *Methods in Molecular Biology, vol. 389:* Pichia *Protocols, Second Edition*
Edited by: J. M. Cregg © Humana Press Inc., Totowa, NJ

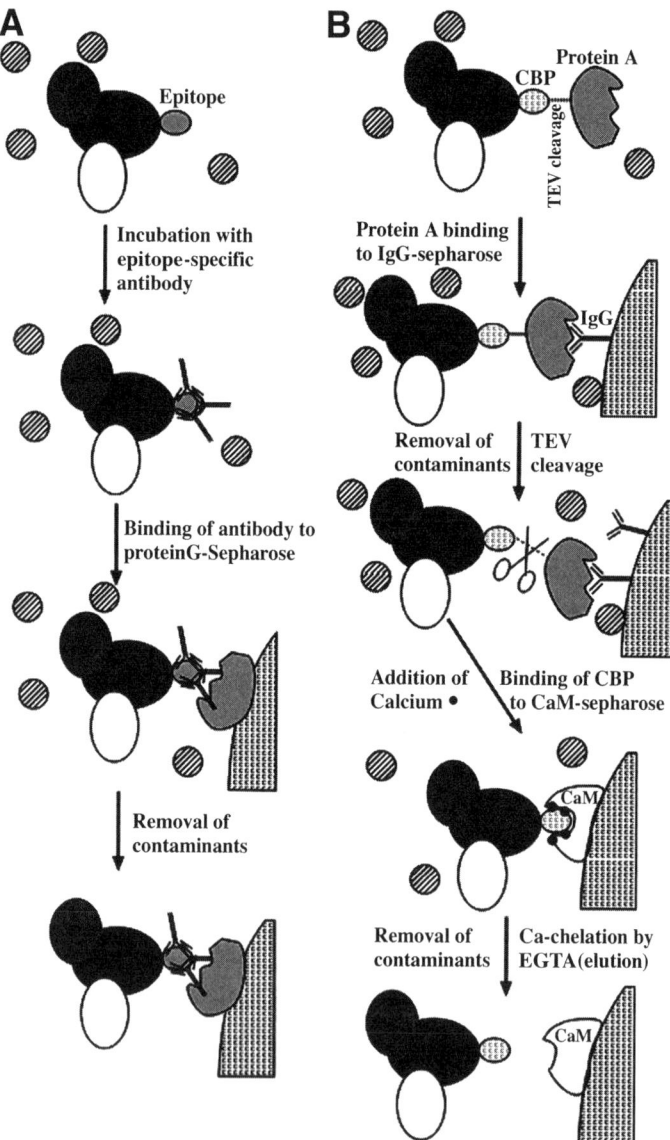

Fig. 1. Biochemical methods to detect protein–protein interactions. (**A**) Immuno-precipitation. The extract containing the protein of interest (black ellipse) is incubated with its corresponding antibody and the mixture is then bound to IgG-sepharose beads. Washes allow the removal of contaminants (hatched circles) and retain intact protein–protein complexes bound to the beads. (**B**) TAP-tag procedure. The extract containing the di-tagged protein is bound to IgG-sepharose beads through interaction with the Protein A moiety of the tag. After several washes, the beads are treated with TEV

Co-IP can be performed on a native protein provided that a good antibody is available, or on an epitope-tagged version of this protein if the tag does not alter the function, location or level of expression of the protein. Depending on the stability of the interaction between the protein of interest and its potential partners, the experiment can be performed with or without the addition of crosslinkers. Both approaches have been used successfully in *P. pastoris* to understand the composition and organization of protein complexes involved in the biogenesis of peroxisomes *(1–3)* (*see* **Fig. 1A**).

The overall protocol includes lysis of cells, solubilization of the protein in the event it is in membranes, incubation of the protein extract with the primary antibody, and then incubation of this mixture with beads (agarose- or sepharose) coupled to Protein A or Protein G depending on the source of the secondary antibody *(4)*. The ternary complex (epitope/primary antibody/secondary antibody-beads) is pelleted, washed, repelleted and the protein content associated with the beads is analyzed by immunoblotting using antibodies against the putative interacting protein(s) (which may or may not be epitope-tagged).

1.1.2. Tandem Affinity Purification

The tandem affinity purification (TAP) tag has been used for the systematic analysis of protein complexes in the proteome of *S. cerevisiae* *(5)*. It can also be used to study the composition of complexes in which a given protein is found *(6)*. The TAP tag consists of two affinity-purification modules separated by a cleavage site for tobacco etch virus (TEV) protease, and fused to the C-terminus of the protein under study *(7)*. The first tag is the calmodulin-binding peptide (CBP) of the kfc (kemptide-factor Xa cleavage site-calmodulin binding peptide) cassette (GenBank X66255) followed by the TEV protease site, and an IgG-binding domain based on Staphylococcal protein A as the second tag. This protein fusion with the dual tag can be introduced into the chromosome at the locus of the wild-type gene in *P. pastoris* using homologous recombination *(8)*. Care should be taken that the addition of this tag does not alter the function, location or level of expression of the protein.

After lysis of the cells, two rounds of affinity purification are performed, as opposed to co-IP which uses only a single purification (*see* **Fig. 1B**). In the first round, the protein is bound to an IgG-sepharose resin via the protein A moiety of the tag. Any contaminating, or non-specifically bound, proteins are left

---

Fig. 1. *(Continued)* protease to release the singly-tagged (CBP) protein. The supernatant is then incubated with calmodulin-sepharose (CaM-sepharose) beads in the presence of calcium. Washes remove other contaminants. The protein and its interacting partners (*dark grey ellipses*), are eluted from the beads by chelation of calcium using EGTA.

behind on the IgG column by releasing the tagged protein from the beads by TEV protease cleavage. This treatment removes the protein A domain on the fusion. The released fusion with its interacting proteins is then passed over a column of calmodulin-sepharose beads in the presence of calcium. The bound proteins are washed to remove the TEV protease and other nonspecific proteins, and the tagged protein is eluted using EGTA to chelate calcium and therefore releasing CBP from calmodulin. This final eluate contains a relatively pure fraction of the protein–CBP fusion together with its interacting partners. This fraction can be analyzed on sodium dodecyl sulfate polyacrylamide gel electrophoresis (SDS-PAGE) (for polypeptide composition), immunoblotting (for identification of predicted interacting proteins) and mass spectrometry (for identification of new proteins).

1.2. Reporter-Based Methods in Heterologous Systems

These methods are in common use and are not performed in *P. pastoris*, so they are reviewed briefly but the methods are not described here.

1.2.1. Yeast Two-Hybrid

Other methods routinely used for the analysis of protein–protein interactions, such as the yeast two-hybrid method (methods recently reviewed in *[9]*) are based on the addition, to the proposed interacting proteins, of tags which will perform a biological function when these tags are brought together spatially as a consequence of protein interactions *(10)*. In a given engineered system, this leads to a scorable phenotype (e.g., growth, resistance) that allows one to conclude whether the proteins interact or not (*see* **Fig. 2A**).

The most common system, referred to as the *GAL4* system, is based on the modular nature of the *S. cerevisiae* Gal4p transcriptional regulator: whereas residues 1–147 have the ability to bind DNA, residues 768–881 can activate transcription of basically any gene provided it is placed under the control of the adequate *cis* regulatory element (Upstream Activating Sequence) bound by Gal4p. The other major yeast two-hybrid system, the LexA system, utilizes in a similar manner the DNA-binding domain of the *Escherichia coli* LexA protein and the transcriptional activation domain of the herpes simplex virus VP16 protein.

Several *S. cerevisiae* host strains differing in the number and nature of the reporter(s) used are available for the two-hybrid interaction test. This technique has allowed tremendous progress in the identification of novel protein components of a given complex, and provided clues regarding the function of proteins following the understanding of the protein networks they are involved in.

1.2.2. Bacterial Two-Hybrid

The bacterial two-hybrid system was recently developed as an alternative method for the study of protein–protein interaction in vivo in a heterologous

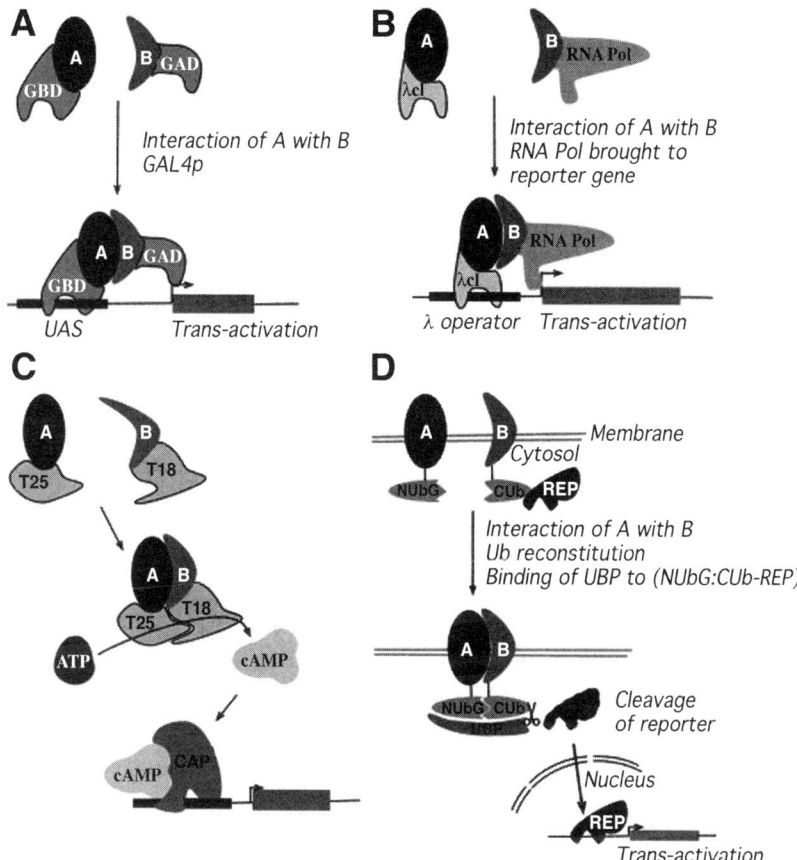

Fig. 2. Study of protein–protein interactions in heterologous systems. (**A**) The yeast two-hybrid GAL4 system is based on the reconstitution of the transcription factor, Gal4p, upon interaction of the proteins under study. (**B,C**) Transcription- and nontranscription-based bacterial two-hybrid assays. (**B**) Interaction of the studied proteins will bring the RNA polymerase to the λ operator and activate transcription at this locus, or (**C**) reconstitute adenylate cyclase activity. (**D**) The split-ubiquitin system is based on the release of a reporter after cleavage by a ubiquitin-specific protease. This occurs only when two proteins, each fused to one half of the ubiquitin molecule, interact and therefore reconstitute ubiquitin.

system (reviewed in *[11]*). These methods are of two types, either based on the reconstitution of (i) a transcription factor (in a similar way as described above for the yeast two-hybrid *[12,13]*), or (ii) of the *Bordetella pertussis* adenylate cyclase protein *(14)* in an engineered *E. coli* strain (*see* **Fig. 2B,C**).

In the first case, the studied protein is fused to the DNA-binding domain of the bacteriophage λ repressor protein (cI), while the second protein is fused to the N-terminal domain of the α subunit of *E. coli* RNA polymerase. The bacterial strain used possesses the *HIS3* reporter gene under the control of the λ operator. Interaction of the two proteins will stabilize the RNA polymerase at the λ operator, thus allowing transcription of the *HIS3* reporter gene. This system is now commercially available under the name BacterioMatch® (trademark of Stratagene, La Jolla, CA).

The nontranscription-based bacterial two-hybrid makes use of the CyaA protein (calmodulin-dependent adenylate cyclase toxin) from *Bordetella pertussis*. The catalytic domain of this protein (residues 1–399, out of 1706) is cleaved into two fragments (namely: T25, residues 1–224, and T18, residues 225–399) that cannot interact with each other. However, upon fusion of each domain to interacting proteins, the catalytic domain is reconstituted and cAMP is synthesized. This is characterized, in an *E. coli cya* strain, by restoration of the cAMP level, which in turn triggers the expression of a reporter gene placed under the control of a cAMP-dependent promoter giving rise to a selectable phenotype.

Although these methods have essentially been used for the study of bacterial proteins, they have been used in several instances for mammalian *(15–18)*, and even viral, proteins *(19)*. This method could therefore be used for the study of in vivo protein–protein interactions. Interestingly, it can be also used as a complementary approach, as reported in a study that compares the traditional yeast two-hybrid to the nontranscription-based bacterial two-hybrid system *(18)*.

1.2.3. Split-Ubiquitin (Ubiquitin-Based Split-Protein Sensor [USPS])

The split-ubiquitin method *(20)* is based on expressing one protein (A) as a fusion to a modified N-terminal domain (residues 1–34 with mutation of I13→G, denoted NUbG) of ubiquitin (Ub), and expressing another protein (B) as a fusion to the Ub C-terminal domain (residues 35–76, CUb) followed by a stable reporter protein. In vivo, NUbG and CUb-reporter have a weak affinity for each other but can assemble and form split-ubiquitin: (NUbG:CUb)-reporter.

In eukaryotes, Ub-protein fusions are cleaved by the action of Ub-specific proteases (UBPs), a process that releases the attached protein. This cleavage step requires Ub to be properly folded (i.e. the N- and the C-terminal regions must interact with each other).

Therefore, upon interaction of the two tagged proteins (A and B) under study, reconstitution of a full-length-equivalent of the Ub molecule occurs, which allows recognition and cleavage by UBP and release of the reporter protein (*see* **Fig. 2C**). The phenotype observed due to protein A-protein B interaction is based on the gain of activity of the released reporter, a transcription factor that activates reporter genes in the nucleus in the suitable strain (methods reviewed in *[21,22]*).

2. Materials

2.1. Co-Immunoprecipitation

Most of the methods for studying protein–protein interactions were developed initially with the analysis of soluble, and often stable rather than transient, protein complexes in mind. However, the application of such procedures to the study of interactions between soluble and membrane proteins, or between membrane proteins, has been more difficult because the procedures for extracting and solubilizing proteins from their membranes of origin require harsher conditions, such as the use of detergents, which might also disrupt the interactions with partner proteins. We present below two different protocols for co-immunoprecipitations. The first is with cleavable crosslinkers to stabilize interactions between membrane proteins prior to their solubilization from membranes *(1)*. This method also has the advantage of being able to capture transient and dynamic interactions between proteins. The second co-IP procedure is without the use of crosslinkers and generally works better for interactions between soluble proteins.

For co-IPs in the presence of crosslinker, the general strategy involves breaking the cells open after spheroplasting, while keeping subcellular organelles intact. The addition of a membrane-permeable crosslinker stabilizes interactions between proteins that are in close contact (determined by the length of the crosslinker). The proteins are then resuspended in a buffer that is partially denaturing, but compatible with antibody–antigen interactions. The protein of interest is immunoprecipitated. After reduction of the crosslinker to release the interacting proteins, the polypeptides are separated by SDS-PAGE and detected by immunoblotting. Controls are necessary to rule out nonspecific crosslinking of proteins, for example, resulting from the use of excessive amounts of the crosslinker.

2.1.1. Materials for Co-Immunoprecipitation of Crosslinked Proteins

1. *P. pastoris* strains PPY12 (*arg4 his4*), SMD1163 (*his4 pep4 prb1*).
2. Yeast culture medium: YPD: 1% yeast extract, 2% peptone, 2% glucose), or peroxisome induction medium such as methanol medium (yeast nitrogen base + 0.7% *[w/v]* ammonium sulfate, 0.05% *[w/v]* yeast extract, 0.5% *[v/v]* methanol, plus required amino acids).
3. Reducing buffer: 100 mM Tris-HCl (pH 7.5), 50 mM ethylene diamine tetraacetic acid (EDTA), 100 mM β-mercaptoethanol.
4. Spheroplasting buffer: 10 mM K_2HPO_4, 10 mM KH_2PO_4, and 1.2 M sorbitol (pH 7.4).
5. Lytic enzyme: Zymolyase 20T (Seikagaku Corp).
6. Lysis buffer: 20 mM K_2HPO_4, 20 mM KH_2PO_4, and 1 mM EDTA (pH 7.5).
7. Protease inhibitors: phenylmethylsulfonylfluoride (PMSF) 0.1 M in ethanol, Protease Inhibitor Cocktail for yeast (Sigma; cat. no. P8215).

8. Dithiobis[succinimidylpropionate] (DSP) (Pierce Chemicals) in dimethyl sulfoxide (DMSO) at concentration of 20 mg/mL. Use freshly prepared solution each time. For mock treatment, add only DMSO to the reaction.
9. Hydroxylamine (Sigma): 1 M solution.
10. TCA (trichloracetic acid) (Fisher Scientific): 100% *(w/v)* solution.
11. Cracking buffer: 50 mM Tris-HCl, 1 mM EDTA (pH 7.5), 1% SDS, and 6M Urea.
12. Tween-20 IP buffer: 50 mM Tris-HCl (pH 7.5), 0.1 mM EDTA, 150 mM NaCl, and 0.% Tween-20.
13. Antiserum against protein of interest.
14. Gamma-bind G Sepharose (Amersham Pharmacia Biotech) or Protein A Sepharose (Amersham Pharmacia Biotech) (*see* **Note 1**).
15. Tween-20 urea buffer: 100 mM Tris-HCl (pH 7.5), 200 mM NaCl, 2 M Urea, and 0.5% Tween-20.
16. TBS buffer: 25 mM Tris-HCl (pH 7.5) and 140 mM NaCl.
17. Urea sample buffer: 6 M Urea, 125 mM Tris (pH 6.8), 6% sodium dodecyl sulfate (SDS), 10% β-mercaptoethanol, and 0.01% bromophenol blue.

2.1.2. Materials for Co-Immunoprecipitation Without the Use of Crosslinker

1. Yeast strain: this protocol has been used with *P. pastoris* strains PPY12 (*arg4 his4*) and SMD1163 (*his4 pep4 prb1*). The strain of interest is either a wild-type (when an antibody against the native form is available [*see* **Note 2**]), or one expressing an epitope-tagged protein. Also, a negative control strain must be used: either a deletion strain or the same strain lacking the tagged protein. This allows the assessment of the specificity of the co-immunoprecipitation for the protein studied, and shows that the presence of an interacting protein in the final extract depends on the presence of the protein studied.
2. Yeast culture medium (*see* **Subheading 2.1.1.**).
3. Gamma-bind G Sepharose (Amersham Pharmacia Biotech) or Protein A-Sepharose (Amersham Pharmacia Biotech) (*see* **Note 1**).
4. Protease inhibitors stock solutions: Leupeptin 1.25 mg/mL, Aprotinin 5 mg/mL, Protease Inhibitor Cocktail solution (Sigma; cat. no. P8215), and PMSF 0.1 M in ethanol.
5. Acid-washed glass beads: diameter 425 µm to 600 µm (Sigma).
6. IP lysis buffer: 50 mM HEPES-KOH (pH 7.5), 0.5 M NaCl, 0.5% *(w/v)* NP-40, 10% *(v/v)* glycerol, and 1 mM EDTA.
7. Wash buffer: 50 mM HEPES-KOH (pH 7.5), 150 mM NaCl, and 1 mM EDTA.
8. Sample buffer, 5X stock: 250 mM Tris-HCl (pH 6.8), 10% *(w/v)* SDS, 0.5 M dithiothreitol (DTT), 0.5% *(w/v)* bromphenol blue, and 50% *(v/v)* glycerol.
9. Nondyed sample buffer: 50 mM Tris-HCl (pH 6.8), 2% *(w/v)* SDS, 0.1M DTT, and 10% *(v/v)* glycerol.
10. Bromphenol blue 100X stock solution: 100 mg/mL.

2.2. Materials for TAP-Tagging

1. Plasmid pMY62 (or pMY63/64 [*see* **Note 3**]).
2. TAP primers: **OMY69:** 5′-TCTGACGCTCAGTGGAACGAA-3′, **OMY70:** 5′-TGCCCCGGAGGATGAGATT-3′.
3. Primers 1, 2, and 3 (*see* **Subheading 3.3.1.** for primer design).
4. *P. pastoris* genomic DNA.
5. LB + zeocin plates: 25 g/L Luria broth (LB) (Sigma; cat. no.L-3522), 15 g/L agar, and 100 µg/mL zeocin.
6. YPD + zeocin plates: 10 g/L yeast extract, 20 g/L peptone, 2% dextrose, 15 g/L agar, and 100 µg/mL zeocin.
7. French Press.
8. Buffer A: 10 mM HEPES-KOH (pH 7.9), 10 mM KCl, 1.5 mM MgCl$_2$, and 0.5 mM DTT, 0.5 mM PMSF, 1 mM leupeptin, protease inhibitor cocktail (use following manufacturer's instructions), and 5 µg/mL aprotinin.
9. Buffer D: 20 mM K-HEPES (pH 7.9), 50 mM KCl, 0.2 mM EDTA, 0.5 mM DTT, 25% *(w/v)* glycerol, 0.5 mM PMSF, 1 mM leupeptin, protease inhibitor cocktail (use as per manufacturer's instructions), and 55 µg/mL aprotinin.
10. Dialysis tubing: Molecular weight cut-off 12–14,000.
11. IgG-agarose beads (Sigma; cat. no. A2909).
12. IPP150 buffer: 10 mM Tris-Cl (pH 8.0), 150 mM NaCl, and 0.1% NP40.
13. TEV cleavage buffer: 10 mM Tris-Cl (pH 8.0), 150 mM NaCl, 0.1% NP40, 0.5 mM EDTA, and 1 mM DTT.
14. TEV protease, recombinant (Invitrogen; cat. no. 10127-017).
15. Calmodulin beads (Stratagene; cat. no. 214303).
16. IPP150 calmodulin-binding buffer: 10 mM Tris-Cl (pH 8.0), 150 mM NaCl, 0.1% NP40, 10 mM β-mercaptoethanol, 1 mM Mg-acetate, 1 mM imidazole, and 2 mM CaCl$_2$.
17. IPP150 calmodulin-elution buffer: 10 mM Tris-Cl (pH 8.0), 150 mM NaCl, 0.1% NP40, 10 mM β-mercaptoethanol, 1 mM Mg-acetate, 1 mM imidazole, and 5 mM EGTA.

3. Methods

3.1. Methods for Co-Immunoprecipitation of Crosslinked Proteins

3.1.1. Lysis of Cells and Crosslinking of Proteins

1. Measure the OD$_{600}$ of a 50 mL yeast culture and when it is approx OD$_{600}$ of 1/mL, collect cells by centrifugation at 2000g for 10 min at room temperature.
2. Resuspend cells in 15 mL of reducing buffer, transfer the suspension into a 50 mL centrifuge tube and incubate them at room temperature for 20 min with gentle shaking.
3. Pellet cells by centrifugation (2000g for 10 min) at room temperature, wash once in the same amount of Spheroplasting buffer.

4. Resuspend pellet in Spheroplasting buffer to give an OD_{600} of 5/mL, add Zymolyase 20T, usually 6 mg/1000 ODs (*see* manufacturer product information for amount to be used).

5. Incubate suspension for 30 to 45 min at 30°C with gentle rotation (*see* **Note 4**).

6. For each cross-linking reaction, transfer 1 mL of spheroplasts to a 1.5 mL reaction tube. For each co-IP, two reactions are required, one with addition of crosslinker agent and one without crosslinker addition as a control. In most cases, detergent-solubilized membrane proteins should not interact without crosslinker.

7. Pellet spheroplasts by centrifugation at low speed (400*g* for 10 s). Resuspend pellet in 1 mL lysis buffer containing protease inhibitors (PMSF to a final concentration of 1 m*M* and 2.5 µL of Protease Inhibitor Cocktail).

8. Immediately add the cross-linking agent, DSP, to a final concentration of 200 µg/mL to stabilize transiently-interacting proteins or detect complexes between membrane proteins (*see* **Note 5**).

9. Incubate at room temperature for 30 min with occasional shaking.

10. Quench the crosslinking reaction by adding 1 *M* hydroxylamine to a final concentration of 20 m*M*.

11. Precipitate protein by adding 100% TCA to a final concentration of 5%. Mix well by inversion, incubate on ice for at least 30 min.

12. Collect TCA precipitates by centrifugation (14,000 rpm; 5 min). Discard supernatant and wash pellet twice with cold acetone (*see* **Note 6**).

13. Dry the washed pellets using a speed-vac concentrator (Savant Instruments).

3.1.2. Co-Immunoprecipitation

1. Dissolve pellets, from **Subheading 3.1.1.**, **step 13** in 100 µL Cracking Buffer using a water-bath sonicator (*see* **Note 7**).

2. Incubate the samples at 65°C for 5 min.

3. Add 1 mL Tween-20 IP Buffer and 10 µL of 100 mg/mL BSA. Mix well by inversion.

4. Centrifuge the mixture at 4°C (14,000*g* for 10 min). Transfer 950 µL of the supernatant to a new 1.5-mL reaction tube.

5. Add antiserum and incubate at 4°C for at least 2 h or overnight with rocking.

6. Add 75 to 100 µL of prehydrated/prewashed Protein A-sepharose 50% slurry for polyclonal or Protein G-sepharose for monoclonal antiserum used for co-IP. Incubate at 4°C for at least 1 h with rocking.

7. Wash beads twice with 1 mL Tween-20 urea buffer, twice with 1 mL Tween-20 IP buffer, and once with 1 mL TBS. Centrifuge at 2000*g* for 10 s each time.

8. Extract co-IP proteins from the sepharose beads by adding 100 µL urea sample buffer and incubating at 65°C for 5 min.

9. Run SDS-PAGE using the samples. The β-mercaptoethanol in the SDS-PAGE sample buffer will reduce the crosslink between the interacting proteins. Use immunoblots with appropriate antibodies to detect the immunoprecipitated protein and other partners it might interact with.

3.2. Method for Co-Immunoprecipitation of Proteins Without Use of Crosslinkers

3.2.1. Cultures

1. Inoculate cells (from a fresh preculture) at $A_{600} = 0.2/mL$. Grow cells to $A_{600} = 0.6–1/mL$ (10–50 mL, depending on the expression level of the protein) in the required medium.
2. Spin down between 6 and 30 ODs of cells, by spinning at $1000g$ for 3 min.

3.2.2. Lysis

1. The following protocol is for 10 mL culture (6 to 10 ODs). Keep all samples in ice, all steps must be done on ice or in a cold room.
2. Give the cells a quick wash with 1.5 mL of cold phosphate buffered saline (PBS) and transfer cells to a 1.5-mL microcentrifuge tube.
3. Centrifuge and remove as much as PBS as possible. Resuspend the cell pellet in 200 μL of IP lysis buffer containing the following volumes of protease inhibitor stock solutions: 2 μL aprotinin, 2 μL leupeptin, 2 μL PMSF (add right before use), 2 μL protease inhibitor cocktail.
4. Add glass beads up to the meniscus and vortex for 1 min in cold room, 10× with 1 min incubation on ice in between.
5. Punch a small hole on the bottom of the tube, small enough to prevents beads from going through. Quickly put this tube into a new, cold microcentrifuge tube, and spin 5 s at $2000g$. Repeat until the beads from the upper tube are dry (i.e., until all the extract has been transferred to the bottom tube).

3.2.3. Solubilization: Clearance

1. Spin down the cell debris. If the protein under study (or predicted interacting proteins) is membrane associated, follow **step a** (solubilization) and then **step b**. If the protein is soluble, go directly to **step b**.
   a. In the case of a membrane protein, first spin down intact yeasts cells and cell debris by centrifuging for1 min at $1000g$ at 4°C. Transfer the supernatant into a clean, cold microcentrifuge tube. Add detergent to solubilize membrane proteins of interest. Proper solubilization of a given protein depends of (i) the nature of the detergent, (ii) its concentration, (iii) the protein/detergent ratio for solubilization, and (iv) time and temperature at which solubilization occurs. This depends mainly on the protein and the characteristics of the membrane and therefore has to be optimized.
   b. Spin down cell debris/unsolubilized material for 10 min at $14,000g$, at 4°C.
2. Transfer the clear extract to a new tube and accurately measure the volume. Determine the volume per OD of cell equivalent (e.g., if the intial 10 ODs end up in a final volume of 200 μL, 1 OD of cells = 20 μL).
3. Remove 1 OD equivalent of cell extract and keep this in a separate tube labeled input. Add 5X sample buffer and store at –20°C.

3.2.4. Immunoprecipitation

1. Add the 1st antibody to the rest of the extract. Dilutions at which the antibody should be used depend on the antibody. For instance, 1 μL of monoclonal α-HA antibody can be used per 200 μL reaction.
2. Mix overnight at 4°C. This time depends on the stability of the interaction and the quality of the antibody. For unstable interactions, this time can be shortened to between 3 and 4 h.
3. Add 100 μL of Protein-A- or Protein G-coupled sepharose beads (previously hydrated and washed in lysis buffer) (*see* **Note 1**).
4. Mix at 4°C for 1 h. A longer incubation (up to 6 h) usually brings more protein, but can also increase the background.
5. Spin beads at 500*g* for 10 s. It is important to spin at low speed to prevent possible membrane vesicles from being pelleted.
6. Transfer 1 OD of cell equivalents to a new cold microcentrifuge tube labeled "unbound." Add 5X sample buffer and keep at –20°C. The rest of the unbound fraction can also be saved.
7. Wash with 1 mL IP lysis buffer: mix at 4°C for 5 min and spin at 500*g* for 10 s, remove supernatant.
8. Repeat twice with 1 mL wash buffer.
9. After the final wash, remove buffer from beads completely. This can be done by aspirating the beads with a syringe and a 30 gage, 1/2 inch needle.
10. Add 50 μL of 1X non-dyed sample buffer to the dried beads and boil for 5 min.
11. Punch a small hole at the bottom of the tube, proceed as in **step 6** (except this time the beads are smaller).
12. Add 100X bromophenol blue stock solution (1X final concentration) to the eluate. Use or freeze at –20°C.

3.2.5. Immunoblot Analysis

1. Denature samples before loading.
2. Load on SDS-PAGE the equivalent of 0.2 OD of "input" and 0.2 OD of "unbound" samples. This can be adjusted in case the protein has a high/low expression. In the unbound fraction, the immunoprecipitated protein is sometimes depleted relative to the input. Usually, the immunoprecipitated protein can be seen when the equivalent of 1 OD of the final extract is loaded. For co-IP proteins, load from 1 to 5–7 OD equivalents depending on the affinity of the antibody, and the strength of the interaction (as long as unrelated marker proteins do not appear).
3. Even after denaturation, there is still a strong reactivity of the antibody present in the extract (and therefore loaded on the gel) with the secondary antibody used in the immunoblotting process. Accordingly, the acrylamide concentration can be adjusted to prevent overlapping of signals (*see* **Note 2**).

3.3. Methods for Long-Homology TAP-Tagging of P. pastoris Proteins

3.3.1. Tagging of the Strain

The TAP-tag sequence has been amplified from CellZome's pBS1479 (*6*), and subcloned into the pPICZ-B (Invitrogen, La Jolla, CA) backbone (leaving behind the bacterial origin of replication, the AOX1 (alcohol oxidase 1) transcription terminator and, most importantly, the prokaryotic/eukaryotic dual functional zeocin resistance cassette), to obtain pMY62, a new TAP-tagging vector with a relatively small size (*see* **Fig. 3A**).

1. Design three primers corresponding to the target gene's coding sequence (*see* **Fig. 3B**). Primer 1 (reverse direction): contains four nucleotides at the 5′-end (for better restriction enzyme digestion), a site for the restriction enzyme *Afl*II, and the region that is complementary to the sequence of the target gene coding sequence preceding, but not including, the stop codon. The reading frame must be maintained with the *Afl*II site CTT.AAG. (If there is an *Afl*II site in the amplification region, use a blunt-end restriction enzyme site instead.) Primer 2 (forward direction): contains four nucleotides at the 5′-end, a site for the restriction enzyme *Cla*I, and the region corresponding to the internal sequence of the gene. Amplify a region longer than 500 bp for better homologous recombination in the later step. (If there is a *Cla*I site in the amplification region, use a blunt-end restriction enzyme site instead.) Primer 1 and Primer 2 should be designed so that the amplified region possesses a unique restriction site located in the middle region of the sequence, which is not present in pMY62, and which will be used for plasmid linearization required before transformation (*see* **Fig. 3A**). Primer 3 (forward direction) is chosen upstream of Primer 2, for verification of the TAP-tag cassette integration at the appropriate locus (*see* **step 4** and **Fig. 3C**).
2. Amplification of the target gene fragment by PCR and insertion of this fragment into pMY62.
   a. Use Primers 1, 2, and the *P. pastoris* genomic DNA (you may use Promega's Wizard Genomic DNA purification kit) as template to amplify the fragment by a high-fidelity thermo-polymerase such as PfuTurbo (Stratagene; cat. no. 600250).
   b. Digest the PCR fragment with *Afl*II and *Cla*I (or *Sca*I/other blunt-end restriction enzyme), clone into appropriately-digested pMY62. Transform *E. coli* and isolate zeocin-resistant colonies on LB + zeocin plate.
   c. The inserted fragment can be sequenced using OMY69 and OMY70 primers to confirm the sequence and the reading frame.
3. Linearize construct to transform *P. pastoris* cells.
   a. Linearize the pMY62 plasmid within the insert by digesting once (as close to the middle of the insert as possible) in the target gene fragment (the vector should not be cut; its sequence is available for computer analysis). Avoid

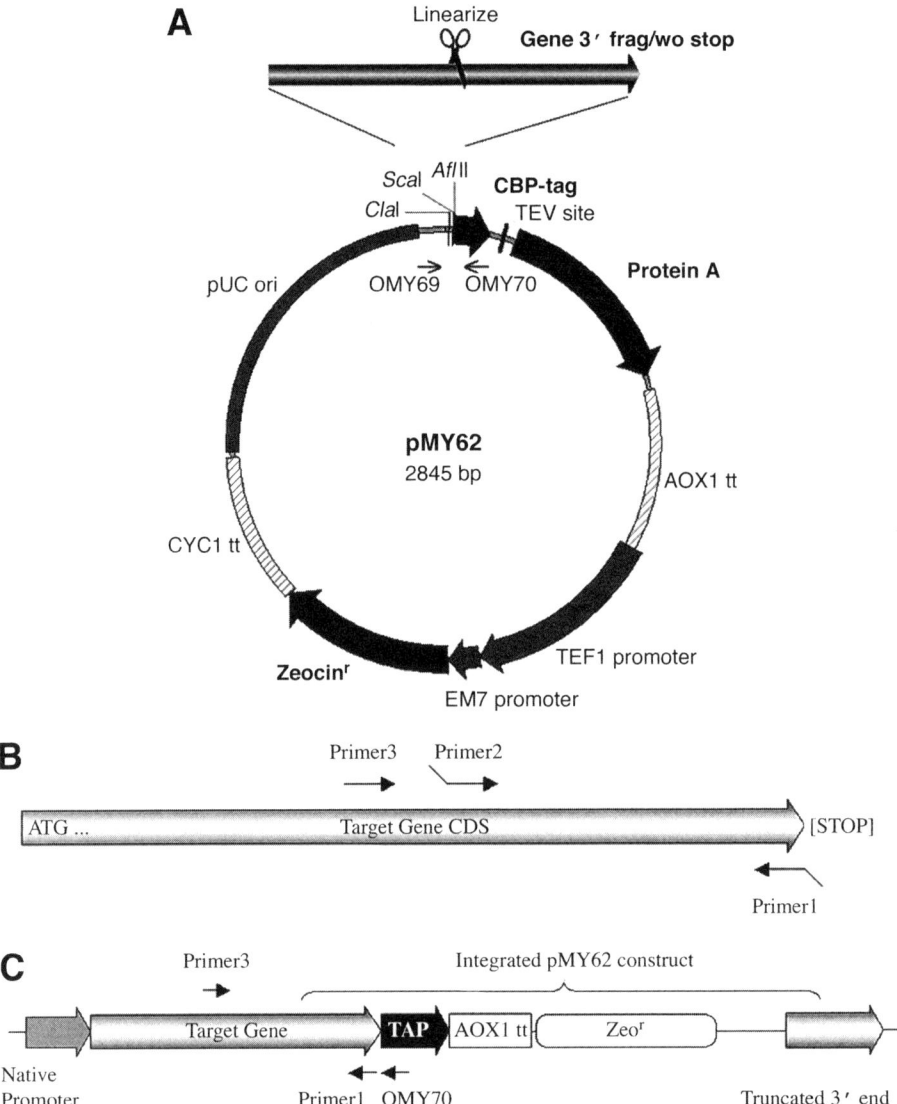

Fig. 3. TAP-tagging strategy in *Pichia pastoris*. (**A**) Map of the TAP-tagging vector (pMY62). The insert is cloned (without its STOP codon for translation) using the restriction sites *Cla*I/*Afl*II or *Sca*I, in-frame with the CBP tag, the TEV protease cleavage site and the Protein A tag. A unique restriction site, in the insert, is used to linearize the vector for transformation by homologous recombination in *P. pastoris*. (**B**) Strategy for primer design (*see* text for details). (**C**) Overall structure of the TAP-tagged locus in its chromosomal context.

regions of less than 100-bp homology with the target gene, on either side of the site of linearization.

b. Transform *P. pastoris* (by electroporation or chemical methods such as LiAc/polyethylene glycol (PEG) *(8)*, plate on YPD + zeocin and isolate zeocin-resistant colonies.

4. PCR identification of correct transformants.

   a. Pick some colonies from the YPD + Zeocin plate and resuspend each of them in 100 µL H_2O.

   b. Prepare raw genomic DNA by taking 50 µL of the resuspended colony and adding Zymolase 20T to a final concentration of 1 mg/mL. Incubate the reaction at 30°C for 30 min. Next, freeze reaction in liquid nitrogen, then thaw in a water bath at room temperature. Repeat freeze-thaw three or more times. Spin down to pellet debris; the raw genomic DNA is in the supernatant.

   c. Run two PCR reactions on each raw genomic DNA sample. A control reaction using Primer3 + Primer1 to check Primer3 and raw genomic DNA quality, and a test reaction using Primer3 + OMY70 to validate the location of the integrated TAP tag. Also do control and test reactions using genomic DNA from a wild-type strain.

   d. PCR results: Wild-type genomic DNA will yield one band in the control reaction and no band in the test reaction. Transformants with incorrect integration should yield the same results. Transformants with the correct integration should produce a band in the control reaction and a band that is slightly larger in the test reaction.

   e. Correct integration of the TAP-tagged segment into the chromosomal locus of the target gene should yield the following layout (*see* **Fig. 3C**).

5. Western blot confirmation of the TAP-fusion protein. The strain may be further confirmed by Western blots with an antibody raised against the target protein (add 20 kDa to the native protein size for the whole TAP-tag, or 5 kDa for the CBP tag remaining after TEV protease cleavage). Alternatively, anti-CBP (Upstate; cat. no. 07-482) at a dilution of 1:5,000) or Protein A antibodies can also be used to detect the fusion protein.

3.3.2. Large-Scale TAP Purification (for Integral-Membrane Peroxins)

(The TAP purification is done essentially following the Séraphin Lab's protocol *[6]*).

1. Inoculate 500 mL of YPD medium in a 2 L flask and incubate overnight at 30°C with vigourous shaking.

2. Harvest the cells by centrifugation at 3000*g* for 10 min and wash them with 500 mL sterile distilled water.

3. Resuspend the cells in 2 L of peroxisome inducing medium (such as methanol medium (*see* **Subheading 2.1.1.**, **step 2**) and incubate overnight at 30°C with vigorous shaking. This step induces peroxisome biogenesis.

4. Harvest the cells by centrifugation at 3000*g* for 10 min and wash them with 500 mL ice-cold water twice, then pellet the cells.
5. Resuspend the cell pellet in 10 mL of buffer A at room temperature.
6. All the following procedures must be carried out at 4°C.
7. Mechanically disrupt the cells using a French Press at 11,000 psi. Add KCl to a final concentration of 0.2 *M* to the lysate.
8. Centrifuge the lysate at 27,000*g* for 30 min at 4°C and add digitonin (A.G. Scientific Inc., cat. no. D-1029) to the supernatant to the desired concentration (0.25–1%) to extract membrane protein complexes. Addition of excessive amount of detergent can destabilize protein complexes. Incubate for 1 h at 4°C.
9. Centrifuge at 100,000*g* for 1 h at 4°C and dialyse the supernatant in buffer D overnight.
10. To the dialysed lysate, add Tris-Cl (pH 8.0) to a final concentration of 10 m*M*, NaCl to 150 m*M*, and NP40 to 0.1%.
11. Wash 200 µL of IgG agarose beads with IPP150 buffer and incubate the dialyzed lysate with the beads for 2 h at 4°C.
12. Wash the beads once with three volumes of IPP150 buffer followed by three volumes of TEV cleavage buffer.
13. Resuspend the IgG agarose beads in 1 mL of TEV cleavage buffer and incubate with 100 units of recombinant TEV protease at 25°C for 90 min.
14. Recover the supernatant and add three volumes of calmodulin-binding buffer and 3 µL of 2 *M* CaCl$_2$/mL of eluate to titrate the EDTA.
15. Wash 200 µL of calmodulin beads with IPP150 calmodulin-binding buffer and incubate the eluate with the beads for 1 h at 4°C.
16. Allow the column to drain by gravity flow and wash the column with 30 mL of IPP150 calmodulin-binding buffer.
17. Elute four fractions of 100 µL each with IPP150 calmodulin-elution buffer.

3.3.3. Analysis of Interacting Proteins

After the two-step TAP procedure, the eluted proteins may be concentrated by TCA precipitation (add TCA to a final concentration of 20%, keep overnight on ice, and centrifuge for 15 min at 14,000*g* at 4°C) and subjected to SDS-PAGE and immunoblotting with antibodies to the target protein and to its putative interacting partners. Alternatively, the proteins in the sample may be subjected to mass spectrometry for identification.

4. Notes

1. It is advisable to use protein G except when the source of the antibody is guinea pig. Neither Protein A nor Protein G bind chicken IgY.
2. Care should be taken in the choice of the epitope and the animal of origin of the primary antibodies used for the detection of interacting protein(s) in the final extract. For instance, if the protein of interest is immunoprecipitated with a monoclonal (mouse) antibody and the interacting protein is also detected using a mouse

antibody, the IgG used for the procedure and present in the extract will react strongly with the secondary anti-mouse antibody used during the immunoblotting (between 50 - and 60 kDa and 20 and 25 kDa). In this case, it is advised to use a different antibody source for (i) immunoprecipitation, or (ii) the detection of interacting proteins. Another alternative consists of using secondary antibody raised against native primary antibodies (TrueBlot™, eBioscience, La Jolla, CA) that will therefore not bind to the denatured primary antibodies present in the SDS-PAGE.

3. TAP-tag vectors with different drug-resistance markers are available. pMY62, as described in the text, has Zeocin resistance; pMY63 and pMY64, carrying neomycin–kanamycin phosphotransferase type I and II, respectively, are essentially the same as pMY62 except that the drugs used for selection are kanamycin in bacteria and G418 in yeast. Their full-length sequences, which are useful for finding an appropriate site for linearization, are available upon request.

4. Spheroplasting is an important step of this procedure. Spheroplasts lyse when diluted in water and this can be easily observed with a light microscope or by the decrease of turbidity at 600 nm after addition of water to spheroplasts.

5. It is important to titrate the amount of cross-linker used so that only specific interactions are detected, but not nonspecific ones. It is a good idea to confirm that the co-IP does not contain non-specific partners due to the addition of too much crosslinker.

6. Resuspend pellet thoroughly for the acetone wash. Use of a water-bath sonicator helps.

7. Solubilize pellet completely in Urea Cracking Buffer until suspension becomes clear.

Acknowledgments

This work was supported by grants DK41737 and DK59844.

References

1. Faber, K. N., Heyman, J. A., and Subramani, S. (1998) Two AAA family peroxins, PpPex1p and PpPex6p, interact with each other in an ATP-dependent manner and are associated with different subcellular membranous structures distinct from peroxisomes. *Mol Cell Biol* **18**, 936–943.

2. Snyder, W. B., Koller, A., Choy, A. J., and Subramani, S. (2000) The peroxin Pex19p interacts with multiple, integral membrane proteins at the peroxisomal membrane. *J Cell Biol* **149**, 1171–1177.

3. Hazra, P. P., Suriapranata, I., Snyder, W. B., and Subramani, S. (2002) Peroxisome remnants in *pex3Δ* cells and the requirement of Pex3p for interactions between the peroxisomal docking and translocation subcomplexes. *Traffic* **3**, 560–74.

4. Graham, T. T. (2004) Unit 8: Protein labeling and immunoprecipitation. 8.5 Immunoprecipitation. In *Short protocols in cell biology: a compendium of methods from Current Protocols in Cell Biology* (ed. Bonifacio, J.S., Dasso, M., Harford, M.,

Lippincott-Schwartz, J., and Yamada, K.M., eds.), pp. 8-17. John Wiley & Sons, Hoboken, NJ, USA, pp. 8–17.

5. Gavin, A. C., Bosche, M., Krause, R., et al. (2002) Functional organization of the yeast proteome by systematic analysis of protein complexes. *Nature* **415,** 141–147.

6. Rigaut, G., Shevchenko, A., Rutz, B., Wilm, M., Mann, M., and Séraphin, B. (1999) A generic protein purification method for protein complex characterization and proteome exploration. *Nat. Biotechnol.* **17,** 1030–1032.

7. Puig, O., Caspary, F., Rigaut, G., et al. (2001) The tandem affinity purification (TAP) method: a general procedure of protein complex purification. *Methods* **24,** 218–229.

8. Higgins, D. R., Busser, K., Comiskey, J., Whittier, P. S., Purcell, T. J., and Hoeffler, J. P. (1998) Small vectors for expression based on dominant drug resistance with direct multicopy selection. *Methods Mol. Biol.* **103,** 41–53.

9. Miller, J. and Stagljar, I. (2004) Using the yeast two-hybrid system to identify interacting proteins. *Methods Mol. Biol.* **261,** 247–262.

10. McAlister-Henn, L., Gibson, N., and Panisko, E. (1999) Applications of the yeast two-hybrid system. *Methods* **19,** 330–337.

11. Hu, J. C., Kornacker, M. G., and Hochschild, A. (2000) *Escherichia coli* one- and two-hybrid systems for the analysis and identification of protein-protein interactions. *Methods* **20,** 80–94.

12. Dove, S. L., Joung, J. K., and Hochschild, A. (1997) Activation of prokaryotic transcription through arbitrary protein-protein contacts. *Nature* **386,** 627–630.

13. Joung, J. K., Ramm, E. I., and Pabo, C. O. (2000) A bacterial two-hybrid selection system for studying protein-DNA and protein-protein interactions. *Proc. Natl. Acad. Sci. USA* **97,** 7382–7387.

14. Karimova, G., Pidoux, J., Ullmann, A., and Ladant, D. (1998) A bacterial two-hybrid system based on a reconstituted signal transduction pathway. *Proc. Natl. Acad. Sci. USA* **95,** 5752–5756.

15. Shaywitz, A. J., Dove, S. L., Kornhauser, J. M., Hochschild, A., and Greenberg, M. E. (2000) Magnitude of the CREB-dependent transcriptional response is determined by the strength of the interaction between the kinase-inducible domain of CREB and the KIX domain of CREB-binding protein. *Mol. Cell. Biol.* **20,** 9409–9422.

16. Jobling, M. G. and Holmes, R. K. (2000) Identification of motifs in cholera toxin A1 polypeptide that are required for its interaction with human ADP-ribosylation factor 6 in a bacterial two-hybrid system. *Proc. Natl. Acad. Sci. USA* **97,** 14,662–14,667.

17. Ghys, K., Fransen, M., Mannaerts, G. P., and Van Veldhoven, P. P. (2002) Functional studies on human Pex7p: subcellular localization and interaction with proteins containing a peroxisome-targeting signal type 2 and other peroxins. *Biochem. J* **365,** 41–50.

18. Fransen, M., Brees, C., Ghys, K., et al. (2002) Analysis of mammalian peroxin interactions using a non-transcription-based bacterial two-hybrid Assay. *Mol. Cell. Proteomics* **1,** 243–252.

19. Dautin, N., Karimova, G., and Ladant, D. (2003) Human immunodeficiency virus (HIV) type 1 transframe protein can restore activity to a dimerization-deficient HIV protease variant. *J. Virol.* **77,** 8216–8226.
20. Johnsson, N. and Varshavsky, A. (1994) Split ubiquitin as a sensor of protein interactions *in vivo*. *Proc. Natl. Acad. Sci. USA* **91,** 10,340–10,344.
21. Fetchko, M. and Stagljar, I. (2004) Application of the split-ubiquitin membrane yeast two-hybrid system to investigate membrane protein interactions. *Methods* **32,** 349–362.
22. Thaminy, S., Miller, J., and Stagljar, I. (2004) The split-ubiquitin membrane-based yeast two-hybrid system. *Methods Mol. Biol.* **261,** 297–312.

17

Localization of Proteins and Organelles Using Fluorescence Microscopy

Jean-Claude Farre, Kanae Shirahama-Noda, Lan Zhang, Keith Booher, and Suresh Subramani

Abstract

This chapter describes the different methods used for localization of proteins and organelles in *Pichia pastoris*. A series of plasmids and a modified immunofluorescence protocol for localization and co-localization of proteins and organelles are described. Also included are protocols for the labeling of different subcellular organelles with vital stains.

Key Words: Protein localization; protein colocalization; fluorescence microscopy; organelle markers; green fluorescent protein variants; peroxisomes.

1. Introduction

In considering the best option for localization of proteins and organelles, it is necessary to consider the advantages and disadvantages of each technique.

Indirect immunofluorescence is a rapid and convenient technique for the detection and localization of endogenous proteins, if an antibody against the protein of interest is available. Different proteins can be localized and colocalized without resorting to plasmid constructions, yeast transformation or having to contend with the availability of suitable selectable markers for the introduction of DNA constructs into the host strain. The antibody specificity is very important in making unambiguous conclusions about the location of a given protein. Additionally, colocalization experiments require primary antibodies from two different species.

In the absence of specific antibodies, the construction of a fusion protein with a fluorescent or epitope tag is the only option to detect the protein, but it requires many considerations and controls. First, the tagged protein must be

From: *Methods in Molecular Biology, vol. 389:* Pichia *Protocols, Second Edition*
Edited by: J. M. Cregg © Humana Press Inc., Totowa, NJ

EcoRI
SacI
KpnI
SmaI
BamHI
SpeI
XhoI
P-GAP
YFP
AOX TT
SspI
SalI
StuI
His4
NcoI
pJCF99
6.2 kb
ORI
Amp
pJCF101
6.0 kb
EcoRI
KpnI
SmaI
BamHI
SpeI
XhoI
PstI
CFP
Arg4
PflMI
HincII
BglII
pJCF146
5.9 kb
pJCF149
5.6 kb
EcoRI SacI* KpnI SmaI BamHI SpeI XhoI PstI*
YFP or CFP
GAA TTC GAG CTC GGT ACC CGG GGA TCC ACT AGT CTC GAG CTG CAG GCA TGC AGC AAG GGC GA
SacI
KpnI
SmaI
BamHI
SpeI
XhoI
SphI
HindIII
pJCF100
6.2 kb
KpnI
SmaI
BamHI
SpeI
XhoI
PstI
SphI
HindIII
pJCF148
5.9 kb
YFP or CFP
SacI* KpnI SmaI BamHI SpeI XhoI PstI* SphI HindIII
AG CTG TAC AAG GAA TTC GAG CTC GGT ACC CGG GGA TCC ACT AGT CTC GAG CTG CAG GCA TGC AAG CTT

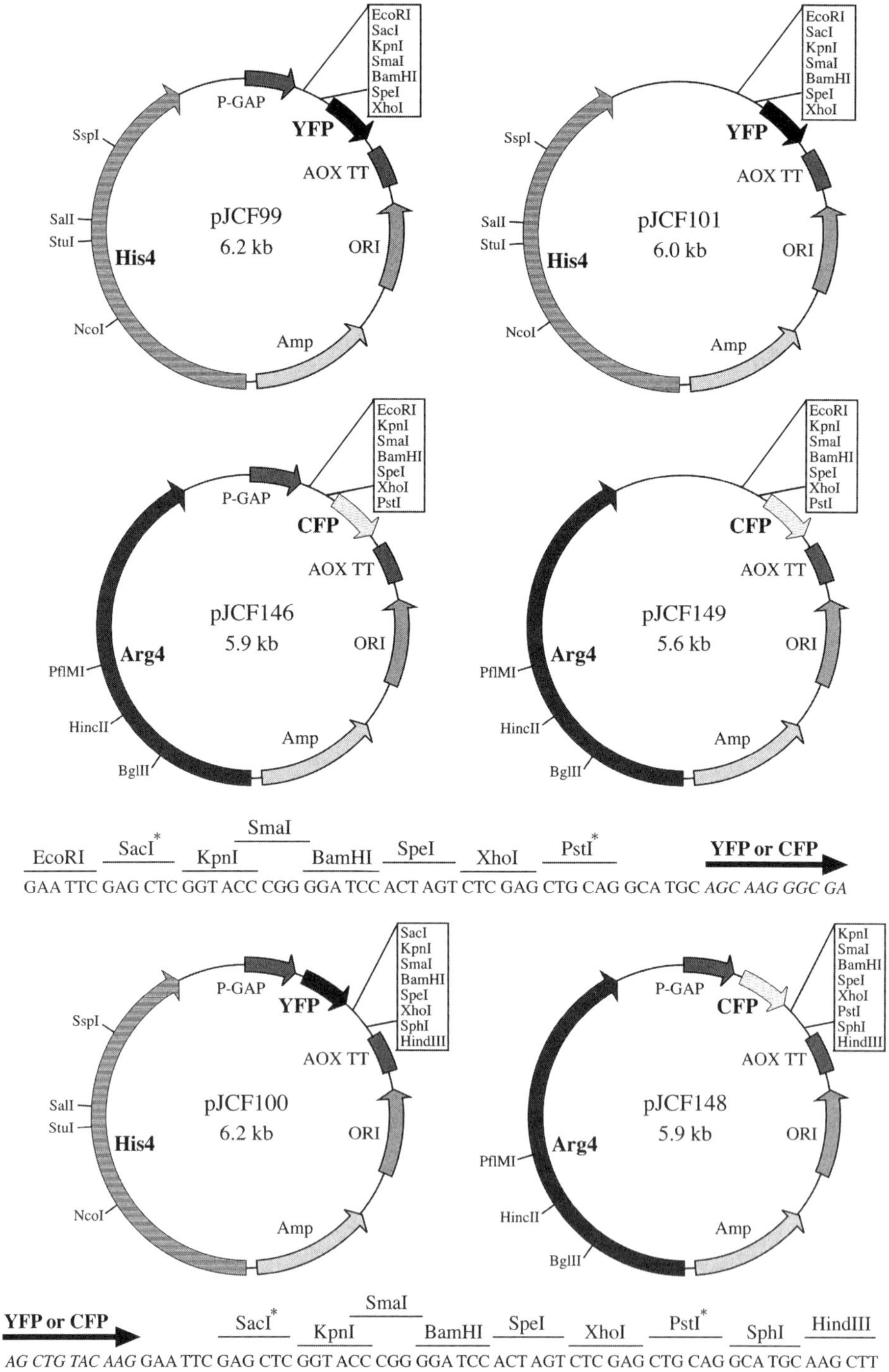

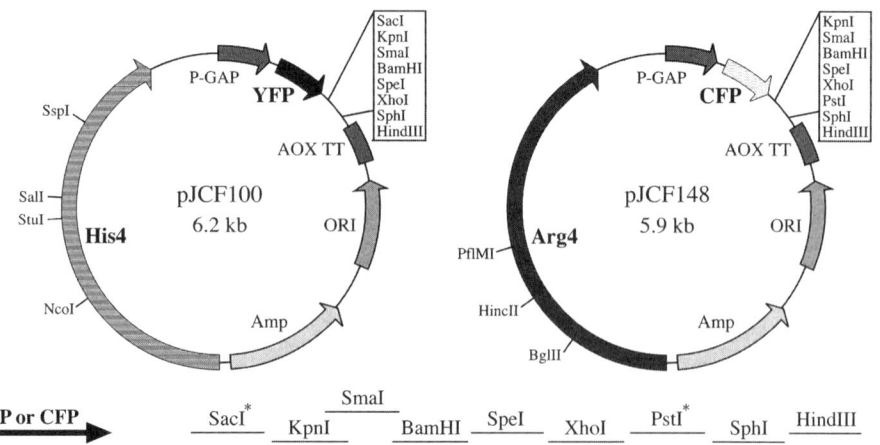

shown to be functional by complementation of a strain lacking the gene for this protein. Second, before the DNA constructs are made for the fusion protein, one needs to consider the selectable markers available in the strain into which the fusion protein is to be introduced. Third, the choice of the appropriate tag and its location in the fusion protein (N-terminal or C-terminal) has to be made. There are many possibilities for this, such as the use of the green fluorescent protein (GFP) from *Aequorea victoria*, or one of the green, blue, cyan, or yellow variants (GFP, BFP, CFP, YFP), or the use of small epitope tags (e.g., HA, Myc, Protein A, Flag, His). GFP and its derivatives offer the possibilities of (i) using live cells, (ii) controlling the level of expression of the fusion protein (e.g., the use of a stronger promoter for detection of a weak signal), (iii) modifying the protein sequence (e.g., use of truncated or mutated forms), and (iv) using CFP-YFP or GFP-BFP combinations in fluorescence resonance energy transfer experiments (FRET: the energy emitted by one fluorophore has the potential to excite a nearby second fluorophore so that the physical proximity of the two proteins can be studied) *(1)*.

1.1. Protein Localization and Colocalization

1.1.1. Fusion Proteins With GFP, BFP, CFP, or YFP

To obtain good images, the first step is to use a bright fluorophore. Many variants of GFP exist. Currently, the brightest GFP molecules are the enhanced GFP (EGFP) and its color variants (EBFP, ECFP, and EYFP). Typically, we fuse these fluorescent proteins to the extreme C termini of proteins whose localization we wish to investigate, and we express these fusions from native (pJCF101 or pJCF149 in **Fig. 1**) or constitutive promoters (e.g., GAP promoter; pJCF99 or pJCF146 in **Fig. 1**). It is important to confirm that the fusion protein complements a deletion strain lacking the protein of interest and that its expression is as close to normal as possible, because overexpression can cause protein mislocalization. If the fusion protein is unable to complement the deletion strain or to be expressed correctly, we resort to fusions at the N termini of the proteins (pJCF100 or pJCF148 in **Fig. 1**).

Usually CFP-YFP or GFP-BFP combinations are correctly separated using appropriate bandpass filters (*see* **Note 1**) and a conventional fluorescence microscope. Maps of the plasmids used in this laboratory for the localization and colocalization experiments are shown in **Fig. 1**. One set of plasmids has the fluorescent protein, YFP, and the *HIS4* marker, and the second set of plasmids has the fluorescent protein, CFP, and the *ARG4* marker.

Fig. 1. *(Opposite page)* Plasmid maps (*indicates that these sites are present only in vectors where they are shown).

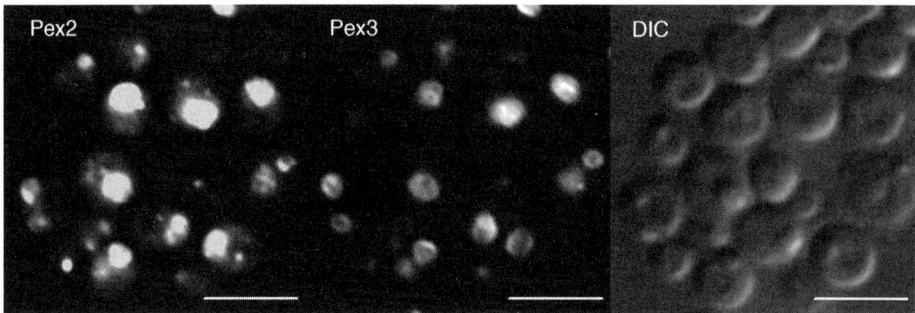

Fig. 2. Colocalization of Pex2 and Pex3 on the peroxisome membrane of *P. pastoris*. The Δ*pex2* cells harboring Pex2-myc were incubated in methanol medium for 6 h and subjected to immunofluorescence. The mouse primary antibody to the myc epitope was visualized using Alexa Fluor 546 goat anti-mouse IgG (Molecular Probes), while the rabbit primary antibody to Pex3 was visualized with Alexa Fluor 488 goat anti-rabbit IgG (Molecular Probes). DIC: Nomarski optics. Bar, 5 μm.

1.1.2. Immunofluorescence

Despite the great developments in the use of protein fusions with fluorescent proteins, indirect immunofluorescence is still a powerful tool to detect the subcellular location(s) of a given protein and to determine its colocalization with other proteins (*see* **Fig. 2**). We have applied the method developed by Rossanese et al. *(2)*, with some modifications, to the localization of peroxisomal membrane proteins. Preservation of the subcellular organelles such as peroxisomes is greatly improved by postfixing the cells with acetone. The acetone treatment might affect the protein structure and might result in some loss of recognition by antibodies. In that case, we recommend trying the method without the acetone fixation. For certain kinds of colocalization experiments, one might want to retain the GFP or red fluorescent protein (RFP) signals during the preparation of samples for fluorescence microscopy using antibodies. The RFP and GFP signals can be preserved during indirect immunofluorescence by fixing the cells with paraformaldehyde instead of formaldehyde (*see* **Note 2**).

1.2. Organelle Localization

1.2.1. Dyes (FM4-64, DAPI, MitoTracker)

Many dyes have been employed to trace organelles (*see* **Fig. 3**). FM4-64 [*N*-(3-triethylammoniumpropyl)-4-(*p*-diethylaminophenyl-hexatrienyl) pyridinium dibromide] is a vital stain for the vacuolar membrane. FM4-64 stains the vacuole red by endocytosis of the dye *(3)*. DAPI (4′, 6-Diamidino-2-phenylindole) stains the nucleus blue by passive diffusion across the membrane and binding to AT regions of the DNA *(4)*. Mitotracker Green FM (Molecular Probes) is a

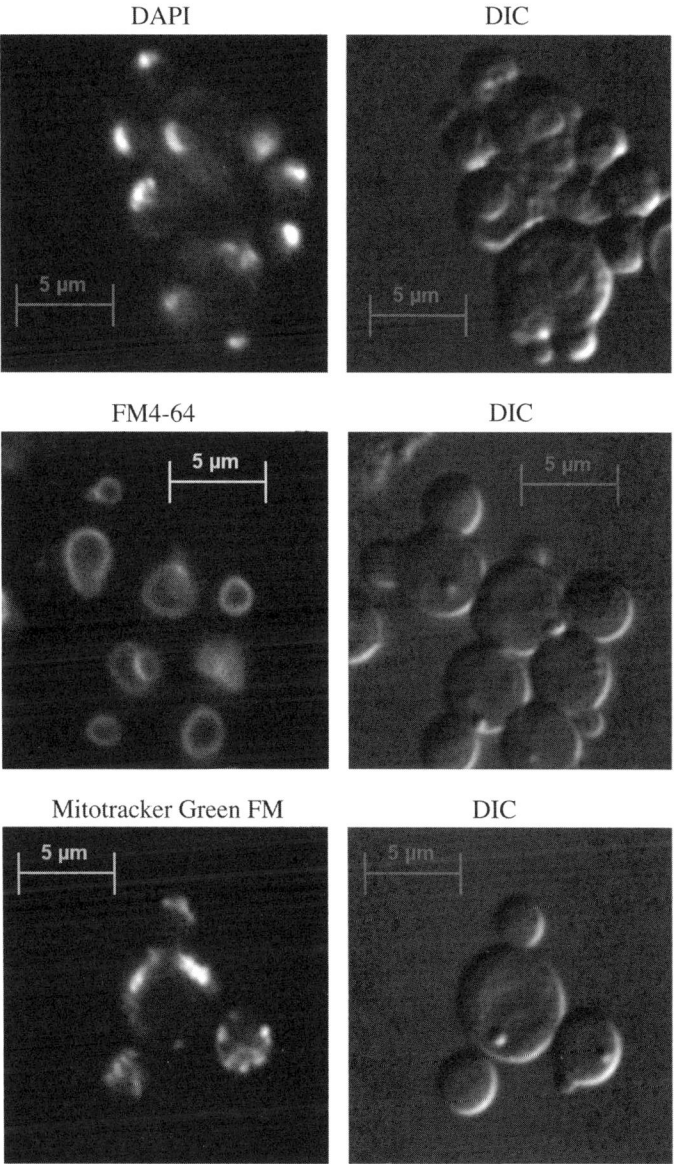

Fig. 3. Localization of the nucleus using DAPI, the vacuole using FM4-64 and the mitochondria using Mitotracker Green FM. DIC: Nomarski optics.

green, fluorescent mitochondrial stain, which appears to localize to mitochondria regardless of mitochondrial membrane potential. The MitoTracker Green FM probe is essentially non-fluorescent in aqueous solutions and only becomes fluorescent once it accumulates in the lipid environment of mitochondria.

1.2.2. Fusion Proteins to Localize Organelles

We have developed a number of fusion proteins in which a peroxisomal targeting signal (PTS) is appended to different fluorescent proteins to follow the biogenesis or the degradation of the peroxisomes. We have used three different targeting signals: (i) Peroxisomal Targeting Signal 1 (PTS1) present at the C terminus of the protein and composed of three amino-acids Ser-Lys-Leu (SKL) *(5)*, (ii) PTS2 present at the N terminus of the protein and whose minimal consensus sequence is (R/K)(L/I/V)-X_5-(H/Q)(L/A) (we add an amino acid sequence of MERLSQLRKHLA to the N terminus of monomeric RFP1, mRFP1) *(6)*, and (iii) Membrane PTS (mPTS) derived from peroxisomal membrane proteins (PMPs). Several targeting signals (mPTSs) for PMPs have been determined, but no consensus sequence has been found. In our studies on peroxisomal membrane localization, the first 40 amino-acids of PpPex3p serve as the mPTS that is fused to GFP variants *(7)*.

These different targeting signals have been used to study each import pathway and to colocalize other proteins with peroxisomes in wild-type or peroxisome biogenesis-deficient *(pex)* cells.

2. Materials

2.1. Strain

The host strain most commonly used is PPY12 (*his4, arg4*) (*see* **Note 3**).

2.2. Dyes

1. FM4-64 (Molecular Probes), dissolved in dimethyl sulfoxide (DMSO) at 1 mg/mL stock.
2. DAPI (Sigma) 1 mg/mL stock in H_2O.
3. MitoTracker Green FM (Molecular Probes) 1 mM stock in DMSO.

2.3. Mounting Solution

There are several commercially available mounting media containing an anti-fading reagent. We use 0.1% *p*-phenylenediamine (Sigma) in 90% glycerol. Dissolve *p*-phenylenediamine in phosphate buffered saline (PBS) to make 1% solution, adjust to pH 9.0, and add glycerol. Store at –80°C in the dark.

2.4. Immunofluorescence

1. Multi-well Teflon-coated glass slide.
2. Cover glass.
3. Fixation buffer: 50 mM potassium phosphate (pH 6.5), 1 mM $MgCl_2$, and 4% formaldehyde.
4. Wash Buffer: 100 mM potassium phosphate (pH 7.5) and 1 mM $MgCl_2$.

5. Zymolyase 20T (Seikagaku-Kogyo): dissolve in H_2O at 10 mg/mL and store at –20°C. Store as aliquots and do not freeze-thaw more than once.
6. 0.1% polylysine (Sigma; cat. no. P 8920).
7. PBS-Block: PBS (137 mM NaCl, 2.7 mM KCl, 1.4 mM KH_2PO_4, 10 mM Na_2HPO_4) (pH 7.4), 1% skim milk, 0.1% bovine serum albumin (BSA), 0.1% n-octyl-β-D-glucopyranoside (Calbiochem; cat. no. 494459).
8. Secondary antibodiesAlexa Fluor 488 and 546 conjugated to goat anti-rabbit IgG, and goat-anti mouse IgG (Molecular Probes).
9. Mounting medium: see Mounting Solution.

2.5. P. pastoris *Vectors*

CFP and YFP were respectively amplified by PCR from pECFP-C1 and pEYFP-C1 (Clontech). pECFP-C1 and pEYFP-C1 encode enhanced cyan and yellow-green variants of the *Aequorea victoria* GFP. Plasmids pJCF99 and pJCF100 were derived from the plasmid pIB2, which expresses genes from the constitutive GAP promoter and has the *HIS4* marker *(8)*. The YFP was located at the *Sph*I site in pJCF99 and at the *Eco*RI site in pJCF100 (*see* **Fig. 1**). pJCF101 was derived from pIB1, which has no promoter and is a useful vector for expressing genes containing endogenous promoters that function in *P. pastoris*. pIB1 also has *HIS4* as a selectable marker *(8)*. The YFP was located at the *Sph*I site in pJCF101 (*see* **Fig. 1**). pJCF135 and pJCF136 (not shown) were derived from pIB1 and pIB2, respectively, by replacement of the *HIS4* marker by *ARG4*. pJCF146 was generated from pJCF136 by insertion of the CFP gene into the *Sph*I site (*see* **Fig. 1**). pJCF148 was generated from pJCF135 by insertion of the CFP fragment into the *Eco*RI site, pJCF149 was derived from pJC135 by cloning the CFP fragment at the *Sph*I site (*see* **Fig. 1**).

3. Methods

3.1. Mounting Live Cells Without Agar

1. Use 1 to 2 mL of log phase cell culture.
2. Centrifuge for 5 min at 2000g at room temperature.
3. Remove the supernatant and resuspend in 250 to 500 μL of fresh media.
4. Mount the cells by spotting 2 μL on a dust-free slide and then pressing the cover glass firmly in place.
5. Seal the cover glass using Cytoseal 60 (Richard-Allan Scientific) (*see* **Note 4**).

3.2. Mounting Live Cells With Agar

1. Make a 1% low-melting agarose (Apex Fine Chemicals) solution in the desired medium (SM, SD, or YPD).
2. Melt the agarose at 55°C and keep liquid at 37°C.
3. Resuspend the cells from **Subheading 3.1., step 3** in 50 to 100 μL of fresh medium.
4. Spread 2 μL of cells on the coverslip.

5. Drop 5 µL of the low melting agarose solution on the slide.
6. Immediately place the cells on the coverslip on top of the drop of agarose on the slide.

3.3. Fixing and Mounting the Cells

1. Centrifuge 1 to 2 mL of log-phase cell culture (OD_{600} of 1–2/mL) as above.
2. Resuspend the cells in 100 µL of 4% paraformaldehyde (electron microscopy grade) in $0.1M$ potassium phosphate buffer (pH 7.5).
3. Incubate at room temperature for 15 min.
4. Centrifuge the cells and remove the supernatant.
5. Wash the cells once in 1 mL of $0.1M$ potassium phosphate buffer (pH 7.5).
6. Resuspend the cells in a small volume of $0.1M$ potassium phosphate buffer (pH 7.5) (*see* **Note 5**).
7. Sonicate the cells briefly using a water-bath sonicator to dissociate them from each other.
8. Mount the cells without agar.

3.4. FM4-64 Staining of the Vacuole in Live Yeast Cells

1. Inoculate 10 mL of YPD with a fresh *P. pastoris* colony.
2. Incubate with a vigorous shaking overnight at 30°C.
3. Use the overnight culture to inoculate a 2 mL culture in the desired medium (SM, SD or YPD [*see* **Note 6**]) in a 50-mL Falcon tube to starting OD_{600} of 0.1/mL.
4. Add 20 µL of 1 mg/mL of FM4-64 and the medium should turn purple when mixed.
5. Incubate with vigorous shaking at 30°C until the culture reaches an OD_{600} of 1 to 2/mL and the medium turns yellow.
6. Centrifuge the cells 5 min at 2000*g* at room temperature.
7. Discard the supernatant, and wash the cells with the desired medium.
8. Centrifuge the cells 5 min at 2000*g* at room temperature.
9. Discard the supernatant, and resuspend the pellets of cells in the desired medium.
10. Observe under the fluorescence microscope using the Rhodamine filter set (*see* **Note 1**).

3.5. DAPI Staining of Nuclei in Live Yeast Cells

1. Inoculate 10 mL of YPD with a fresh *P. pastoris* colony.
2. Incubate with vigorous shaking overnight at 30°C.
3. Use the overnight culture to inoculate a 2 mL culture in the desired medium (SM, SD or YPD [*see* **Note 7**]) in a 50-mL Falcon tube to starting OD_{600} of 0.1/mL.
4. Grow the cells up to an OD_{600} of 1.0/m.
5. Add 5 µL DAPI to the final concentration of 2.5 µg/mL.
6. Grow the cells for 30 min.
7. Centrifuge the cells 5 min at 2000*g* at room temperature.
8. Resuspend cells and wash with 1X PBS.
9. Centrifuge the cells 5 min at 2000*g* at room temperature.
10. Finally resuspend the cells in 1X PBS and observe under the fluorescence microscope using the DAPI filter set.

3.6. DAPI Staining of Fixed Cells

1. After the antibody incubation for immunofluorescence (*see* **Subheading 3.8.3., step 7**) add 20 µL of 1 µg/mL DAPI in 1X PBS over the cells (*see* **Note 8**).
2. Incubate for 2 min.
3. Wash 3× with 1X PBS.
4. Add mounting solution and place the coverslip over the cells (eliminate excess).
5. Seal the edge of the coverslip with Cytoseal 60.

3.7. MitoTracker Green FM Staining of Mitochondria (see Note 9)

1. Inoculate 10 mL of YPD with a fresh *P. pastoris* colony.
2. Incubate with vigorous shaking overnight at 30°C.
3. Use the overnight culture to inoculate a 2 mL in the desired medium (SM, SD, or YPD [*see* **Note 7**]) in a 50-mL Falcon tube to starting OD_{600} of 0.1/mL.
4. Grow the cells up to an OD_{600} 1.0/mL.
5. Add 0.2 µL MitoTracker Green FM to a final concentration of 100 nM.
6. Incubate for 1 h at 30°C.
7. Wash cells 2× with fresh medium.
8. Examine the cells under the fluorescence microscope using the FITC or GFP filter set.

3.8. Immunofluorescence

3.8.1. Fixing and Spheroplasting of Cells

1. Collect the cells of 10-mL culture at an OD_{600} of 0.25 to 1.0/mL by centrifugation at room temperature.
2. Resuspend the cells in 5 mL of freshly prepared fixation buffer.
3. Fix the cells for 1 h at room temperature in a 15-mL tube on a rotator.
4. Collect the cells by centrifugation for 3 min at 1000g.
5. Aspirate the supernatant completely.
6. Resuspend the cells in 5 mL of freshly prepared wash buffer and centrifuge again as above. Repeat the wash again.
7. Resuspend the cells in wash buffer to an OD_{600} of 10/mL (*see* **Note 10**).
8. Add 0.6 µL of 2-mercaptoethanol to 100 µL of cell suspension.
9. Add 10 µL of 10 mg/mL Zymolyase 20T to the cell suspension (*see* **Note 11**).
10. Incubate the cell suspension for 10 to 20 min with mixing end-over-end or on a rotator (*see* **Note 12**) at room temperature.
11. Centrifuge the spheroplasts for 2 min at 400g.
12. Gently resuspend the spheroplasts in 100 µL of wash buffer and centrifuge again.
13. Resuspend the spheroplasts in 200 µL of wash buffer (*see* **Note 10**).

3.8.2. Adhesion and Permeabilization of Cells

1. Add 20 µL of 0.1% polylysine (Sigma) to each well of multi-well slide glass (*see* **Note 13**).
2. Remove the solution by aspiration after 5 min.

3. Wash each well three times with water and air dry. (Alternatively, wash the slides with running water and air dry).
4. Add 20 μL of the spheroplast suspension to each well.
5. After 3 min, blot off excess liquid with a piece of filter paper.
6. Postfix the spheroplasts by immersing the dried slide glass in 40 mL of pre-chilled 100% acetone in a 50-mL Falcon tube for 5 min at –20°C.
7. Remove the slide glass, blot off excess solvent, and allow dry. You can store the slides at 4°C (*see* **Note 10**).

3.8.3. Antibody Incubation

1. Incubate each well to re-hydrate and block with a drop of PBS-Block for 30 min.
2. Aspirate the solution and add 15 μL of primary antibodies mixture in PBS-Block.
3. Incubate in a humid chamber for 1 to 2 h at room temperature or overnight at 4°C.
4. Wash each well eight times with PBS-Block.
5. Add 15 μL of secondary antibodies mixture in PBS-Block.
6. Incubate the slide in a humid chamber for 30 to 60 min. Keep the slide glass in the dark.
7. Wash each well eight times with PBS-Block and completely remove the liquid by aspiration after the final wash.
8. Add suitable mounting medium to each well, cover the slide glass with a cover glass, and seal with nail polish or equivalent (e.g., Cytoseal 60).
9. Observe under a fluorescence microscope or store the slide at 4°C (*see* **Note 10**). The signals are stable for at least 1 wk.

4. Notes

1. The bandpass filters we use are the Endow GFP filter set (41017), Blue GFP filter set (31021), Yellow GFP BP filter set (41028), Cyan GFP V2 filter set (31044 V2), and Rhodamine filter set (C2915) from Chroma Technology Corp.
2. We fix the cells with the fixation buffer containing 4% paraformaldehyde (EM grade), and convert the cells to spheroplasts in the wash buffer containing 1.2 M sorbitol. The spheroplasts are permeabilized by treating the cells with 0.5% Triton X-100 in the wash buffer for 5 to 15 min and applied to the wells. The later steps are essentially the same as above except the sealing. Organic solvents affect the structure of the fluorescent proteins so that one should avoid using acetone or methanol to fix the cells, and sealants that contain organic solvents.
3. In some cases, autofluorescence may be a problems. *ade*⁻ strains are particularly fluorescent. Also, dead cells yield strong fluorescence with the GFP and Rhodamine filters.
4. The sample is good for about 1 h until carbon dioxide production causes the coverslip to bow outward.
5. If you keep the suspension at 4°C, the signals of fluorescent proteins are stable for a few weeks.
6. pH problems: the medium should turn purple with the FM4-64. If it turns yellow, it is because the pH of the medium is too acidic (less than pH 5.5). Adjust the pH of the medium to between 6.0 and 6.8 with KOH.

7. Avoid growing the cells in YPD for the fluorescence microscopy experiment, because it gives a high background.
8. You can add DAPI directly in the mounting solution (100 ng/mL final concentration), but the background could be higher.
9. The dye will stain live cells but is not well-retained after aldehyde fixation. MitoTracker Green FM can be used to stain mitochondria in fixed cells as well.
10. You can interrupt the process at this point.
11. The original method used recombinant yeast lytic enzyme (ICN Biomedicals), and most labs use Oxalyticase or Zymolyase 100T for immunofluorescence, but Zymolyase 20T also works.
12. We suggest doing this process as a time course at least for the first time because it is the most critical part of the procedure. In our hands, 10 to 20 min incubation gives good results.
13. We usually spin the poly-lysine and water for 10 min at 14,000g before the preparation of the slides to remove any dust in the solution, and clean the slides with 100% ethanol or acetone.

Acknowledgments

This work was supported by NIH grants DK41737 and DK59844 to SS. JCF was supported by an EMBO postdoctoral fellowship.

References

1. Miyawaki, A., Llopis, J., Heim, R., et al. (1997) Fluorescent indicators for Ca^{2+} based on green fluorescent proteins and calmodulin. *Nature* **388,** 882–887.
2. Rossanese, O. W., Soderholm, J., Bevis, B. J., et al. (1999) Golgi structure correlates with transitional endoplasmic reticulum organization in *Pichia pastoris* and *Saccharomyces cerevisiae. J. Cell Biol.* **145,** 69–81.
3. Vida, T. A. and Emr, S. D. (1995) A new vital stain for visualizing vacuolar membrane dynamics and endocytosis in yeast. *J. Cell Biol.* **128,** 779–792.
4. Dann, O., Bergen, G., Demant, E., and Voltz, G. (1971) Trypanocide Diamidine des 2-Phenyl-benofurans, 2-Phenyl-indens und 2-Phenyl-indols. *Liebigs Ann. Chemi.* **749,** 68–89.
5. Gould, S. J., Keller, G. A., and Subramani, S. (1987) Identification of a peroxisomal targeting signal at the carboxy terminus of firefly luciferase. *J. Cell Biol.* **105,** 2923–2931.
6. Campbell, R. E., Tour, O., Palmer, A. E., et al. (2002) A monomeric red fluorescent protein. *Proc. Natl. Acad. Sci. USA* **99,** 7877–7882.
7. Wiemer, E. A., Luers, G. H., Faber, K. N., Wenzel, T., Veenhuis, M., and Subramani, S. (1996) Isolation and characterization of Pas2p, a peroxisomal membrane protein essential for peroxisome biogenesis in the methylotrophic yeast *Pichia pastoris. J. Biol. Chem.* **271,** 18,973–18,980.
8. Sears, I. B., O'Connor, J., Rossanese, O. W., and Glick, B. S. (1998) A versatile set of vectors for constitutive and regulated gene expression in *Pichia pastoris. Yeast* **14,** 783–790.

18

Fluorescence Microscopy and Thin-Section Electron Microscopy

Benjamin S. Glick

Abstract

Intracellular structures in *Pichia pastoris* can be visualized by the complementary methods of fluorescence microscopy and electron microscopy. An improved immunofluorescence protocol yields better optics and more reliable antigen preservation than conventional methods. As an alternative to immunofluorescence, if a protein of interest is fused to GFP or another fluorescent tag, the cells can be fixed and viewed directly. For higher-resolution studies of organelle morphology, thin-section electron microscopy of permanganate-fixed cells yields good preservation of intracellular membranes.

Key Words: Immunofluorescence; fluorescence microscopy; electron microscopy; yeast; *Pichia pastoris*; fluorescent proteins; permanganate fixation.

1. Introduction

A major advantage of *Pichia pastoris* as a cell biological system is that it can be analyzed using techniques originally developed for *Saccharomyces cerevisiae*. For localizing proteins within a yeast cell, the traditional method is immunofluorescence microscopy *(1)*. More recently, fluorescent protein tags have been used to study the localization and dynamics of intracellular components *(2)*. For examining the morphology of organelles, the simplest approach is thin-section electron microscopy (EM) of aldehyde-fixed cells *(3,4)*. More advanced EM methods include the localization of antigens by immunoelectron microscopy, and the three-dimensional analysis of intracellular structures using electron tomography *(5–7)*.

Here we describe two protocols for fluorescence microscopy and one protocol for thin-section EM of budding yeast cells. The immunofluorescence microscopy protocol is somewhat more elaborate than standard methods, but

From: *Methods in Molecular Biology, vol. 389:* Pichia *Protocols, Second Edition*
Edited by: J. M. Cregg © Humana Press Inc., Totowa, NJ

yields better and more consistent images. The protocol for fixing fluorescently tagged cells is very simple. The protocol for thin-section EM allows for clear and reproducible visualization of intracellular membranes. These methods work equally well for *P. pastoris* and *S. cerevisiae*.

2. Materials

2.1. Stock Solutions for Fluorescence and Electron Microscopy

1. Formaldehyde, 16% solution, EM grade ampules (Ted Pella; cat. no. 18505).
2. Glutaraldehyde, 25% solution, EM grade ampules (Ted Pella; cat. no. 18426).
3. 10X stock solution 1 M KPO_4 (pH 7.5). Autoclave, and store at room temperature.
4. 5X stock solution 0.2 M KPO_4 (pH 6.8). Autoclave, and store at room temperature.
5. 5X stock solution 0.25 M KPO_4 (pH 6.5). Autoclave, and store at room temperature.
6. 2000X stock solution 2.0 M $MgCl_2$. Autoclave, and store at room temperature.
7. 2000X stock solution phenylmethylsulfonylfluoride (PMSF)/pepstatin: 0.5 M PMSF, 2 mM pepstatin. Dissolved in water-free dimethyl sulfoxide (DMSO), freeze in liquid nitrogen, and store in aliquots at –80°C.
8. Yeast lytic enzyme: recombinant, protease-free (ICN; cat. no. 153529) (*see* **Note 1**). Dissolve in 0.1 M KPO_4 (pH 7.5), 1 mM $MgCl_2$ to 20,000 U/mL. Dispense into 50 μL aliquots, freeze in liquid nitrogen, and store at –80°C.
9. Poly-L-lysine: molecular weight > 300,000 (Sigma; cat. no. P-1524). Dissolve to 0.1% in deionized H_2O, and store at –20°C.
10. Phosphate-buffered saline (PBS) (pH 7.4). Autoclave, and store at room temperature.
11. 500X stock solution Hoechst 33258: 0.1% in H_2O. Store in the dark at –20°C.
12. β-mercaptoethanol (Sigma; cat. no. M-6520).
13. Mounting solution. Work at very low light levels while preparing this solution. Dissolve 1 g of phenylene diamine in 100 mL PBS. Adjust the pH to 9.0 with a few drops of 0.5 M Na_2CO_3. Immediately transfer 10 mL of this solution to a fresh 150-mL beaker. While stirring, gradually add 90 mL glycerol. Continue stirring until well mixed. Store the mounting solution in 0.2-mL aliquots at –80°C (*see* **Note 2**).

2.2. Working Solutions for Immunofluorescence Microscopy

1. Fixing Solution: 50 mM KPO_4 (pH 6.5), 1 mM $MgCl_2$, 4% formaldehyde. Make fresh.
2. PM: 0.1 M KPO_4 (pH 7.5), 1 mM $MgCl_2$. Make fresh. Just before use, add PMSF/pepstatin and sonicate in a bath sonicator to dissolve the PMSF.
3. PBS-Block: PBS, 1% dried milk, 0.1% bovine serum albumin (BSA) (globulin-free) (Sigma; cat. no. A-7638), and 0.1% octyl glucoside. Make fresh.
4. Methanol or acetone prechilled to –20°C.

2.3. Working Solution for Fixing Fluorescently Tagged Cells

1. 5 mL 2X Fixative. 2 mL of 0.25M KPO_4 (pH 6.5), 5 μL of 2.0 M $MgCl_2$, 2.5 mL of 16% formaldehyde, 100 μL of 25% glutaraldehyde, and 0.5 mL of deionized H_2O. Make fresh just before use.

2.4. Working Solutions for Electron Microscopy

1. Fixing Solution: 50 m*M* KPO$_4$ (pH 6.8), 1 m*M* MgCl$_2$, and 2% glutaraldehyde. Make fresh.
2. Wash Solution: 50 m*M* KPO$_4$ (pH 6.8). Make fresh.
3. 4% KMnO$_4$. Make fresh. The crystals take a while to dissolve with vigorous vortexing.
4. 2% uranyl acetate. Sterile filter this solution and store it at room temperature in the dark. Discard the solution if a precipitate forms.

3. Methods

3.1. Immunofluorescence Microscopy

The basic procedures for immunofluorescence microscopy of budding yeast were developed in the 1980s *(1)*. Compared with those early methods, the protocol described below *(8)* yields improved optics because the cells are attached to the coverslip instead of the microscope slide, thereby avoiding the distortion that is caused by imaging through a layer of mounting medium. This protocol also gives more reliable preservation of cellular structures because protease-free recombinant lyticase is used instead of the conventional Zymolyase. In principle, any antigen can be localized by immunofluorescence as long as suitable antibodies are available. Most researchers perform so-called indirect immunofluorescence by incubating with an unlabeled primary antibody followed by a fluorescently-conjugated secondary antibody. If desired, two different antigens can be localized simultaneously using a combination of a mouse monoclonal antibody and a rabbit polyclonal antibody, in combination with secondary antibodies that are typically conjugated to red or green fluorophores. Nuclear and mitochondrial DNA can also be visualized with blue fluorescent DNA stains such as Hoechst or 4,6-diamidino-2-phenylindole (DAPI).

It must be noted that immunofluorescence has significant limitations. The preservation of cellular morphology is variable. In many cases, especially with some mutant strains, a subset of the cells remain unpermeabilized after lyticase treatment. Certain antigens, including many peripheral membrane proteins, can lose their normal localization during formaldehyde fixation. The cells typically exhibit a finely punctate background fluorescence that may obscure a genuine signal. Some antibodies, particularly polyclonal antibodies, can bind to structures in addition to the antigen of interest. The issue of background staining can be addressed by modifying the protein of interest with an epitope tag, and then testing whether the signal obtained with an anti-tag antibody is specific to strains containing the tagged protein. Many rabbit sera contain antibodies against yeast α-1,6-mannose carbohydrate linkages, and therefore stain compartments of the secretory and endocytic pathways; this staining can be eliminated by preincubating the antibody with yeast mannan *(8)*.

1. Inoculate a yeast culture in the appropriate medium and let it grow overnight. Aim to have an OD_{600} at harvesting time of between 0.25 and 1.0.

2. Working in a hood, pipet 8 mL of the culture onto a 150-mL Nalgene bottle-top filter (45-mm diameter, 0.2-μm pores, SFCA membrane). Apply vacuum (*see* **Note 3**). As soon as the medium has passed through the filter, break the vacuum and add 5 mL of fixing solution. Swirl to resuspend the cells, and transfer the suspension to a 15-mL Falcon tube. Incubate at room temperature for 2 h, vortexing once briefly after 1 h.

3. During the fixation, prepare the primary and secondary antibody mixtures and precool the organic solvent.

4. Spin for 3 min at 1000*g* in a low-speed centrifuge. Remove the supernatant carefully and completely with a Pasteur pipet. From this point until the organic solvent fixation step, it is important to work quickly.

5. Resuspend the cells in 5 mL PM, then spin for 3 min at 1000*g* and remove the supernatant completely.

6. Resuspend the cells in PM to a final OD_{600} of 10. (For example, if the original culture was grown to an OD_{600} of 0.1, resuspend the cells in 400 μL PM.) Transfer 100 μL of the cell suspension to an 0.5-mL: Eppendorf tube.

7. Add 0.6 μL: β-mercaptoethanol followed by 20 μL (400 units) of yeast lytic enzyme. Mix end-over-end at room temperature for 15 min.

8. Harvest the spheroplasts by spinning 2 min in a microfuge at 1000*g* (about 3500 rpm). Resuspend the spheroplasts by gentle pipetting in 100 μl PM. Repeat this spin and resuspension step.

9. Dilute the primary and secondary antibodies to the appropriate concentration (*see* **Note 4**) in uncentrifuged PBS-Block (typical volume: 100 μL). Wait at least 10 min, then spin for 5 min at top speed in a microfuge to remove particulate matter. Save the supernatants.

10. Prepare wells as follows on the surface of a number 1.5 glass coverslip (*see* **Note 5**). For 1 to 4 samples use a 22 × 22 mm coverslip, and for 5 to 10 samples use a 22 × 50 mm coverslip. To create wells, cut out a suitable rectangular piece from the adhesive backing of a plastic document pouch for a postal courier (e.g., FedEx), and use a leather punch on the largest setting to create holes. Attach the perforated plastic to the surface of the coverslip, ensuring that the plastic around the wells is smooth and flat (*see* **Note 6**).

11. Add 10 μL polylysine to each well. Leave for 30 s and then aspirate off the polylysine. Let dry. Wash each well three times by adding a drop of deionized H_2O, then aspirating off the drop. Let the slide air-dry.

12. Centrifuge the PBS-Block solution for 10 min at 2000*g* in a low-speed centrifuge. Use this centrifuged solution for all steps except antibody dilution.

13. Add 10 μl of spheroplasts to each polylysine-coated well of the coverslip. After 3 min, gently blot off the excess liquid with a cotton swab (Q-tip). Let the coverslip dry completely.

14. Immerse the coverslip for 5 min in a 50-mL Falcon tube containing 40 mL of methanol or acetone cooled to –20°C (*see* **Note 7**). Remove the coverslip and immediately invert it onto a paper towel to blot the excess organic solvent from the wells. Allow any remaining organic solvent to evaporate.

15. Optional: create hydrophobic barriers around each well by drawing a thick circle around it with a Sanford Expo dry-erase marker.

16. Add a drop of PBS-Block to each well. Incubate at least 30 min. From now until adding the mounting solution, keep the coverslip in a humid chamber (i.e., a large Petri dish or other closed chamber containing a wet paper towel).

17. Aspirate to remove the PBS-Block (*see* **Note 8**). Cover the cells with 10 μL of a primary antibody that has been diluted in PBS-Block. Incubate 1 h at room temperature.

18. Wash each well eight times by aspirating to remove the liquid and then immediately adding a drop of PBS-Block. It is important to remove as much liquid as possible during the washes. Wash one well after the other so that each well sits in the wash solution for a while.

19. Add the fluorophore-conjugated secondary antibody diluted in PBS-Block. If DNA staining is desired, supplement this antibody solution with a 1:500 dilution of 0.1% Hoechst dye. Incubate for 30 min at room temperature in the dark.

20. Wash 8 × with PBS-Block as above. After the final wash, be sure to remove the PBS-Block completely. Allow all traces of liquid to evaporate.

21. Add 5 μL of mounting solution to each well, avoiding bubbles. Invert the coverslip carefully onto a glass slide. Press down firmly. Seal the edges of the coverslip with clear nail polish.

22. View immediately with a fluorescence microscope equipped with suitable filters, using a ×100 objective. For preservation, store the slide at –20°C in the dark.

3.2. Fixing Fluorescently Tagged Cells

The direct viewing of fixed, fluorescently tagged cells has the great advantage of being easy and fast. Fluorescence patterns in the fixed cells tend to be stable for at least 1 wk at 4°C. Compared to immunofluorescence microscopy, the preservation of cellular structure is better and the background fluorescence is lower. However, this method has caveats. As with immunofluorescence, the aldehyde fixation step can disrupt the localization of certain proteins. Thus, an important control experiment is to examine living cells from log-phase cultures to ensure that the fixation procedure does not alter the fluorescence pattern. More generally, care must be taken to ensure that the fluorescent tag does not perturb the function or localization of a protein. The green fluorescent protein (GFP) tag or other fluorescent tag should be truly monomeric (*9*). If possible, the relevant gene should be replaced with the tagged gene to avoid problems with overexpression, and to confirm that the tagged protein functions normally.

1. Working in a hood, place 300 μL of 2X fixative in an Eppendorf tube, and cool the solution on ice.

2. To the 2X fixative, add 300 µL from a growing yeast culture in rich or minimal medium. Add the culture solution while vortexing the tube. The final 1X fixative concentration is: 50 mM KPO$_4$ (pH 6.5), 1 mM MgCl$_2$, 4% formaldehyde, and 0.25% glutaraldehyde. Allow fixation to proceed for 1 h on ice.
3. Spin in a microfuge at 1500g for 2 min. Remove the supernatant, and gently resuspend the cell pellet by pipetting in 0.6 mL PBS.
4. Repeat the centrifugation and resuspension step.
5. Optional: to visualize DNA, spin again and resuspend the cells in 0.6 mL PBS containing a 1:500 dilution of 0.1% Hoechst dye. Stain for 30 min on ice, and then wash 3 × with PBS by spinning and resuspending.
6. Finally, spin once again and resuspend the cell pellet in 50 µL PBS.
7. Place 1 to 2 µL of the fixed cell suspension on a glass slide and cover with a number 1.5 coverslip. View with a fluorescence microscope equipped with suitable filters, using a ×100 objective.
8. Store the fixed cell suspension at 4°C in the dark.

3.3. Thin-Section Electron Microscopy

The procedure described here (*4,8,10*) is relatively quick and reliable, and it yields excellent contrast of intracellular membranes, partly because ribosomes are not visible (*see* **Fig. 1**). Oxidation of the cell wall with KMnO$_4$ facilitates penetration of the resin. However, KMnO$_4$ also extracts chromatin and other cellular material, and may alter the detailed structure of organelles, so caution should be used when interpreting the micrographs. If optimal quality and fidelity of preservation are needed, the reader should consider the more advanced technology of rapid freezing followed by freeze substitution (*11*).

1. Grow a 50-mL yeast culture to an OD$_{600}$ of about 0.5.
2. Filter the cells slowly on a 0.22 µM bottle-top filter down to a volume of about 5 mL. Do not dry the cells completely.
3. Rapidly add 40 mL of ice-cold fixing solution. Allow the fixation to proceed for 1 h on ice (*see* **Note 9**).
4. Spin the fixed cells 3 min at 1000g in a low-speed centrifuge at 4°C. Pour off the supernatant, and resuspend the pellet by vortexing in 25 mL ice-cold wash solution.
5. Repeat **step 4** 2 ×, for a total of 3 washes.
6. Spin again after the last resuspension. This time resuspend the pellet in 1 mL wash solution, and transfer the cell mixture to an Eppendorf tube.
7. The remaining steps until resin polymerization are all done at room temperature. Spin for 3 min in a microfuge at 2000g (about 5000 rpm) (*see* **Note 10**). Resuspend the cells in 0.75 mL 4% KMnO$_4$. Mix end-over-end for 30 min (*see* **Note 11**).
8. Spin for 3 min at 2000g. Resuspend the cells in 0.75 mL H$_2$O. Repeat twice more.
9. Spin as above, and resuspend the cells in 0.75 mL 2% uranyl acetate. Mix end-over-end for 1 h.

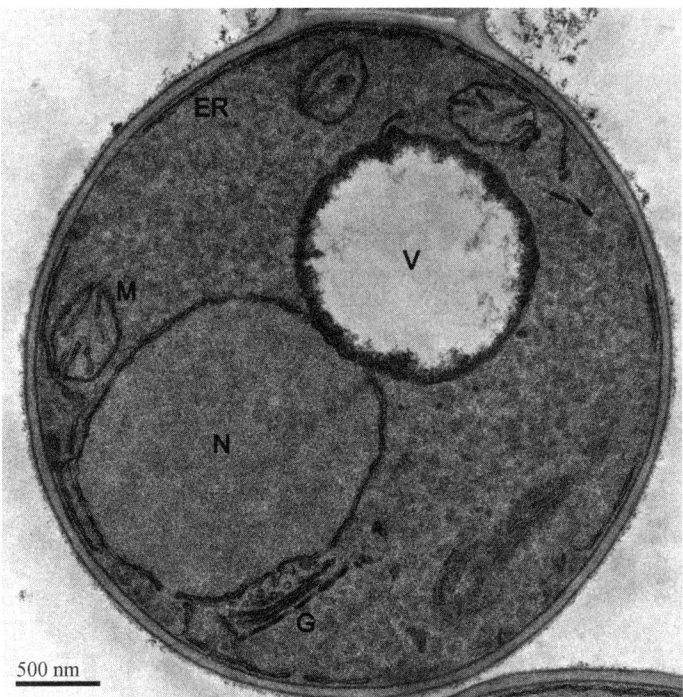

Fig. 1. Thin-section electron micrograph of a permanganate-fixed *P. pastoris* cell. Visible are the nucleus (N), endoplasmic reticulum (ER), mitochondria (M), vacuole (V), and Golgi apparatus (G). Note that unlike *S. cerevisiae*, *P. pastoris* contains stacked Golgi cisternae *(8)*. The scale bar is 500 nm. Micrograph courtesy of Stephanie Levi.

10. Wash 4 × with H_2O as in **step 8**.
11. Prepare normal Spurrs resin according to the formula given in the kit instructions. Work in the hood and wear gloves, as these compounds are extremely toxic. It is convenient to mix the resin components in a disposable plastic beaker placed on a balance in the hood. Mixing can be done with a glass rod.
12. Dehydrate the cells in the following graded series of EtOH solutions. After each spin (3 min at 2000g) and resuspension, mix the cells end-over-end for 5 min. Starting with a fresh, unopened bottle of water-free EtOH, make the following 10 dilutions of increasing EtOH percentage: 50, 70, 80, 85, 90, 95, 100, 100, 100, and 100%.
13. Resuspend the cells in 0.75 mL of a 3:1 EtOH:Spurrs mixture. Mix end-over-end for 1 h.
14. Spin for 3 min at 2000g. Resuspend the cells in 0.75 mL of a 1:1 EtOH:Spurrs mixture. Mix end-over-end overnight.
15. Spin for 3 min at 2000g. Resuspend the cells in 0.75 mL of a 1:3 EtOH:Spurrs mixture. Mix end-over-end for 1 h.

16. Spin for 3 min in a microfuge at full speed. Resuspend the cells in 0.75 mL of pure Spurrs. Mix end-over-end for 1 h.
17. Repeat **step 16**. The results may be improved if the time for this second incubation in pure Spurrs is overnight instead of 1 h.
18. Spin for 3 min in a microfuge at full speed. Resuspend the cells in 300 µL fresh Spurrs. Transfer 100 µL of the cell suspension into each of three BEEM capsules sitting in a metal rack, making sure to avoid trapping any air at the bottom of the capsule. Fill each capsule to the brim with Spurrs. Insert a small square of paper containing a printed sample number. Leave the capsules uncapped throughout the procedure, including during polymerization.
19. Place the rack with the capsules in a vacuum dessicator, and degas for 15 min.
20. Place the rack on a block of styrofoam in an accurate temperature-controlled oven set at 68°C. Allow the resin to polymerize for 36 to 48 h. The samples are now ready for sectioning (*see* **Note 12**) and staining with lead citrate.

4. Notes

1. Recombinant lytic enzyme is expensive. As an alternative, lytic enzyme can be purified from bacteria engineered to produce this protein *(12)*.
2. Some researchers use commercial antifade mounting solutions such as Vectashield (Vector Labs).
3. This vacuum filtration method is an excellent alternative to centrifugation, which takes longer and imposes cellular stress that can potentially alter the cytoskeleton.
4. Antibody concentrations that give a good signal-to-noise ratio need to be determined empirically. For monoclonal antibodies, typical final concentrations range from 1 to 5 µg/mL. For polyclonal antibodies, typical dilutions range from 1:50 to 1:5000. For fluorescent secondary antibodies, typical dilutions range from 1:100 to 1:250.
5. Most modern fluorescence microscopes are designed to give optimal images with a coverslip thickness of 1.5.
6. Coverslips with pre-made wells are now commercially available from Grace Bio-Labs under the name "Chambered Coverglass."
7. Some antibodies work better with methanol fixation, others with acetone fixation. As a general rule, methanol works better for antibodies that recognize denatured epitopes whereas acetone works better for antibodies that recognize native epitopes.
8. Aspirate using a Pasteur pipet linked to a vacuum. Always aspirate from the edge of the well rather than the center because a strong vacuum will damage the cells. It is advisable to draw out the Pasteur pipet over a flame to make a smaller opening, thereby ensuring a gentle suction. First touch the pipet tip to the outer edge of the well, then remove remaining liquid by touching the pipet tip to the inner rim of the well. After aspirating liquid from a well, always add the subsequent solution within a few seconds.
9. Glutaraldehyde is toxic. Work in the hood, wear gloves and safety glasses, and dispose of all glutaraldehyde-containing solutions in an organic liquids waste container.

10. For each of the microfuge-based solution changes in this EM protocol, it is very important to remove the supernatant entirely after the centrifugation. If some liquid remains after aspiration, an additional 1-min spin at 2000*g* can be used to restore a tight pellet and facilitate complete removal of the supernatant.

11. The $KMnO_4$ solution is a potent oxidizing agent, so use caution. Wear gloves and safety glasses, and dispose of the waste in a caustic liquids container.

12. Because of the cell wall, yeasts are more difficult to section than many other cell types. Sectioning should be done with a well-sharpened diamond knife.

References

1. Pringle, J. R., Adams, A. E. M., Drubin, D. G., and Haarer, B. K. (1991) Immunofluorescence methods for yeast. *Methods Enzymol.* **194,** 565–602.

2. Tatchell, K. and Robinson, L. C. (2002) Use of green fluorescent protein in living yeast cells. *Methods Enzymol.* **351,** 661–683.

3. Byers, B. and Goetsch, L. (1991) Preparation of yeast cells for thin-section electron microscopy. *Methods Enzymol.* **194,** 602–608.

4. Kaiser, C. A. and Schekman, R. (1990) Distinct sets of *SEC* genes govern transport vesicle formation and fusion early in the secretory pathway. *Cell* **61,** 723–733.

5. Mulholland, J. and Botstein, D. (2002) Immunoelectron microscopy of aldehyde-fixed yeast cells. *Methods Enzymol.* **351,** 50–81.

6. O'Toole, E., Winey, M., McIntosh, J. R., and Mastronarde, D. N. (2002) Electron tomography of yeast cells. *Methods Enzymol.* **351,** 81–95.

7. Mogelsvang, S., Gomez-Ospina, N., Soderholm, J., Glick, B. S., and Staehelin, L. A. (2003) Tomographic evidence for continuous turnover of Golgi cisternae in *Pichia pastoris. Mol. Biol. Cell* **14,** 2277–2291.

8. Rossanese, O. W., Soderholm, J., Bevis, B. J., et al. (1999) Golgi structure correlates with transitional endoplasmic reticulum organization in *Pichia pastoris* and *Saccharomyces cerevisiae. J. Cell Biol.* **145,** 69–81.

9. Zacharias, D. A., Violin, J. D., Newton, A. C., and Tsien, R. Y. (2002) Partitioning of lipid-modified monomeric GFPs into membrane microdomains of live cells. *Science* **296,** 913–916.

10. Gould, S. J., McCollum, D., Spong, A. P., Heyman, J. A., and Subramani, S. (1992) Development of the yeast *Pichia pastoris* as a model organism for a genetic and molecular analysis of peroxisome assembly. *Yeast* **8,** 613–628.

11. McDonald, K. and Müller-Reichert, T. (2002) Cryomethods for thin section electron microscopy. *Methods Enzymol.* **351,** 96–123.

12. Shen, S.-H., Chrétien, P., Bastien, L., and Slilalty, S. N. (1991) Primary sequence of the glucanase gene from *Oerskovia xanthineolytica*. Expression and purification of the enzyme from *Escherichia coli. J. Biol. Chem.* **266,** 1058–1063.

Index